Microbe-Induced Abiotic Stress Alleviation in Plants

Microbe-Induced Abiotic Stress Alleviation in Plants

Topic Editors

Ying Ma
Christopher Rensing

MDPI

Basel • Beijing • Wuhan • Barcelona • Belgrade • Novi Sad • Cluj • Manchester

Topic Editors
Ying Ma
Department of Life Sciences
University of Coimbra
Coimbra
Portugal

Christopher Rensing
College of Resources and Environment
Fujian Agriculture and Forestry University
Fuzhou
China

Editorial Office
MDPI AG
Grosspeteranlage 5
4052 Basel, Switzerland

This is a reprint of the Topic, published open access by the journals *Agriculture* (ISSN 2077-0472), *International Journal of Plant Biology* (2037-0164), *Microorganisms* (ISSN 2076-2607), *Plants* (ISSN 2223-7747) and *Soil Systems* (ISSN 2571-8789), freely accessible at: https://www.mdpi.com/topics/6K0E199945.

For citation purposes, cite each article independently as indicated on the article page online and as indicated below:

Lastname, A.A.; Lastname, B.B. Article Title. *Journal Name* **Year**, *Volume Number*, Page Range.

ISBN 978-3-7258-7991-5 (Hbk)
ISBN 978-3-7258-7992-2 (PDF)
https://doi.org/10.3390/books978-3-7258-7992-2

Contents

plants

Editorial

Microbe-Induced Abiotic Stress Alleviation in Plants: From Hidden Partners to Central Drivers of Resilience

Ying Ma

Centre for Functional Ecology, Associate Laboratory TERRA, Department of Life Sciences, University of Coimbra, Calçada Martim de Freitas, 3000-456 Coimbra, Portugal; cathymaying@gmail.com

1. Introduction

Our understanding of plant stress biology has shifted with debate toward a more integrative, microbiome-centered perspective. What was once primarily viewed as an intrinsic, plant-centered physiological problem is now often interpreted as a system-level outcome shaped by the interactions between plants and their associated microbiomes. Microorganisms, long considered background inhabitants of soil and plant tissues, are now thought to play a more decisive role in determining plant fitness under environmental stress.

Abiotic stress imposes one of the most severe constraints on global food production and ecosystem stability [1]. Drought, salinity, extreme temperatures, and heavy metal contamination together cause a large proportion of crop yield losses worldwide and are likely to intensify under climate change [2]. However, plants do not face these stresses alone. Evidence accumulated since the mid-2010s suggests that rhizosphere microbes, endophytes, and phyllosphere communities strongly influence the responses of plants to stress by modulating nutrient acquisition, hormone signaling, redox balance, and ion transport [3]. Microbial functions such as phosphate solubilization, nitrogen fixation, exopolysaccharide (EPS) production, siderophore release, and metal immobilization or volatilization are frequently linked to increases in the tolerance of plants to stress [3].

Recent studies have increasingly emphasized the importance of microbial contributions; however, this perspective is not fully supported, especially when moving from controlled environments to field conditions [4]. As climate models now integrate biosphere–atmosphere feedbacks, modern plant stress biology must incorporate the microbiome as a coregulator of plant adaptation [5]. Despite this recognition, our conceptual and experimental frameworks do not fully capture the ecological reality in which plants experience multiple, interacting stresses in microbially complex environments.

The overall picture, however, hides important mechanistic and institutional blind spots. Much of what we know about microbe-mediated stress alleviation has been derived from simplified laboratory systems using single microbial strains and single stress factors [3]. These approaches have been invaluable in identifying the beneficial traits of microbes, but they inaccurately represent real agricultural soils, where plants simultaneously encounter fluctuating water availability, salinity, nutrient limitation, and toxic metal conditions while interacting with thousands of microbial taxa [6]. As such, the performance of microbial inoculants identified through these approaches has been inconsistent in the field, underscoring the need to transition from strain-based to community-based and system-level perspectives.

Plants are not passive in this process. Under stress, plants can actively reshape their associated microbiomes through changes in root exudation and immune signaling. Stress-induced metabolites and signaling compounds may selectively enrich the abundance of microbes capable of buffering osmotic stress, detoxifying harmful elements, or enhancing antioxidant defenses [7]. However, the mechanisms underlying these recruitment processes remain only partly understood. In particular, the direct links between microbial genes, metabolites, and plant stress response pathways have been difficult to establish.

Stress is also highly heterogeneous in space, although this aspect has often been overlooked in research. Soil structure, moisture, redox conditions, and nutrient gradients create microhabitats that strongly influence microbial activity [8]. In saline or heavy-metal-contaminated soils, microbial EPS and biofilm formation can modify soil aggregation, ion mobility, and metal bioavailability, thereby altering the stress environment experienced by plant roots [9].

Climate change further complicates these interactions. Rising temperatures, shifting precipitation patterns, and more frequent extreme climate events affect plant physiology as well as microbial community composition and stability [10]. Under drought, for example, shifts in microbial carbon metabolism and phytohormone production may influence whether plants successfully acclimate [11]. Likewise, warming and soil acidification can alter metal speciation and microbial detoxification processes, with consequences for both crop performance and phytoremediation outcomes [12].

In this context, the MDPI Topic **"Microbe-Induced Abiotic Stress Alleviation in Plants"** provides a platform for integrating insights from molecular biology, microbial ecology, plant physiology, and soil science. The contributions provide diverse lines of evidence, although the research does not yet converge on a single, coherent framework.

2. Evidence Emerging from the Topic "Microbe-Induced Abiotic Stress Alleviation in Plants"

The contributions to this Topic highlight the multidimensional nature of microbial stress mitigation. Rather than pointing to a single dominant mechanism, the studies collectively suggest that multiple microbial functions operate across different stress contexts. Traits such as EPS production, phytohormone modulation (e.g., indole-3-acetic acid), metal chelation, and volatile organic compound (VOC) emission have been implicated in alleviating osmotic imbalance, oxidative damage, and ion toxicity in plants [13–15]. However, their effectiveness appears to be variable and often depends on environmental conditions and host–microbe compatibility. This variability is particularly evident in the case of microbial VOCs. Unlike root-associated interactions, VOC-mediated effects can occur without direct physical contact, suggesting a more diffuse mode of influence on plant physiology [14]. The reproducibility of these VOC-mediated effects remains uncertain under field conditions. More broadly, a synthesis study indicates that microorganisms can enhance plant tolerance to drought and salinity by influencing stress signaling pathways, antioxidant systems, and metabolic processes [16]. In cadmium (Cd)-contaminated systems, inoculation with EPS-producing, Cd-tolerant bacterium enhances plant antioxidant activity and growth while reducing Cd bioavailability through adsorption, chelation, and mineral precipitation. These effects often coincide with shifts in soil aggregation and rhizosphere communities, potentially limiting Cd uptake and increasing phytoremediation efficiency [17].

Evidence from this Topic suggests that microbial consortia can, in some cases, outperform single inoculants, particularly when complementary functional traits are involved. Interactions between plant growth–promoting rhizobacteria and arbuscular mycorrhizal fungi, for example, may enable the coordinated regulation of nutrient acquisition, stress signaling, and antioxidant responses [18]. However, these synergistic effects are not universal and often depend on environmental conditions, host specificity, and the stability of microbial interactions over time.

https://doi.org/10.3390/plants15101427

This context dependency becomes more apparent when considering broader biological interactions. Studies on plant viral coinfections show that outcomes can shift between antagonism and synergism depending on host tissue and physiological state, effectively and highly dynamically reshaping metabolic and defense responses [19]. Likewise, emerging metagenomic evidence indicates that aboveground endophytic fungi indirectly restructure rhizosphere microbial communities by altering functional gene abundance linked to carbon fixation and nitrogen cycling [20]. Such findings reinforce a conceptual shift from inoculating individual "beneficial strains" toward designing stress responsive microbial consortia capable of sustained functional stability across heterogeneous environments.

A further set of contributions links microbial activity to soil structure, aggregation, and biogeochemical cycling, demonstrating that microorganisms modify not only plant physiology but also the physical and chemical milieu in which roots function. By regulating soil aggregation, pore architecture, and mineral–organic interfaces, microbes modulate water retention, ion transport, and metal bioavailability. These soil-mediated pathways are particularly critical under salinity and heavy metal stress, where the spatial distribution and chemical speciation of ions ultimately determine toxicity and plant exposure. For instance, inoculation with *Enterobacter* sp. significantly increases plant biomass, Cd accumulation, and translocation in *Hybrid Pennisetum*, while reconfiguring the rhizosphere microbial community structure, increasing soil nutrient availability, and reshaping metabolic profiles, ultimately enhancing phytoremediation efficiency [21]. Metagenomic analyses indicate that the presence of symbiotic *Epichloë* endophytic fungi alters the functional potential of the rhizosphere microbiome by increasing the abundance of microbial genes associated with the reductive tricarboxylic acid carbon fixation pathway and nitrification, while promoting the accumulation of genes conferring resistance to heavy metals in the rhizosphere [20]. Complementary studies on plant–pathogen interactions reveal that biotic stress profoundly disrupts the photosynthetic machinery and cellular redox balance, as multiple viral in fections suppress the expression of photosynthesis-related gene while simultaneously inducing antioxidant defenses, thereby reshaping plant metabolic homeostasis under stress [22]. These findings illustrate how microbial processes operating from molecular and physiological responses to microbial community dynamics and soil structural collectively shape plant resilience under environmental stress.

Rather than converging on a single narrative, these studies point to a more context-dependent picture: the effects of microbes on plant stress tolerance are shaped by soil properties, climate variability, and plant genotype. Interactions within microbial communities can either enhance or constrain these effects. Consistent improvements in plant performance will remain difficult to achieve unless these processes are collectively considered.

3. Future Perspectives: Predictive and Climate-Smart Plant–Microbiome Engineering

What, then, should define the next generation of research in microbe-induced stress alleviation?

The evidence assembled in this Topic points toward several priorities: (1) developing predictive models that integrate plant physiology, microbial community dynamics, and soil–climate interactions; (2) identifying keystone taxa and functional guilds that govern resilience under multiple-stress conditions; (3) linking microbial metabolites and signaling molecules to plant stress response pathways; (4) scaling up from controlled experiments to long-term, multisite field trials; (5) embedding microbial strategies within One Health and climate-smart agriculture frameworks.

Ultimately, the future of sustainable crop production under climate change will depend on our ability to harness the adaptive capacity of plant-associated microbiomes. Microbes

https://doi.org/10.3390/plants15101427

are not simply inputs to be applied but partners to be managed, designed, and integrated into agroecosystems. By advancing a systems-level understanding of plant–microbe–soil interactions, this Topic seeks to contribute to a new paradigm in which microbial processes become central to environmental resilience, food security, and ecological sustainability.

Funding: This work was supported by Fundação para a Ciência e Tecnologia (FCT), I.P., under Scientific Employment Stimulus—Individual Call (2024.10481.CEECIND), by the Project UID/04004/2025—Centre for Functional Ecology—Science for the People & the Planet (DOI: 10.54499/UID/04004/2025), and by Associate Laboratory TERRA (LA/P/0092/2020; DOI: 10.54499/LA/P/0092/2020).

Conflicts of Interest: The author declares no conflicts of interest.

References

1. Ma, Y. Abiotic Stress Responses and Microbe-Mediated Mitigation in Plants. *Agronomy* **2023**, *13*, 1844.
2. Khan, M.S.U.; Afridi, N.; Ritu, S.A.; Shipar, S.I.; Zaman, S.B.; Hasan, N.T.; Uddin, S.; Hasan, M.; Rahimi, M.; Sumi, M.J.; et al. Enhancing Root Resilience through Sustainable Agriculture to Mitigate Heavy Metal Pollution and Abiotic Stresses in a Changing Climate. *Rhizosphere* **2026**, *37*, 101251. [CrossRef]
3. Cao, M.Y.; Narayanan, M.; Shi, X.J.; Chen, X.; Li, Z.L.; Ma, Y. Optimistic Contributions of Plant Growth-Promoting Bacteria for Sustainable Agriculture and Climate Stress Alleviation. *Environ. Res.* **2023**, *217*, 114924. [CrossRef] [PubMed]
4. Iqbal, B.; Li, G.; Alabbosh, K.F.; Hussain, H.; Khan, I.; Tariq, M.; Javed, Q.; Naeem, M.; Ahmad, N. Advancing Environmental Sustainability through Microbial Reprogramming in Growth Improvement, Stress Alleviation, and Phytoremediation. *Plant Stress* **2023**, *10*, 100283. [CrossRef]
5. Ma, Y.; Li, X.; Luo, Y.M. Microbiome-driven Innovations for Climate-Resilient Crop Production. *Nat. Food* **2026**, *7*, 304–308. [CrossRef]
6. Ma, Y.; Dias, M.C.; Freitas, H. Drought and Salinity Stress Responses and Microbe-Induced Tolerance in Plants. *Front. Plant Sci.* **2020**, *11*, 591911. [CrossRef]
7. dos Santos, M.S.; Rodrigues, S.P.; Hemerly, A.S.; Fernandes, A.A.R.; Fernandes, P.M.B. Harnessing Plant–Microorganism Interactions to Mitigate Biotic and Abiotic Stresses for Sustainable Crops. *Plants* **2026**, *15*, 647. [CrossRef]
8. Zhang, N.; Zheng, G.; Han, X.; Chen, X.; Lu, X.; Yan, J.; Zou, W. Microbial Resource Limitation in Soil Aggregates Effects Soil Organic Carbon Storage under Straw Incorporation. *Soil Till. Res.* **2026**, *256*, 106874. [CrossRef]
9. Li, Y.; Narayanan, M.; Shi, X.; Chen, X.; Li, Z.; Ma, Y. Biofilms Formation in Plant Growth-Promoting Bacteria for Alleviating Agro-Environmental Stress. *Sci. Total Environ.* **2024**, *907*, 167774. [CrossRef]
10. Afkhami, M.E.; Classen, A.T.; Dice, C.G.; Hernandez, D.J.; Li, V.W.; Rawstern, A.H.; Rudgers, J.A.; Stinchcombe, J.R.; Crawford, K.M. Unraveling Complexity in Climate Change Effects on Beneficial Plant–Microbe Interactions: Mechanisms, Resilience, and Future Directions. *New Phytol.* **2026**, *249*, 93–113. [CrossRef]
11. Sayer, E.J.; Crawford, J.A.; Edgerley, J.; Askew, A.P.; Hahn, C.Z.; Whitlock, R.; Dodd, I.C. Adaptation to Chronic Drought Modifies Soil Microbial Community Responses to Phytohormones. *Commun. Biol.* **2021**, *4*, 516. [CrossRef]
12. Drabesch, S.; Lechtenfeld, O.J.; Bibaj, E.; Ninin, J.M.L.; Pachecco, J.L.; Fendorf, S.; Planer-Friedrich, B.; Kappler, A.; Muehe, E.M. Climate Induced Microbiome Alterations Increase Cadmium Bioavailability in Agricultural Soils with pH Below 7. *Commun. Earth Environ.* **2024**, *5*, 637. [CrossRef]
13. Zhang, H.; Wang, K.; Liu, X.; Yao, L.; Chen, Z.; Han, H. Exopolysaccharide-Producing Bacteria Regulate Soil Aggregates and Bacterial Communities to Inhibit the Uptake of Cadmium and Lead by Lettuce. *Microorganisms* **2024**, *12*, 2112. [CrossRef] [PubMed]
14. Salazar-Ramírez, M.T.; Palacio-Rodríguez, R.; Quezada-Rivera, J.J.; Velásquez-Chávez, T.E.; Muro-Pérez, G.; Ortega-Reyes, H.I.; Orozco-Vidal, J.A.; Yescas-Coronado, A.G.; Verástegui-Hernández, G.A.; Sáenz-Mata, J. Volatile Organic Compounds from Candelilla-Associated PGPR Enhance *Arabidopsis thaliana* Seedling Growth Under Salinity Stress. *Soil Syst.* **2025**, *9*, 135. [CrossRef]
15. Huang, Z.-Y.; Li, Y.; Xiong, H.-A.; Naz, M.; Yan, M.-T.; Zhang, R.-K.; Liu, J.-Z.; Ren, X.-T.; Ren, G.-Q.; Dai, Z.-C.; et al. Rhizobacteria's Effects on the Growth and Competitiveness of *Solidago canadensis* Under Nutrient Limitation. *Agriculture* **2025**, *15*, 1646. [CrossRef]
16. Feng, D.; Li, W.; Huang, P.; Gu, M.; Tang, G.; Ding, Y.; Cao, G.; Xu, W. Mechanisms of Microorganisms Alleviating Drought and Salt Stresses in Plants. *Microorganisms* **2025**, *13*, 2565. [CrossRef]
17. Li, J.; Xu, X.; Song, L.; Na, M.; Xu, S.; Zhang, J.; Huang, Y.; Li, X.; Zheng, X.; Zhou, J. Investigating the Mechanism of Cadmium-Tolerant Bacterium *Cellulosimicrobium* and Ryegrass Combined Remediation of Cadmium-Contaminated Soil. *Plants* **2024**, *13*, 1657. [CrossRef]

18. Savastano, N.; Bais, H. Synergism or Antagonism: Do Arbuscular Mycorrhizal Fungi and Plant Growth-Promoting Rhizobacteria Work Together to Benefit Plants? *Int. J. Plant Biol.* **2024**, *15*, 944–958. [CrossRef]
19. Niraula, P.M.; Baldrich, P.; Cheema, J.A.; Cheema, H.A.; Gaiter, D.S.; Meyers, B.C.; Fondong, V.N. Antagonism and Synergism Characterize the Interactions between Four North American Potato Virus Y Strains. *Int. J. Plant Biol.* **2024**, *15*, 412–428. [CrossRef]
20. Wang, C.; Shi, C.; Huang, W.; Zhang, M.; He, J. The Impact of Aboveground *Epichloë* Endophytic Fungi on the Rhizosphere Microbial Functions of the Host *Melica transsilvanica*. *Microorganisms* **2024**, *12*, 956. [CrossRef]
21. Chen, Z.-J.; Li, M.-L.; Gao, S.-S.; Sun, Y.-B.; Han, H.; Li, B.-L.; Li, Y.-Y. Plant Growth-Promoting Bacteria Influence Microbial Community Composition and Metabolic Function to Enhance the Efficiency of *Hybrid pennisetum* Remediation in Cadmium-Contaminated Soil. *Microorganisms* **2024**, *12*, 870. [CrossRef]
22. Neliana, I.R.; Soleha, W.; Suherman; Darsono, N.; Harmoko, R.; Sawitri, W.D.; Sugiharto, B. Alteration of Photosynthetic and Antioxidant Gene Expression in Sugarcane Infected by Multiple Mosaic Viruses. *Int. J. Plant Biol.* **2024**, *15*, 757–768. [CrossRef]

https://doi.org/10.3390/plants15101427

microorganisms

MDPI

Article

Exopolysaccharide-Producing Bacteria Regulate Soil Aggregates and Bacterial Communities to Inhibit the Uptake of Cadmium and Lead by Lettuce

Heyun Zhang [1], Ke Wang [1], Xinru Liu [1], Lunguang Yao [2], Zhaojin Chen [1,*] and Hui Han [1,2,*]

[1] Collaborative Innovation Center of Water Security for Water Source Region of Mid-route Project of South-North Water Diversion of Henan Province, Nanyang Normal University, Nanyang 473061, China; zhangheyun66@163.com (H.Z.); m13080182307@163.com (K.W.); 17651962022@163.com (X.L.)

[2] Henan Field Observation and Research Station of Headwork Wetland Ecosystem of the Central Route of South-to-North Water Diversion Project, Nanyang Normal University, Nanyang 473061, China; lunguangyao@nynu.edu.cn

* Correspondence: zhaojin_chen@163.com (Z.C.); hanhui2018@nuynu.edu.cn (H.H.)

Abstract: The accumulation of heavy metals in the soil not only causes serious damage to the soil ecosystem, but also threatens human health through the food chain. Exopolysaccharides have the functions of adsorbing and chelating heavy metals and reducing their bioavailability in the soil. In our study, exopolysaccharide-producing bacteria with a high efficiency in adsorbing cadmium (Cd) and lead (Pb) were screened from heavy metal-contaminated farmland. Through pot experiments, the influence of functional strains on the size distribution, heavy metal content, and bacterial community structure of soil aggregates in lettuce was studied using high-throughput sequencing technology. The results show that 11 strains secreting exopolysaccharides were initially screened from heavy metal-contaminated soil. Among them, strain Z23 had a removal rate of 88.6% for Cd and 93.2% for Pb. The rate at which Cd was removed by strain Z39 was 92.3%, and the rate at which Pb was removed was 94.4%. Both strains belong to *Bacillus* sp. Strains Z23 and Z39 induced the formation of $Fe_2Pb(PO_4)_2$, $Cd_2(PO_4)_2$, and Pb_2O_3 in the solution. The pot experiments showed that strains Z23 and Z39 increased (19.1~23.9%) the dry weight and antioxidant enzyme activity of lettuce roots and leaves, while reducing (40.1~61.7%) the content of Cd and Pb. Strains Z23 and Z39 increased the proportion of microaggregates (<0.25 mm) and the content of exopolysaccharides in rhizosphere soil and reduced (38.4–59.7%) the contents of available Cd and Pb in microaggregates, thus inhibiting the absorption of heavy metals by lettuce. In addition, the exopolysaccharide content and the bacterial community associated with heavy metal resistance and nitrogen (N) cycling (*Patescibacteria*, *Saccharimonadales*, *Microvirga*, and *Pseudomonas*) in microaggregates were key factors affecting the available heavy metal content in soil. These results show that the exopolysaccharide-producing bacteria Z23 and Z39 reduced the absorption of Cd and Pb by lettuce tissues, thus providing strain resources for the safe utilization of soils that exceed heavy metal standards for farmland and for reducing the heavy metal content in vegetables.

Keywords: immobilization; exopolysaccharides; microaggregates; heavy metal; antioxidant enzyme; nitrogen cycling

1. Introduction

Due to industrial production activities, such as mining and gold plating or the improper disposal of agricultural waste, heavy metal pollution has occurred in some farmland [1]. Cadmium (Cd) and lead (Pb) are among the most toxic environmental pollutants, are easily enriched in agricultural soils, have strong migration capabilities, and cannot be degraded by microorganisms [2,3]. They are absorbed by crops and ultimately enter the human body through the food chain and pose potential risks to human health [4]. For example, chronic Cd poisoning causes kidney damage, mainly involving renal tubule lesions. In severe cases, chronic renal failure, osteomalacia, and osteoporosis may occur [5]. Currently, there are several methods to prevent crops from absorbing heavy metals from the soil, such as screening crop varieties, in situ passivation remediation, agronomic measures (deep dive and water and fertilizer management), and foliar fertilizer spraying [6]. Among them, in situ passivation remediation is a feasible and effective method. The most important aspect of in situ passivation remediation technology is the screening of passivators, which involves compost, straw, lime, clay minerals, and functional microorganisms [7,8]. Compared to chemical passivators, such as lime and clay minerals, microorganism passivators have inevitably lived in heavy metal-contaminated soil for a long time, thus demonstrating strong resistance to heavy metals and exhibiting corresponding detoxification mechanisms [9]. At the same time, they also play a certain role in controlling the interaction with crop roots [10]. These microorganisms immobilize heavy metals in the soil, reduce their toxicity, and secrete substances such as indole acetic acid (IAA), siderophores, polyamines, and arginine to promote plant growth, improve crop quality, and improve soil quality [11,12].

Exopolysaccharides are metabolic products secreted by microorganisms during growth and metabolism and are easily separated from the bacterial cell wall [13]. They are a type of macromolecular substance that has good adsorption and passivation effects on heavy metals. Exopolysaccharides adsorb organic nutrients in the external environment, provide energy for cell metabolism, and adsorb inorganic ions, including heavy metal ions, effectively promoting the formation of exopolysaccharide gel-like substances and minerals [14]. The mechanisms by which microorganisms use exopolysaccharides to adsorb heavy metals mainly involve surface complexation and ion exchange [15]. The exopolysaccharides produced by biofilms can also be used to remove aluminum (Al), Pb, zinc (Zn), and nickel (Ni) [16]. Copper (Cu) binds to hydroxyl and phosphoryl groups and Al binds to amino and carbonyl groups in exopolysaccharides [17]. Pb^{2+} can be removed because it chelates with carboxyl, hydroxyl, and C-O-C groups in exopolysaccharides [18].

In nature, there are many kinds of microorganisms that can synthesize exopolysaccharides, such as bacteria, fungi, and actinomycetes [19]. According to statistics, 76 species of microorganisms belonging to 49 genera have been found to produce exopolysaccharides, among which the main exopolysaccharide-producing bacteria are *Bacillus*, *Pseudomonas*, *Streptococcus*, and *Lactobacillus* [20,21]. Due to the important role of exopolysaccharides in the field of heavy metal pollution bioremediation, it has become a research focus at home and abroad. *Halomonas* sp. produces exopolysaccharides under arsenic (As) stress, which oxidizes toxic As^{3+} to As^{6+} of lower toxicity, thereby detoxifying or bioremediating As in the soil and promoting the growth of rice under As stress [22]. The mechanisms by which exopolysaccharide-producing bacteria adsorb heavy metals mainly include complexation, ion exchange, biological precipitation, and oxidation–reduction [23,24]. Although some scholars have reported the use of exopolysaccharide-producing bacteria to immobilize heavy metals, the mechanisms used by these bacteria in immobilizing Cd and Pb are not well understood. In addition, there are few studies on the effects of exopolysaccharide-producing bacteria isolated from heavy metal-contaminated soil on the absorption of Cd and Pb by lettuce [25].

It is assumed that exopolysaccharide-producing bacteria can immobilize Cd and Pb and inhibit the absorption of heavy metals by vegetables. However, the soil and microbiological mechanisms used by exopolysaccharide-producing bacteria in inhibiting the absorption of Cd and Pb by lettuce are still unclear. Therefore, rhizosphere soils were collected from Cd- and Pb-contaminated fields and exopolysaccharide-producing strains were screened according to the exopolysaccharide production medium. The mechanisms of Cd and Pb adsorption used by functional strains were studied by means of scanning electron microscopy (SEM) and X-ray photoelectron spectroscopy (XPS). Moreover, the effect of functional strains on the growth and heavy metal absorption in lettuce and its soil and microbial mechanisms were studied through pot experiments. The research results are expected to provide strain resources and technical support for implementing heavy metal-contaminated farmland bioremediation and ensuring crop safety.

2. Materials and Methods

2.1. Screening of Exopolysaccharide-Producing Bacteria

Cd-contaminated soil was collected from vegetable roots in the suburbs of Jiyuan city, Henan Province, China (35°03′ N, 112°61′ E), and brought back to the laboratory. Then, the rhizosphere soil (2 g) was weighed and placed into a shaker containing 50 mL sterile deionized water, and shaken at 30 °C for 2 h. The culture solution (1 mL) was placed into a test tube containing 9 mL of sterile deionized water, and a pipette was used to blow the evenly mixed solution into 10^{-1} diluent. A pipette was used to draw 1 mL of 10^{-1} diluent, which was then poured into another test tube containing 9 mL of sterile deionized water to form 10^{-2} diluent, and finally to form 10^{-3}, 10^{-4}, and 10^{-5} soil diluents. Then, 0.5 mL diluents of different gradients were absorbed and evenly coated on solid nitrogen medium (sucrose 10.0 g, K_2HPO_4 2.0 g, $MgSO_4$ $7H_2O$ 0.5 g, yeast powder 0.5 g, NaCl 0.1 g, agar powder 20.0 g, deionized water 1000 mL, pH 7.0, Shanghai McLean Biochemical Technology Co., Ltd., Shanghai, China) on a plate and incubated at 30 °C for 3 days in a biochemical incubator (Shengwei-SPXD300, Shanghai, China). A sterilized toothpick was used to pick out the colonies with a large amount of mucus on the surface, and a single colony was obtained through four zoning lines and labeled.

2.2. Determination of the Ability of Exopolysaccharide-Producing Bacteria to Adsorb Heavy Metals

A shake flask containing 50 mL of nitrogen-containing liquid medium was filled with Cd and Pb mother liquor, resulting in a Cd concentration of 5 mg L^{-1} and a Pb concentration of 10 mg L^{-1} in the flask. The bacterial suspension of the selected strains was inoculated at a ratio of 1% and cultured for 3 days at 28 °C with shaking. Each strain experiment was repeated three times [26]. After the culture was completed, 5 mL of culture solution was taken, centrifuged at 7000 r min^{-1} for 10 min in a constant temperature shaker (PY180S, Huicheng, China), and the supernatant was collected for digestion on an electric hot plate. The concentrations of Cd and Pb in the supernatant were measured using an inductively coupled plasma optical emission spectrometer (ICP-OES, Shimadau-ICPE 9820, Shanghai, China), and the removal rate of Cd and Pb by the strain was calculated. The content of polysaccharides in the culture solution was determined using sulfuric acid–anthrone colorimetry. Briefly, 1 mL of the aforementioned supernatant was added to 3 mL of absolute ethanol, and the liquid in the centrifuge tube was tapped with a pipette until white flocculent precipitate appeared. The centrifuge tube was covered and placed in a 4 °C refrigerator for 24 h, and then centrifuged at 7000 r min^{-1} for 10 min. The supernatant was decanted, dried in a 37 °C oven, and added to 1 mL of deionized water and 0.5 mL of Sevage reagent (chloroform/n-butanol = 4:1). The mixture was inverted and mixed evenly, and then left to stand for 1 h at 37 °C, followed by centrifugation at 7000 r min^{-1} for 10 min. The supernatant was decanted and added to 5 mL of sulfuric acid anthrone, heated in a 100 °C water bath for 10 min, and then the absorbance at 620 nm was measured using a spectrophotometer (UV762PC, Nanjing, Xuxi, China). The polysaccharide content was calculated using a standard curve [27]. The strains with the highest ability to remove Cd and Pb and secrete polysaccharides were the target strains for further study.

2.3. Determination of Biological Characteristics of Functional Strains

The DNA of the strains was extracted and genus identification was performed using 16S rDNA sequencing [28]. Solid nitrogen-containing culture plates containing different concentrations of Cd (100–800 mg L^{-1}) and Pb (120–2000 mg L^{-1}) were prepared to determine the resistance of the target strains to Cd and Pb. The indoleacetic acid (IAA) produced by the strains was quantified using the methods described by Jiang et al. [29]. The siderophores produced by the strains were quantified using the chrome azurol-s (CAS) analytical method [30]. The ability of the target strains to produce 1-aminocyclopropane-1-carboxylate

(ACC) deaminase was determined according to the method used by Belimov et al. [31]. Gram staining was used to identify the positive and negative strains.

2.4. Adsorption of Cd and Pb by Exopolysaccharide-Producing Bacteria

The mother solution of Cd and Pb was added to a shake flask containing 50 mL of nitrogen-containing liquid medium, resulting in a Cd concentration of 5 mg L^{-1} and a Pb concentration of 10 mg L^{-1} in the flask. The target bacterial suspension was inoculated at a ratio of 1% into the flask, and the flask was incubated at 28 °C for 7 days. There were three treatments: the solution containing Cd and Pb (CK), the solution containing both bacterial strains and heavy metals, and the solution containing only bacterial strains. Each treatment had three replicates. On days 0, 1, 3, 5, and 7, a portion of the culture broth was taken, centrifuged at 7000 r min^{-1} for 10 min, and the supernatant was collected. The concentrations of Cd and Pb in the solution were determined using ICP-OES, the pH of the supernatant was measured using a pH meter (PHS-3CT, Jingsheng, Shanghai, China), and the polysaccharide content in the supernatant was determined using sulfuric acid–anthrone colorimetry.

2.5. Mechanism of Exopolysaccharide-Producing Bacteria Adsorbing Cd and Pb

The cell pellets of the 7th day of the adsorption experiment were collected and then mixed with 15 mL of 2.5% (v/v) glutaraldehyde. They were left to stand at 4 °C for 4 h and then centrifuged at 5000 r min^{-1} for 10 min. The supernatant was removed. Following this, 50%, 75%, 90%, and 100% ethanol solutions were added in order, and the solution was left to stand for 10 min at room temperature. Then, the solution was centrifuged at 5000 r min^{-1} for 10 min, and the pellets were placed in a freeze dryer for 48 h [32]. The morphology of the samples was characterized using a field emission scanning electron microscope–energy spectrometer (SEM-EDS, Hitachi I-SU8010, Dongjing, Japan). The surface functional groups of the samples were characterized using a Fourier-transform infrared spectrometer (FIRT, America Thermo Fisher Scientific-iS50, Shanghai, China). The composition of the sediment on the surface of the cells was determined using an X-ray diffractometer (XRD, D-MAX2500, Japan Corporation Co., Ltd., Tokyo, Japan). The typical elemental valence state changes on the surface of the cells were analyzed using an X-ray photoelectron spectrometer (XPS, America Thermo Fisher Scientific K-Alpha, Shanghai, China).

2.6. Pot Experiment with Lettuce

Soil samples were collected (35°03′ N, 112°61′ E), naturally air-dried, and placed in pots, with each pot containing 3 kg of soil (Cd 2.45 mg kg^{-1}; Pb 179 mg kg^{-1}; pH 6.92; organic matter 27.9 g kg^{-1}, total N 1.74 g kg^{-1}, available P 0.48 g kg^{-1}, total P 1.41 g kg^{-1}, total K 1.32 g kg^{-1}, and available K 1.11 g kg^{-1}). Lettuce seeds were soaked in 75% ethanol for 3 min, washed with deionized water 3 times, evenly sprinkled in the pots, and covered with fine soil. After the lettuce sprouted, the seedlings were gradually thinned out, with 5 seedlings per pot. A total of 3 treatments were set up in the experiment: the control group (CK), the Z23 treatment group (Z23), and the Z39 treatment group (Z39), with 4 replicates for each treatment. Among all of the screened strains, Z23 and Z39 had the strongest ability to remove Cd and Pb, which were used for this part of the experiment. When the lettuce grew into 4 small leaves, the strains were inoculated. The strains Z23 and Z39 were activated and prepared into a bacterial suspension (10^8 CFU mL^{-1}). A small ditch was drawn around the lettuce, the bacterial solution was poured in at an amount of 25 mL (10^8 CFU mL^{-1}) per pot, and the control was replaced with sterile deionized water. After inoculating the strains, the cultivation was continued for 30 days, and then the harvesting process was carried out.

2.7. Effects of Exopolysaccharide-Producing Bacteria on Lettuce Growth and Cd Uptake

The roots and leaves of lettuce were separated using scissors, and the roots were soaked in a 0.01 mol L^{-1} EDTA-2Na solution for 10 min and rinsed with deionized water.

Fresh lettuce roots and leaves were taken, and the activities of superoxide dismutase (SOD) in the samples were determined using a SOD assay kit (A001-1-1, Nanjing Jiancheng Bioengineering Institute, Nanjing, China). The activities of peroxidase (POD) in the samples were determined using a POD assay kit (A084-3-1, Nanjing Jiancheng Bioengineering Institute, Nanjing, China). Three randomly selected lettuce leaves and roots were placed in a file bag and labeled, and then dried in an oven at 80 °C until constant weight and measured for dry weight. The dried lettuce leaves and roots were cut with scissors; weighed to 0.1 g; added to a crucible with 3 mL of HNO_3 (65%), 1 mL of HCl (36%), 2 mL of $HClO_4$ (70%), and 2 mL of HF (42%); and left for 18 h before digestion. Finally, the contents of Cd and Pb in the digestion solution were determined using ICP-OES [33], and the contents of Cd and Pb in the lettuce tissue were calculated.

2.8. Effects of Exopolysaccharide-Producing Bacteria on the Structure of Rhizosphere Soil Aggregates

Rhizosphere soil was collected from the rhizosphere of the lettuce, allowed to dry naturally without grinding, and then placed directly on a sieve (with a pore size of 0.25 mm) to collect soil aggregates of two particle sizes: large aggregates (>0.25 mm) and microaggregates (<0.25 mm) [34]. The content of DTPA-Cd and DTPA-Pb in the large aggregates and microaggregates was determined using ICP-OES [33]. The content of polysaccharides was determined using sulfuric acid–anthrone colorimetry. The total potassium (TK), total phosphorus (TP), ammoniacal nitrogen (NO_4^+), and nitric nitrogen (NO_3^-) contents in the rhizosphere soils of the lettuce were detected via the methods described by Han et al. [35].

2.9. Determination of Bacterial Community Structure in Soil Aggregates

The total DNA of rhizosphere soil large aggregates and microaggregates in the CK, Z23, and Z39 treatment groups was extracted. Then, PCR amplification of the V3–V4 variable region of the total DNA was performed [36]. A PE 2*300 library was constructed using the purified amplification fragments according to the standard operating procedures of the Illumina MiSeq platform (Illumina, San Diego, CA, USA). For each cluster file, the diversity (Sobs, Chao1, Simpson, and Shannon and PD indices) was calculated with MOTHUR. The analysis project was assisted by the Shanghai Meiji Company, Shanghai, China (https://www.majorbio.com/, accessed on 8 August 2023).

2.10. Data Analysis

Excel 2010, SPSS 17.0, Avantage 6.6, and Omnic 8.2 software were used for data analysis and statistics. The mathematically processed results are presented in the form of M ± SE, where M is the arithmetic mean and SE is the standard error (n = 4). All the data were subjected to homogeneity testing before Tukey's multiple comparison. One-way ANOVA and Tukey's test ($p < 0.05$) were used to compare the heavy metal removal rate, pH, and polysaccharide contents of the solution; the Cd and Pb contents in the lettuce plants; and the SOD and POD activities in the fresh lettuce. Two-factor ANOVA was used to compare dry weights of lettuce plants, the DTPA-extractable Cd and Pb contents in the rhizosphere soils of the lettuce, and the polysaccharide contents in the soils of the lettuce. Redundancy analysis (RDA) was performed using CANOCO 4.5 to examine the relationship between the soil quality and the DTPA-extractable Cd and Pb contents.

3. Results

3.1. Screening of Exopolysaccharide-Producing Bacteria with Heavy Metal Adsorption Capacity

The exopolysaccharide-producing bacteria were initially screened from heavy metal-contaminated rhizosphere soil using an exopolysaccharide-producing medium, and 11 stable exopolysaccharide-producing bacterial strains were finally selected. The removal rates of Cd for these 11 strains ranged from 38.6% to 92.3%, and the removal rates for Pb ranged from 49.2% to 94.4%. Among them, strain Z23 had a removal rate of 88.6% for Cd and 93.2% for Pb. Strain Z39 had a removal rate of 92.3% for Cd and 94.4% for Pb (Figure 1).

Subsequently, the exopolysaccharide-producing abilities of these 11 bacterial strains were determined, and it was found that the polysaccharide content ranged from 64.7 mg L^{-1} to 157 mg L^{-1}, with strain Z23 producing 134 mg L^{-1} of polysaccharide and strain Z39 producing 157 mg L^{-1} of polysaccharide (Figure 1). Comprehensive analysis showed that strains Z23 and Z39 had high adsorption and removal abilities for Cd and Pb as well as polysaccharide production and could be used as test strains for further research.

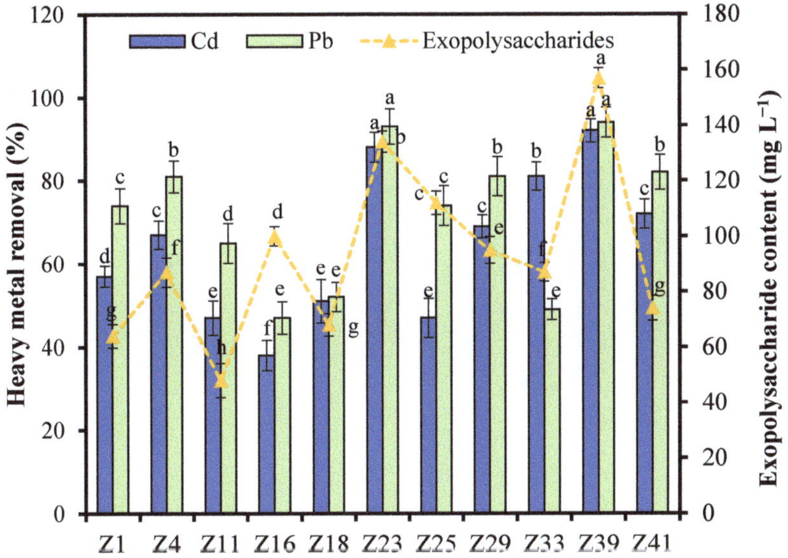

Figure 1. The ability of strains to remove Cd and Pb and produce exopolysaccharides. Error bars are ±standard error (n = 4). Different lowercase letters (a–h) for the same test indicator indicate significant ($p < 0.05$) differences between treatments according to Tukey's test.

3.2. Biological Characteristics of the Target Bacterial Strain

The total DNA of the tested strains Z23 and Z39 was used as a template for 16S rDNA amplification and sequencing. The 16S rDNA sequence of strain Z23 was 100% similar to that of *Bacillus paramobilis* BML-BC017, and it was preliminarily identified as *Bacillus* sp. (Figure S1). The 16S rDNA sequence of strain Z39 was 100% similar to that of *Bacillus altitudinis* 41KF2b, and it was preliminarily identified as *Bacillus* sp. (Figure S1). In general, strains Z23 and Z39 belong to the genus *Bacillus*. Strains Z23 and Z39 had the ability to secrete high concentrations of IAA, siderophores, and ACC polyaminase, indicating that both strains had the ability to promote crop growth (Table S1). The lethal concentration of Cd for strain Z23 was 400 mg L^{-1}, and the lethal concentration of Pb was 1800 mg L^{-1}. The lethal concentration of Cd for strain Z39 was 500 mg L^{-1}, and the lethal concentration of Pb was 2000 mg L^{-1} (Table S1). These results indicated that both strains had high resistance to heavy metals and survived well in contaminated environments, providing possibilities for the subsequent remediation of heavy metal-contaminated soil.

3.3. Adsorption of Cd and Pb by Functional Strains

The optical density in 600 nm (OD_{600}) of strains Z23 and Z39 in solutions with Cd (5 mg L^{-1}) and Pb (10 mg L^{-1}) gradually increased with the increase in the culture time (Figure 2a). When strains Z23 and Z39 were cultured for 7 days, the OD_{600} in the solution reached 1.65 and 1.87, respectively, indicating that the viable strains Z23 and Z39 were able to grow normally under the stress of Cd and Pb. The concentration of Cd in the CK solution remained between 4.79 mg L^{-1} and 4.86 mg L^{-1}, while that in the Z23 and Z39 solutions gradually decreased with an increase in the culture time. The Cd removal rate reached 76.2% and 85.6% on day 7 of culture, respectively (Figure 2b), indicating that strains Z23 and Z39 had strong adsorption capacity for Cd. As the culture days increased, the concentration of Pb in the CK solution remained between 7.25 mg L^{-1} and 7.35 mg L^{-1}. The concentration of Pb in the Z23 and Z39 solutions gradually decreased, with the Pb reaching 2.75 mg L^{-1} in the Z23 solution and 1.35 mg L^{-1} in the Z39 solution on day 7 of culture (Figure 2c). The pH in the CK solution remained unchanged (6.48), while both Z23 and Z39 significantly increased the pH of the solution (Figure 2d). On day 7 of culture, the pH in the Z23 solution was 8.23 and in the Z39 solution was 8.51, indicating that strains Z23 and Z39 increased the pH of the solution and thereby reduced the concentration of Cd and Pb.

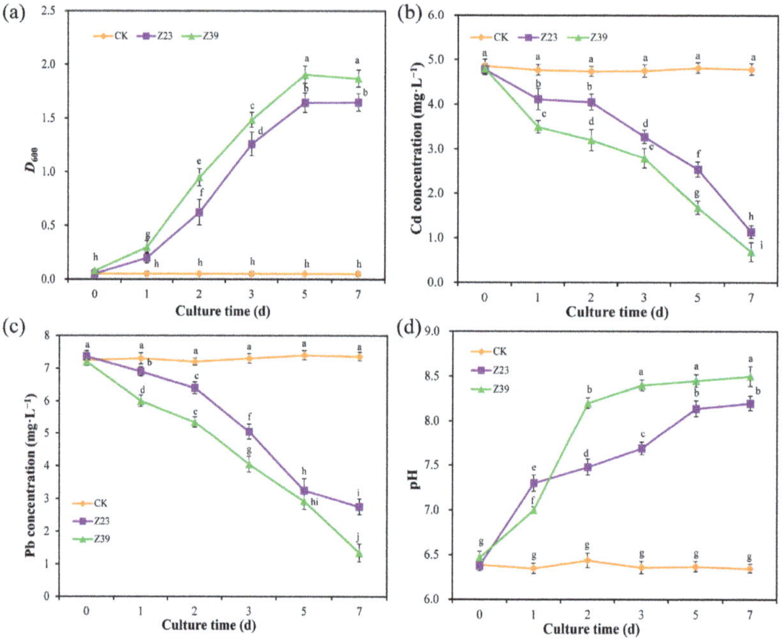

Figure 2. Effects of functional strains on the concentration of heavy metals in the culture solution. (**a**) OD_{600}; (**b**) Cd concentration (mg L^{-1}); (**c**) Pb concentration (mg L^{-1}); (**d**) pH. Error bars are ±standard error (n = 4). Bars indicated by the same letter were not significantly ($p > 0.05$) different according to Tukey's test.

3.4. Mechanism of Functional Strains Adsorbing Cd and Pb

The cell pellets of strains Z23 and Z39 were collected, and the SEM images showed that there were many precipitate particles on the cell walls of strains Z23 and Z39, while EDS analysis showed that these precipitate particles contained Cd and Pb (Figure 3), indicating that the cell walls of strains Z23 and Z39 immobilized Cd and Pb. A comparison with standard cards revealed that some compounds containing Cd and Pb were observed on the cell walls of the tested strains during the process of Z23 and Z39's immobilization of heavy metals, including iron(II) lead(II) phosphate ($Fe_2Pb(PO_4)_2$), cadmium phosphate ($Cd_3(PO_4)_2$), and lead sesquioxide (Pb_2O_3) (Figure 3). Moreover, the synthesized compounds varied among the different strains. In strain Z23, $Fe_2Pb(PO_4)_2$ and $Cd_3(PO_4)_2$ were detected, while in strain Z39, $Fe_2Pb(PO_4)_2$, $Cd_3(PO_4)_2$, and Pb_2O_3 were detected. Therefore, strains Z23 and Z39 induced the formation of precipitates of Cd and Pb, thereby reducing the content of Cd and Pb in the solution. When synthesizing Cd and Pb precipitates, the types of precipitates synthesized on the cell surface of strain Z39 were more diverse than those of strain Z23, which was consistent with the adsorption rate results of Cd and Pb for the strains. The adsorption and removal effect of Cd and Pb by strain Z39 was slightly better than that of strain Z23.

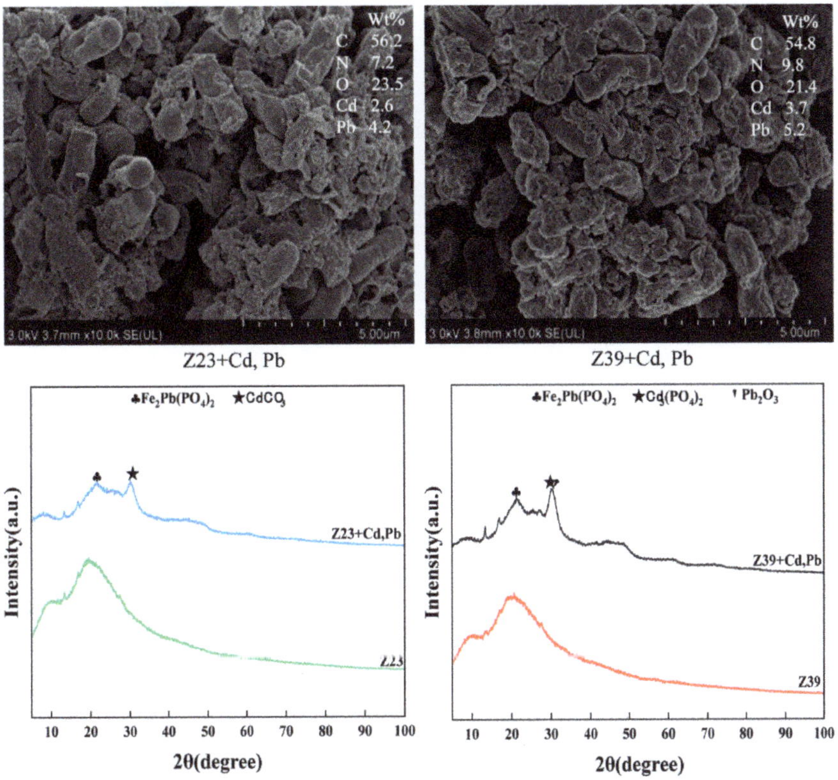

Figure 3. SEM-EDS and XRD images of functional strains adsorbing heavy metals.

3.5. XPS Analysis

The XPS spectrum analysis of the samples showed changes in the main elements carbon (C), oxygen (O), nitrogen (N), Cd, and Pb (Figure 4). It was found that the peak intensities of C and O elements were higher, indicating that EPS was mainly composed of C and O elements. C showed three types of peaks: carbon–carbon single bond/carbon–hydrogen bond (C-C/C-H) at 284.8 eV, carbon–oxygen single bond/carbon–nitrogen single bond (C-O/C-N) at 286.3–286.54 eV, and carbon–oxygen double bond (C=O) at 287.9–288 eV. Under the stress of Cd and Pb, the content of C-(C, H) on the surface of the bacteria decreased, while the content of C-(N, O) and C=O increased, indicating that C-(C, H) fracture occurred in the C element in EPS. C combined with O or N to form C-N or C=O and existed in proteins, indicating that the strain can produce protein-rich EPS after heavy metal stress, and the increase in O or N functional groups with a C structure improved the heavy metal removal ability of EPS [37]. The content of C-O-C decreased, while the content of C=O increased, and a new O-Cd peak appeared (533.6–533.8 eV), indicating that part of the C-O was converted to C=O. The increase in C=O was conducive to the adsorption of metals by EPS [38]. There were two types of N in the EPS of the strains without Cd and Pb: C-NH_2/C_2-NH at 399.8–400.1 eV, and R-NH^{3+} at 401.7–402.86 eV. Under Cd and Pb stress, the contents of C-NH_2/C_2-NH and R-NH^{3+} in the EPS decreased and R-NO^{2-} was produced, indicating that C-NH_2/C_2-NH was converted into R-NO^2 groups, which could bind to heavy metals [39]. In addition, new strong characteristic peaks of Cd, Cd3$d_{3/2}$,

$Cd3d_{5/2}$, $Pb4f_{5/2}$, and $Pb4f_{7/2}$ appeared on the cell surface, indicating that heavy metals had been successfully adsorbed on the cell surfaces of strains Z39 and Z23.

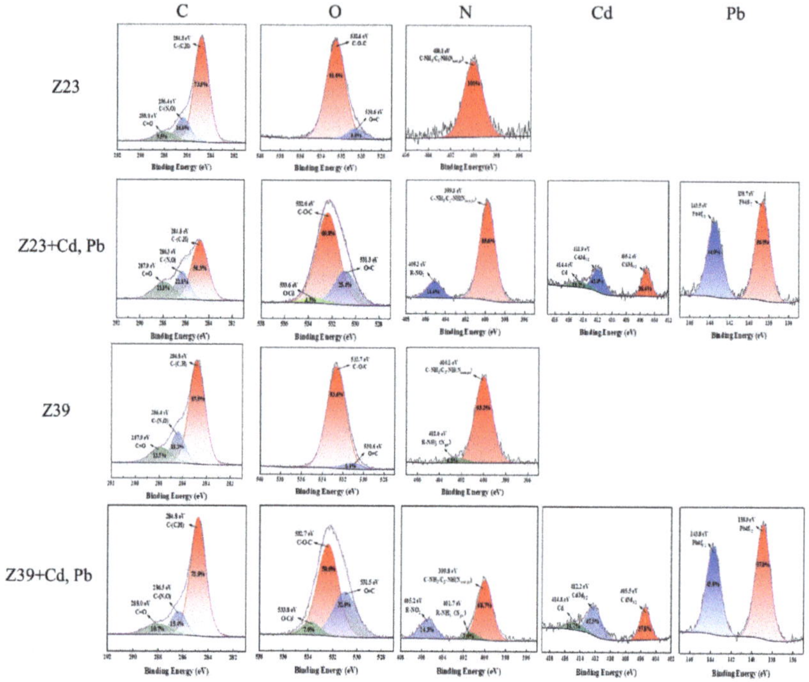

Figure 4. XPS images of functional strains adsorbing heavy metals.

3.6. Dry Weight, Heavy Metal Content, and Antioxidant Enzyme Activity in Lettuce

Compared with the control, inoculation with strains Z23 and Z39 significantly increased the dry weight of lettuce roots and leaves by 19.1% to 23.9% (Figure 5a). The contents of Cd and Pb in the lettuce roots in the CK treatment were 0.53 mg kg^{-1} and 0.75 mg kg^{-1}, respectively (Figure 5b). However, the inoculation of strain Z23 decreased the contents of Cd and Pb in the lettuce roots by 52.8% and 42.7%, respectively. The inoculation of strain Z39 decreased the contents of Cd and Pb in the lettuce roots by 49.1% and 40.1%, respectively (Figure 5b). Moreover, the contents of Cd and Pb in the lettuce leaves were 0.32 mg kg^{-1} and 0.57 mg kg^{-1} in the CK treatment. Strain Z23 decreased the contents of Cd and Pb in the lettuce leaves by 52.8% and 42.7%, respectively. Strain Z39 decreased the contents of Cd and Pb in the lettuce leaves by 52.8% and 42.7%, respectively (Figure 5b). The POD activity values in the lettuce roots and leaves were 42.6 U g^{-1} and 18.9 U g^{-1}, respectively, in the CK treatment; however, the inoculation of strains Z23 and Z39 decreased the POD activity in the lettuce roots and leaves by 17.1% to 34.9% (Figure 5c), indicating that these strains reduced the toxicity of Cd and Pb in lettuce, leading to a decrease in POD activity in lettuce tissues. Similarly, the SOD activity values in lettuce roots and leaves in the CK treatment were 85.6 U g^{-1} and 47.3 U g^{-1}, respectively. The inoculation of strains Z23 and Z39 also significantly decreased the SOD activity in lettuce roots by 16.5% to 29.1% (Figure 5d), indicating that the inoculation of strains Z23 and Z39 reduces the toxicity of Cd and Pb in lettuce, leading to a decrease in SOD activity in lettuce tissues.

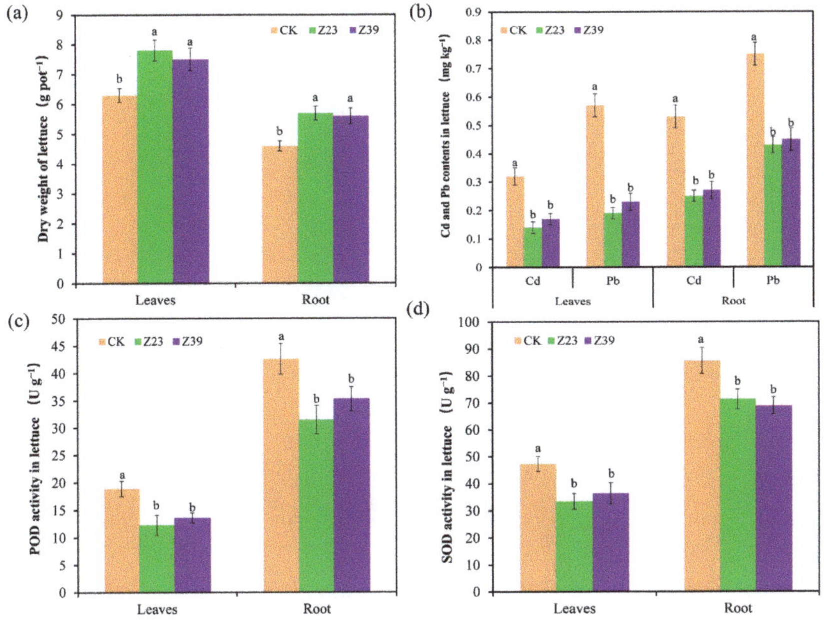

Figure 5. Effects of different treatments on the dry weight, Cd and Pb contents, and antioxidant enzymes of lettuce. (**a**) Dry weight of lettuce; (**b**) Cd and Pb contents in lettuce; (**c**) POD activity; (**d**) SOD activity. Error bars are ±standard error (n = 4). Bars indicated by the same letter were not significantly ($p > 0.05$) different according to Tukey's test.

3.7. Particle Size Distribution and Polysaccharide Content in Soil Aggregates

The polysaccharide content in soil can affect the distribution of aggregate size and the mobility of heavy metals [40]. In the CK control group, the proportion of large aggregates (>0.25 mm) in the rhizosphere soil was 64.2%, and the proportion of microaggregates (<0.25 mm) was 35.8% (Figure 6a). The proportion of large aggregates in the rhizosphere soil was 42.5%, and the proportion of microaggregates was 57.5% in the Z23 treatment. In the treatment group inoculated with Z39, the proportion of large aggregates in the rhizosphere soil was 47.6%, and the proportion of microaggregates was 52.4% (Figure 6a). This indicated that strains Z23 and Z39 increased the proportion of microaggregates in the lettuce rhizosphere soil, thereby enhancing the ability of the soil to adsorb and immobilize Cd and Pb [39]. Further research showed that both Z23 and Z39 significantly increased (20.6–48.8%) the polysaccharide content in the rhizosphere soil compared to the CK control, especially in microaggregates (Figure 6b), indicating that microaggregates play an important role in immobilizing heavy metals. The content of DTPA-Cd in large aggregates was 0.65 mg kg^{-1} and the content of DTPA-Pb was 1.23 mg kg^{-1} in the CK group (Figure 6c). However, the contents of DTPA-Cd and DTPA-Pb in large aggregates in the treatment groups of strains Z23 and Z39 significantly decreased (22.8–44.6%) (Figure 6c). Similarly, in the lettuce rhizosphere soil of the CK group, the content of DTPA-Cd in microaggregates was 0.62 mg kg^{-1} and the content of DTPA-Pb was 1.12 mg kg^{-1}. However, the content of DTPA-Cd and DTPA-Pb in microaggregates in the treatment groups of strains Z23 and Z39 also significantly decreased (38.4–59.7%) (Figure 6c). These results showed that inoculating the polysaccharide-producing bacteria Z23 and Z39 reduced the content of available heavy metals in the rhizosphere soil, thereby preventing the lettuce from absorbing heavy metals.

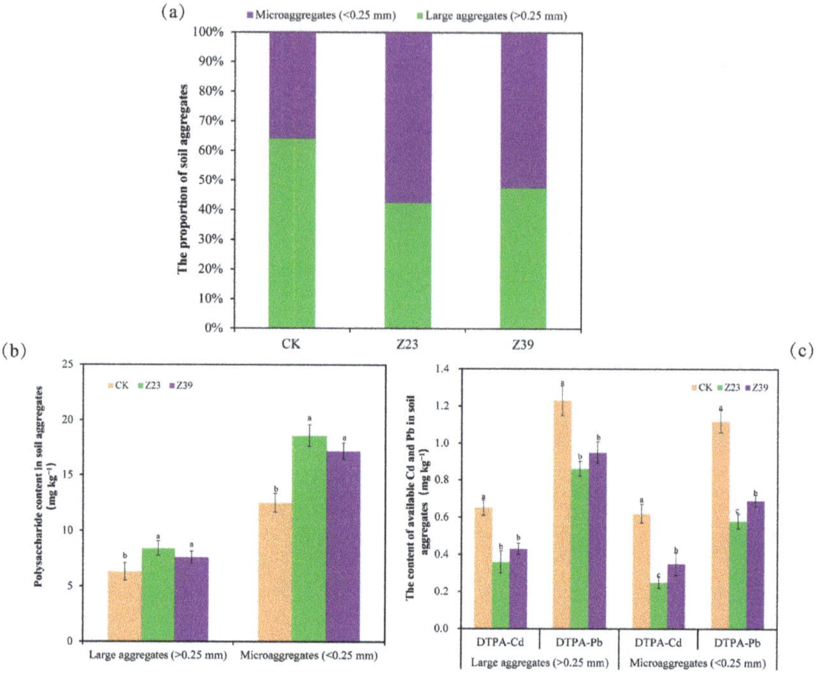

Figure 6. Effects of exopolysaccharide-producing bacteria on the particle size distribution and exopolysaccharide content of soil aggregates in the rhizosphere soil of lettuce. (**a**) Proportion of different particle size aggregates. (**b**) Exopolysaccharide content. (**c**) Contents of available heavy metals in lettuce rhizosphere soil. Error bars are ±standard error (n = 4). Bars indicated by the same letter were not significantly ($p > 0.05$) different according to Tukey's test.

3.8. Bacterial Community Structure and Diversity of Soil Aggregates

The changes in bacterial community structure were analyzed for large aggregate soil and microaggregate soil in the CK treatment group (CK-B and CK-S), the Z23 treatment group (Z23-B and Z23-S), and the Z39 treatment group (Z39-B and Z39-S). The PCoA results showed that the samples treated differently were dispersed, large aggregate samples were clustered together, and microaggregate samples were clustered together (Figure 7a), indicating that the bacterial community structure of the large aggregate samples was different from that of the microaggregate samples. The Alpha diversity results also indicated that inoculating strains Z23 and Z39 could alter the bacterial community diversity in rhizosphere soil aggregates, significantly improving the bacterial community diversity in microaggregates in particular (Table S2). The results of the β NTI further indicated that the changes in bacterial communities in the microaggregates in the Z23 and Z39 treatment groups were homogeneously dispersed (Figure 7b), meaning that the influence of the exogenous inoculation of functional strains on the bacterial community structure was significant and this influence was a dominant random process. For all treatments, Actinobacteria, Proteobacteria, Chloroflexi, Acidobacteriota, Firmicutes, Myxococcota, Bacteroidota, Gemmatimonadota, and Planctomycetota were the dominant phyla, with a total relative abundance of more than 98% (Figure 7c). Compared with the CK treatment, inoculation with Z23 and Z39 increased the relative abundances of Proteobacteria, Firmicutes and Bacteroidota, and decreased those of Actinobacteria and Chloroflexi in the microaggregates (Figure 7c). Moreover, strains Z23 and Z39 also increased the relative abundances of *Rhizobiales*, *Gaiellales*, *Bacillales*, and *Burkholderiales* in the microaggregates (Figure 7d).

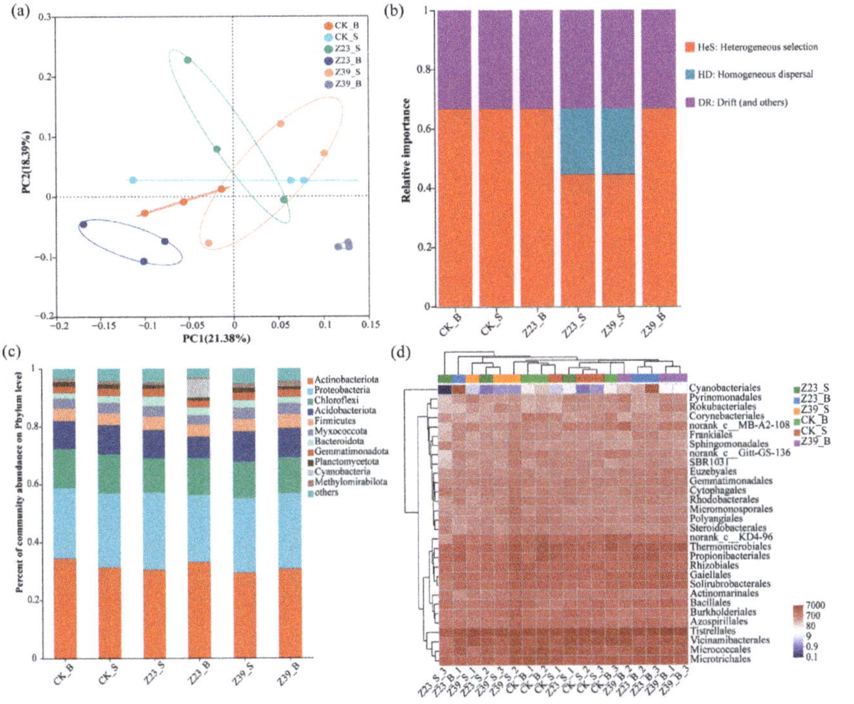

Figure 7. Effects of strains Z23 and Z39 on the bacterial community diversity of rhizospheric soil aggregates. (**a**) PCoA. (**b**) Analysis of βNTI community structure. (**c**) Relative abundances of sequences at the phylum level. (**d**) Relative abundances of sequences at the order level.

3.9. Functional Bacteria in Soil Aggregates

The exogenous inoculation of bacterial strains may have an impact on the indigenous functional bacterial community in the soil. The cladogram (Figure 8) showed that the key bacterial groups mainly included p__Bacteroidota, c__Acidobacteriae, g__Bryobacter, f__Sphingomonadaceae, o__Sphingomonadales, and o__Chitinophagales in the Z23-S treatment and c__Cyanobacteriia, o__Azospirillales, g__Rubellimicrobium, and g__Microcoleus__PCC-7113 in the Z23-B treatment. The cladogram also showed that the key bacterial groups mainly included f__Phycisphaeraceae, c__Saccharimonadia, g__Luteolibacter, o__Saccharimonadales, p__Patescibacteria, and o__Phycisphaerales in the Z39-S treatment and c__Thermoleophilia, o__Solirubrobacterales, f__Methyloligellaceae, and g__Rhodococcus in the Z39-B treatment (Figure 8). In addition, compared to the macroaggregates, the microaggregates contained a greater variety of key and functional bacteria, indicating that the exogenous inoculation of EPS-producing bacteria had a more significant impact on the bacterial community structure in the microaggregates and could mobilize more types of bacteria to participate in the detoxification of heavy metals. According to the literature, *Phycisphaeraceae* and *Patescibacteria* have the ability to immobilize heavy metals [41,42]. *Saccharimonadales* participates in glucose metabolism and N cycling [43]. *Chitinophagales* has the ability to dissolve P and promote plant growth [44]. *Bryobasterales* mediates iron (Fe) cycling and Cd fixation in rhizosphere soil [45,46]. These results indicated that inoculation with strains Z23 and Z39 altered the structure of the functional bacterial community and enriched strains with heavy metal-resistant, plant growth-promoting, and heavy metal-immobilizing properties in the rhizosphere soil, thereby inhibiting the absorption of Cd by lettuce.

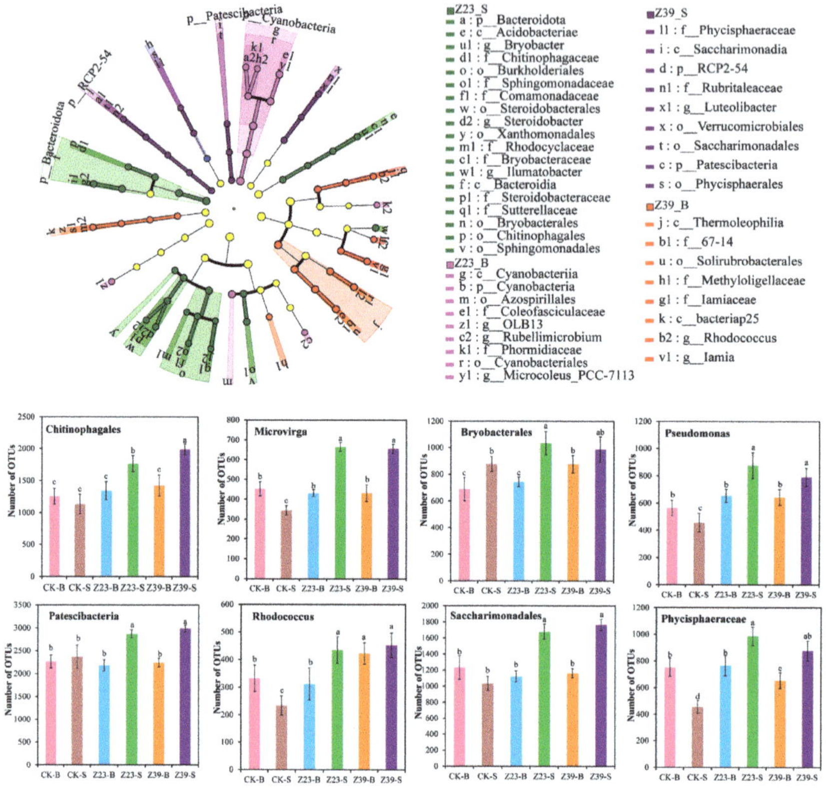

Figure 8. LEfSe analysis based on OTUs between the six treatments and the number of OTUs of typical strains. Bars indicated by the same letter were not significantly ($p > 0.05$) different according to Tukey's test.

3.10. Correlation Analysis

Spearman correlation analysis and RDA revealed the key factors affecting the available heavy metal content in the rhizosphere soil (Figure 9). The effects of strains Z23 and Z39 on the total potassium (TK), total phosphorus (TP), ammoniacal nitrogen (NO_4^+), nitric nitrogen (NO_3^-), and EPS contents in the rhizosphere soils of the lettuce are detailed in Table S3. The results showed that the DTPA-Cd and Pb contents were strongly negatively correlated with the pH, NO_4^+, NO_3^-, and EPS levels (r = −0.534 to −0.905, $p < 0.05$) in the rhizosphere soil (Table S4). Furthermore, the DTPA-Cd and Pb contents presented a strong negative correlation with the abundance of *Patescibacteria*, *Saccharimonadales*, *Microvirga*, and *Pseudomonas* (r = −0.754 and −0.846, $p < 0.05$) (Table S4). The results indicated that the EPS content and the bacterial community associated with heavy metal resistance and N cycling (*Patescibacteria*, *Saccharimonadales*, *Microvirga*, and *Pseudomonas*) in microaggregates were key factors affecting the available heavy metal content in soil.

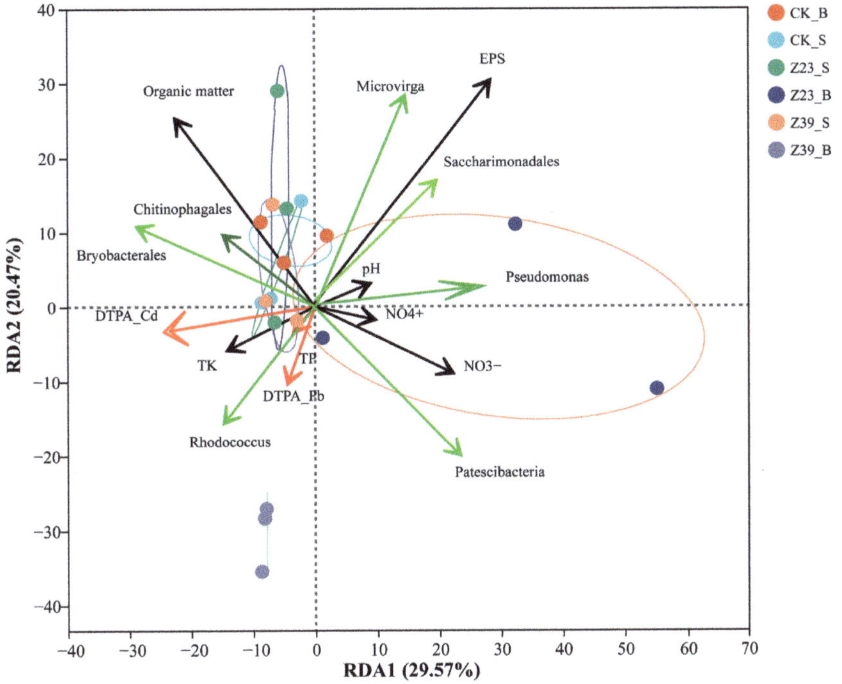

Figure 9. RDA of DTPA-extractable Cd and Pb contents and soil quality in the rhizosphere soils of lettuce.

4. Discussion

Heavy metal-immobilizing bacteria are types of microorganisms that have been extensively studied for their ability to fix heavy metals and inhibit their absorption by crops [47,48]. The mechanisms of heavy metal immobilization used by these bacteria mainly include surface adsorption, intracellular enrichment, extracellular adsorption and chelation, biological precipitation, and oxidation–reduction [49,50]. There are multiple pathways for bacteria to repair heavy metals in the soil, among which the pathway of fixing and precipitating heavy metals through exopolysaccharides has been shown to have great potential [51]. Exopolysaccharide-producing bacteria immobilize heavy metals in solution and soil by secreting exopolysaccharides, thereby reducing the mobility of heavy metals [52]. In our study, two exopolysaccharide-producing bacteria, *Bacillus* sp. Z23 and Z39, were isolated from heavy metal-contaminated soil and exhibited high efficiency in adsorbing Cd and Pb and secreting exopolysaccharides. In solutions with a Cd concentration of 5 mg L^{-1} and a Pb concentration of 10 mg L^{-1}, the adsorption and removal rates for strains Z23 and Z39 were above 90% for both, indicating that these two strains can efficiently immobilize Cd and Pb. Karthik et al. [53] isolated two exopolysaccharide-producing bacteria, AR6 and AR8, from the rhizosphere of kidney beans and demonstrated their good biosorption of chromium (Cr), reducing the accumulation of Cr in the roots and shoots. Cd, Cr, and Cu stimulated *Bacillus* sp. S3 to produce exopolysaccharides and enhanced its ability to adsorb and detoxify heavy metals [54]. Microbial adsorption occurs passively and in an independent metabolic manner, and both live and dead bacteria can absorb heavy metals through several physical and chemical mechanisms, such as ion exchange, complexation, precipitation, and chelation [55]. Even under low-metal-ion conditions, their ability to absorb ions is enhanced due to their excellent sensitivity, accumulation rate, and extent.

Negatively charged cell surfaces and exopolysaccharides bind to positively charged metal ions to fix heavy metals [56]. For example, the acetyl amino (-NH_2) group of chitin; the structural polysaccharide of fungi; the amine (-NH_3), sulfhydryl (-SH), and carboxylic (-COOH) groups in proteins; the phosphodiester (-COOP), phosphate (PO_4^{3-}), and hydroxyl (-OH) groups in polysaccharides; and other abundant active functional groups and non-carbohydrate substituents give the polymer an overall negative charge [57]. The XRD and XPS results indicated that strains Z23 and Z39 not only induced the mineralization and precipitation of Cd and Pb to form $Fe_2Pb(PO_4)_2$, $Cd_2(PO_4)_2$, and Pb_2O_3, but also played an important role in the adsorption and immobilization of Cd and Pb through the secretion of exopolysaccharides. Iyer et al. [58] studied the tolerance of the marine bacterium *Enterobacter cloaceae* (which produces extracellular polysaccharides) to Cr(VI), and found that the strain not only has tolerance to Cr(VI), but also strong Cr removal ability. Studies have shown that polysaccharide-producing bacteria reduce the bioavailability of heavy metals in soil, alleviate stress conditions, and promote plant growth and health [20]. *Agrobacterium tumefaciens* F2 was shown to have high adsorption ability for heavy metals such as Pb^{2+}, Cd^{2+}, and nickel (Ni^{2+}), with the best adsorption effect for Pb^{2+}. The adsorption mechanism was mainly due to the combination of heavy metals with functional groups such as C=O in carboxyl groups and C-O-C in sugar derivatives. Amino-acid-based proteins in EPS also adsorb heavy metals to some extent [59]. The protein components in extracellular polysaccharides have abundant -COOH, -OH, NH_3^-, and PO_4^{3-} groups, with the highest adsorption capacity for Pb^{2+} [60]. Two polysaccharide-producing bacteria, *Bacillus gibsonii* PM11 and *Bacillus xiamenensis* PM14, were found to improve the nutrient utilization rate of flax and promoted its growth by alleviating metal stress [61]. In this study, strains Z23 and Z39 increased the dry weight of the lettuce roots and leaves, indicating that these strains have the ability to alleviate the toxicity of Cd and Pb in lettuce and promote the growth of various lettuce tissues. The polysaccharide-producing bacteria not only increased the content of extracellular polysaccharides and proteins in the soil, but also had the function of passivating plants from absorbing heavy metals [62]. Morillo Pérez et al. [63] showed that polysaccharides secreted by *Paenibaillus jamilae* could adsorb Cu(II), Ni(II), and Pb(II), and found a maximum adsorption capacity of 303 mg g^{-1} for Pb(II). The polysaccharide-producing microorganisms produced extracellular polysaccharides, provided the necessary C and N sources for plants, and also showed good application prospects in the remediation of heavy metal pollution in soil. The exopolysaccharides secreted by exopolysaccharide-producing bacteria released phosphate ions and induced heavy metals to form precipitates, thus reducing the availability of heavy metals [64]. In addition, these bacteria also secreted some plant growth hormones, promoted plant growth, and prevented plants from absorbing heavy metals [14]. In this study, the polysaccharide-producing bacteria Z23 and Z39 reduced the content of Cd and Pb in wheat roots and leaves, indicating that these strains have the ability to inhibit the absorption of heavy metals by lettuce and can be used as heavy metal passivators in further research.

Soil aggregates are the basic components of the soil structure. Due to the different composition and properties of aggregates with different particle sizes, their characteristics in terms of the adsorption of environmental pollutants are also different [65]. The behavior of heavy metals in soil is largely restricted by the distribution of aggregates, which affect their bioavailability and migration ability. Research has shown that there is a significant relationship between the enrichment of heavy metals and the size of aggregates, with smaller aggregates having a larger specific surface area and higher organic matter content, which is beneficial for the enrichment of heavy metals [66]. Regulating the size of the soil aggregates is important for studying the distribution and bioavailability of heavy metals. By adding exopolysaccharide-producing bacteria to the soil, the size of the soil aggregates can be changed, thus affecting the absorption of heavy metals by crops [67]. Wu et al. [68] found that *Bacillus amyloliquefaciens* HYD-B17, *Bacillus licheniformis* HYTAPB18, and *Bacillus subtilis* RMPB44 with high exopolysaccharide production improved the stability of soil aggregates. Our study showed that the polysaccharide-producing bacteria Z23 and Z39 increased the proportion of microaggregates in the rhizosphere soil of lettuce, thereby enhancing the ability of the soil to adsorb and fix Cd and Pb. In addition, exopolysaccharide-producing bacteria were introduced into the soil to bind the soil particles together and form stable aggregates. The stability of soil aggregates is a symbolic factor for excellent soil structure. Therefore, improving the stability of aggregates improves the retention of soil water and fertilizer [67]. Exopolysaccharide-secreting bacteria are of great significance for the formation and stability of soil aggregates and the improvement of the soil structure [69]. In recent decades, scholars have conducted a significant amount of in-depth research on the mechanism of soil microorganisms in the formation and stability of soil aggregates [70]. Current studies have mainly focused on the influence of fungi on the formation of soil aggregates, because fungi can physically intertwine soil particles in aggregates through mycelium and connect them through the organic cementing of metabolites [71]. However, because the mycelium is easily decomposed by other microbial activities and forms an unstable aggregate structure, it does not last long in the soil. Therefore, more scholars began to study the effects of bacteria and their exopolysaccharides on the formation of soil aggregates [72]. The results showed that extracellular polysaccharides produced by bacteria are more beneficial to the formation and stability of soil aggregates than bacteria. In our study, inoculating Z23 and Z39 significantly increased the content of polysaccharides in the soil aggregate. The content of polysaccharides in microaggregates was significantly higher than that in large aggregates, indicating that microaggregates play an important role in immobilizing heavy metals.

Soil aggregates are the main environment in which soil microorganisms are active. Soil aggregates provide ecological niches for different microbial communities with specific classifications, functions, and ecological adaptability to environmental disturbances [73]. Therefore, soil aggregates can determine the resources available to soil microorganisms through oxygen diffusion, water flow, the accessibility of organic matter, and nutrient availability [74]. The chemical composition, organic matter decomposition ability, cation exchange capacity, surface reactivity, and adsorption characteristics of different particle size aggregates also vary; therefore, there are differences in the soil microbial communities living in them. Soil microorganisms can form and stabilize soil aggregates by binding clay particles together through their own activities and extracellular polysaccharides produced by certain fungal hyphae. In our study, inoculation with polysaccharide-producing bacteria increased the proportion of microaggregates in the rhizosphere soil of lettuce. More importantly, compared to macroaggregates, microaggregates had a more significant impact on the bacterial community structure. The vast majority (90%) of soil microorganisms are related to microaggregates, with the vast majority (70%) living within microaggregates and very few microorganisms in direct contact with each other. Only a small portion (<10%) of microorganisms are located on the surface of large aggregates exposed to large pores. Some scholars have found that fungi in soil, especially arbuscular mycorrhizal fungi, produce hyphae that are mainly related to the formation and stability of large aggregates [75]. Some sand particles in aggregates are only connected by fungal hyphae, while soil bacteria are mainly involved in the formation and stability of microaggregates [76]. In our study, inoculation with strains Z23 and Z39 altered the structure of the functional bacterial community and enriched strains with heavy metal-resistant, plant growth-promoting, and heavy metal-immobilizing properties in the rhizosphere soil, thereby inhibiting the absorption of Cd and Pb by lettuce. Thus, strains Z23 and Z39 have shown great potential for the remediation of excess Cd and Pb in farmland, collaborating with indigenous microbial communities in the soil to regulate the structure of rhizosphere soil aggregates, allowing heavy metals to precipitate into soil microaggregates, and reducing the mobility and available content of Cd and Pb, thereby avoiding the crop root absorption of heavy metals and ensuring food safety. However, functional strains are susceptible to external environmental influences and their functions may be unstable. Further research on the stability of strain functions and the validation of field effects is needed.

5. Conclusions

Two bacterial strains, Z23 and Z39, with high exopolysaccharide production were screened from heavy metal-contaminated soil. Both strains belong to the genus *Bacillus* sp. Strain Z23 had a removal rate of 88.6% for Cd and 93.2% for Pb. Strain Z39 had a removal rate of 92.3% for Cd and 94.4% for Pb. Strains Z23 and Z39 reduced the content of Cd and Pb in solution, inducing the formation of precipitates such as $Fe_2Pb(PO_4)_2$, $Cd_2(PO_4)_2$, and Pb_2O_3. Moreover, the polysaccharides secreted by the strains also participate in the adsorption of heavy metals. Pot experiments showed that strains Z23 and Z39 increased the dry weight and antioxidant enzyme activity of lettuce roots and leaves, while reducing the content of Cd and Pb by 40.1–61.7%. The strains increased the proportion of microaggregates (<0.25 mm) and exopolysaccharide content in the rhizosphere soil, reduced the content of available Cd and Pb, and prevented the absorption of heavy metals by lettuce. Moreover, the inoculation with strains Z23 and Z39 altered the structure of the functional bacterial community in the microaggregates, thereby inhibiting the absorption of Cd and Pb by lettuce. Our results provide strain resources and a theoretical basis for microbial remediation in heavy metal-contaminated fields.

Supplementary Materials: The following supporting information can be downloaded at https://www.mdpi.com/article/10.3390/microorganisms12112112/s1: Figure S1: Phylogenetic tree constructed using the Neighbor Joining method on the basis of 16S rDNA sequences of strains. Table S1: Promoting characteristics and heavy metal resistance of functional strains. Table S2: Estimation of bacterial community diversity. Table S3. Determination of physical and chemical indicators of lettuce rhizosphere soil. Table S4: Spearman correlation analysis of physical and chemical properties of lettuce rhizosphere soil.

Author Contributions: Conceptualization, Z.C. and H.H.; methodology, H.Z. and K.W.; software, H.Z. and X.L.; investigation, H.Z., K.W. and X.L.; writing—original draft preparation, H.Z., L.Y. and K.W.; writing—review and editing, Z.C. and H.H.; project administration, H.H. and H.Z.; funding acquisition, H.H. All authors have read and agreed to the published version of the manuscript.

Funding: This research was funded by the National Natural Science Foundation of China (Grant No. 42377039) and the National Natural Science Foundation of Henan (Grant No. 242300421149).

Data Availability Statement: All data are included in the article and Supplementary Materials.

Conflicts of Interest: The authors declare no conflicts of interest.

References

1. Zhou, S.; Cheng, Q.; Weindorf, D.C.; Yang, B.; Gong, Z.; Yuan, Z. Multiple approaches for heavy metal contamination characterization and source identification of farmland soils in a metal mine impacted area. *Appl. Geochem.* **2024**, *174*, 106125. [CrossRef]
2. Dong, S.; Li, L.; Chen, W.; Chen, Z.; Wang, Y.; Wang, S. Evaluation of heavy metal speciation distribution in soil and the accumulation characteristics in wild plants: A study on naturally aged abandoned farmland adjacent to tailings. *Sci. Total Environ.* **2024**, *917*, 170594. [CrossRef] [PubMed]
3. Yang, H.; Li, R.; Li, J.; Guo, Y.; Gao, T.; Guo, D.; Zhang, Q. Changes of heavy metal concentrations in farmland soils affected by non-ferrous metal smelting in China: A meta-analysis. *Environ. Pollut.* **2023**, *336*, 122442. [CrossRef] [PubMed]
4. León-Cañedo, J.A.; Alarcón-Silvas, S.G.; Fierro-Sañudo, J.F.; Rodríguez-Montes de Oca, G.A.; Partida-Ruvalcaba, L.; Díaz-Valdés, T.; Páez-Osuna, F. Mercury and other trace metals in lettuce (*Lactuca sativa*) grown with two low-salinity shrimp effluents: Accumulation and human health risk assessment. *Sci. Total Environ.* **2019**, *650*, 2535–2544. [CrossRef] [PubMed]
5. Suhani, I.; Sahab, S.; Srivastava, V.; Singh, R.P. Impact of cadmium pollution on food safety and human health. *Curr. Opin. Toxicol.* **2021**, *27*, 1–7. [CrossRef]
6. Han, H.; Wu, X.; Yao, L.; Chen, Z. Heavy metal-immobilizing bacteria combined with calcium polypeptides reduced the uptake of Cd in wheat and shifted the rhizosphere bacterial communities. *Environ. Pollut.* **2020**, *267*, 115432. [CrossRef]
7. Pan, W.; Lu, Q.; Xu, Q.-R.; Zhang, R.-R.; Li, H.-Y.; Yang, Y.-H.; Liu, H.-J.; Du, S.-T. Abscisic acid-generating bacteria can reduce Cd concentration in pakchoi grown in Cd-contaminated soil. *Ecotoxicol. Environ. Saf.* **2019**, *177*, 100–107. [CrossRef]
8. Siedt, M.; Schäffer, A.; Smith, K.E.C.; Nabel, M.; Roß-Nickoll, M.; van Dongen, J.T. Comparing straw, compost, and biochar regarding their suitability as agricultural soil amendments to affect soil structure, nutrient leaching, microbial communities, and the fate of pesticides. *Sci. Total Environ.* **2021**, *751*, 141607. [CrossRef]

9. Zhang, Y.; Zhou, Q.; Gao, C.; Lu, Y.; Sheng, Y.; Xiao, M.; Yun, Y.; Selvaraj, J.N.; Zhang, X.; Li, Y.; et al. Endophytic bacteria for Cd remediation in rice: Unraveling the Cd tolerance mechanisms of *Cupriavidus metallidurans* CML2. *J. Hazard. Mater.* **2024**, *469*, 133846. [CrossRef]
10. Rawat, P.; Sharma, A.; Shankhdhar, D.; Shankhdhar, S.C. Improvement of phosphorus uptake, phosphorus use efficiency, and grain yield of upland rice (*Oryza sativa* L.) in response to phosphate-solubilizing bacteria blended with phosphorus fertilizer. *Pedosphere* **2022**, *32*, 752–763. [CrossRef]
11. Disi, Z.A.; Attia, E.; Ahmad, M.I.; Zouari, N. Immobilization of heavy metals by microbially induced carbonate precipitation using hydrocarbon-degrading ureolytic bacteria. *Biotechnol. Rep.* **2022**, *35*, e00747. [CrossRef] [PubMed]
12. Wang, D.; Chen, H.; Xin, C.; Yuan, Y.; Sun, Q.; Cao, C.; Chao, H.; Wu, T.; Zheng, S. Insight into adsorption of Pb(II) with wild resistant bacteria TJ6 immobilized on biochar composite: Roles of bacterial cell and biochar. *Sep. Purif. Technol.* **2024**, *331*, 125660. [CrossRef]
13. Priyadarshanee, M.; Das, S. Spectra metrology for interaction of heavy metals with extracellular polymeric substances (EPS) of *Pseudomonas aeruginosa* OMCS-1 reveals static quenching and complexation dynamics of EPS with heavy metals. *J. Hazard. Mater.* **2024**, *466*, 133617. [CrossRef] [PubMed]
14. Reddy, K.V.; Ranjit, P.; Priyanka, E.; Maddela, N.R.; Prasad, R. Bioremediation of heavy metals-contaminated sites by microbial extracellular polymeric substances—A critical view. *Environ. Chem. Ecotoxicol.* **2024**, *6*, 408–421. [CrossRef]
15. Li, X.; Yu, Y.; Li, Y.; Yin, Z.; Liu, X.; Lian, B. Molecular mechanism of increasing extracellular polysaccharide production of *Paenibacillus mucilaginosus* K02 by adding mineral powders. *Int. Biodeterior. Biodegrad.* **2022**, *167*, 105340. [CrossRef]
16. Xu, X.; Zhang, L.; Han, Y.; Zhou, Z. Biosorption of Pb^{2+} and Zn^{2+} by Ca-alginate immobilized and free extracellular polysaccharides produced by *Leuconostoc citreum* B-2. *Int. J. Biol. Macromol.* **2021**, *193*, 2365–2373. [CrossRef]
17. Lahiri, D.; Nag, M.; Dutta, B.; Dey, A.; Ray, R.R. Chapter 1—Bacterial extracellular polysaccharides in biofilm formation and function. In *Application of Biofilms in Applied Microbiology*; Shah, M.P., Ed.; Academic Press: Cambridge, MA, USA, 2022; pp. 1–23.
18. Ajao, V.; Nam, K.; Chatzopoulos, P.; Spruijt, E.; Bruning, H.; Rijnaarts, H.; Temmink, H. Regeneration and reuse of microbial extracellular polymers immobilised on a bed column for heavy metal recovery. *Water Res.* **2020**, *171*, 408–421. [CrossRef]
19. Ali, M.; Cybulska, J.; Frąc, M.; Zdunek, A. Application of polysaccharides for the encapsulation of beneficial microorganisms for agricultural purposes: A review. *Int. J. Biol. Macromol.* **2023**, *244*, 125366. [CrossRef]
20. Ghosh, P.K.; Maiti, T.K. Structure of Extracellular Polysaccharides (EPS) Produced by Rhizobia and their Functions in Legume–Bacteria Symbiosis:—A Review. *Achiev. Life Sci.* **2016**, *10*, 136–143. [CrossRef]
21. Yan, C.; Ji, S.; Wu, R.; Li, M.; He, K.; Shi, H.; Wang, C.; Yang, H.; Guo, J.; Wu, J. Structural properties and biological activities of the extracellular polysaccharide of *Bacillus subtilis* LZ13-4. *Int. J. Biol. Macromol.* **2024**, *259*, 129176. [CrossRef]
22. Mukherjee, P.; Mitra, A.; Roy, M. Halomonas Rhizobacteria of Avicennia marina of Indian Sundarbans Promote Rice Growth Under Saline and Heavy Metal Stresses Through Exopolysaccharide Production. *Front. Microbiol.* **2019**, *10*, 1207. [CrossRef] [PubMed]
23. Mota, R.; Rossi, F.; Andrenelli, L.; Pereira, S.B.; De Philippis, R.; Tamagnini, P. Released polysaccharides (RPS) from *Cyanothece* sp. CCY 0110 as biosorbent for heavy metals bioremediation: Interactions between metals and RPS binding sites. *Appl. Microbiol. Biotechnol.* **2016**, *100*, 7765–7775. [CrossRef] [PubMed]
24. Zhou, Y.; Xia, S.; Zhang, J.; Nguyen, B.T.; Zhang, Z. Insight into the influences of pH value on Pb(II) removal by the biopolymer extracted from activated sludge. *Chem. Eng. J.* **2017**, *308*, 1098–1104. [CrossRef]
25. Shahid, M.; Altaf, M.; Ali, S.; Tyagi, A. Isolation and assessment of the beneficial effect of exopolysaccharide-producing PGPR in *Triticum aestivum* (L.) plants grown under NaCl and Cd -stressed conditions. *Plant Physiol. Biochem.* **2024**, *215*, 108973. [CrossRef] [PubMed]
26. Chen, L.; He, L.-Y.; Wang, Q.; Sheng, X.-F. Synergistic effects of plant growth-promoting *Neorhizobium huautlense* T1-17 and immobilizers on the growth and heavy metal accumulation of edible tissues of hot pepper. *J. Hazard. Mater.* **2016**, *312*, 123–131. [CrossRef]
27. Wang, Y.; Jing, Y.; Leng, F.; Wang, S.; Wang, F.; Zhuang, Y.; Liu, X.; Wang, X.; Ma, X. Establishment and Application of a Method for Rapid Determination of Total Sugar Content Based on Colorimetric Microplate. *Sugar Tech* **2016**, *19*, 424–431. [CrossRef]
28. Teng, Z.; Shao, W.; Zhang, K.; Huo, Y.; Li, M. Characterization of phosphate solubilizing bacteria isolated from heavy metal contaminated soils and their potential for lead immobilization. *J. Environ. Manag.* **2019**, *231*, 189–197. [CrossRef] [PubMed]
29. Jiang, C.-y.; Sheng, X.-f.; Qian, M.; Wang, Q.-y. Isolation and characterization of a heavy metal-resistant *Burkholderia* sp. from heavy metal-contaminated paddy field soil and its potential in promoting plant growth and heavy metal accumulation in metal-polluted soil. *Chemosphere* **2008**, *72*, 157–164. [CrossRef]
30. Rajkumar, M.; Nagendran, R.; Lee, K.J.; Lee, W.H.; Kim, S.Z. Influence of plant growth promoting bacteria and Cr^{6+} on the growth of Indian mustard. *Chemosphere* **2006**, *62*, 741–748. [CrossRef]
31. Belimov, A.A.; Hontzeas, N.; Safronova, V.I.; Demchinskaya, S.V.; Piluzza, G.; Bullitta, S.; Glick, B.R. Cadmium-tolerant plant growth-promoting bacteria associated with the roots of Indian mustard (*Brassica juncea* L. Czern.). *Soil Biol. Biochem.* **2005**, *37*, 241–250. [CrossRef]
32. Han, H.; Wang, Q.; He, L.-y.; Sheng, X.-f. Increased biomass and reduced rapeseed Cd accumulation of oilseed rape in the presence of Cd-immobilizing and polyamine-producing bacteria. *J. Hazard. Mater.* **2018**, *353*, 280–289. [CrossRef] [PubMed]

33. Feng, Y.; Yang, J.; Liu, W.; Yan, Y.; Wang, Y. Hydroxyapatite as a passivator for safe wheat production and its impacts on soil microbial communities in a Cd-contaminated alkaline soil. *J. Hazard. Mater.* **2021**, *404*, 124005. [CrossRef] [PubMed]
34. Wang, Y.; Xu, Y.; Li, L.; Yang, Y.; Xu, C.; Luo, Y.; Wang, Y.; Liang, X.; Sun, J. Immobilization of Cd by mercapto-palygorskite in typical calcareous and acidic soil aggregates: Performance and differences. *Chemosphere* **2023**, *323*, 138223. [CrossRef] [PubMed]
35. Han, H.; Cai, H.; Wang, X.; Hu, X.; Chen, Z.; Yao, L. Heavy metal-immobilizing bacteria increase the biomass and reduce the Cd and Pb uptake by pakchoi (*Brassica chinensis* L.) in heavy metal-contaminated soil. *Ecotoxicol. Environ. Saf.* **2020**, *195*, 110375. [CrossRef] [PubMed]
36. Fisher, K.A.; Yarwood, S.A.; James, B.R. Soil urease activity and bacterial ureC gene copy numbers: Effect of pH. *Geoderma* **2017**, *285*, 1–8. [CrossRef]
37. Yang, Y.; Wikieł, A.J.; Dall'Agnol, L.T.; Eloy, P.; Genet, M.J.; Moura, J.J.G.; Sand, W.; Dupont-Gillain, C.C.; Rouxhet, P.G. Proteins dominate in the surface layers formed on materials exposed to extracellular polymeric substances from bacterial cultures. *Biofouling* **2016**, *32*, 95–108. [CrossRef]
38. Ye, S.; Ma, Z.; Liu, Z.; Liu, Y.; Zhang, M.; Wang, J. Effects of carbohydrate sources on biosorption properties of the novel exopolysaccharides produced by Arthrobacter ps-5. *Carbohydr. Polym.* **2014**, *112*, 615–621. [CrossRef]
39. Lian, Z.; Yang, Z.; Song, W.; Sun, M.; Gan, Y.; Bai, X. Characteristics of EPS from *Pseudomonas aeruginosa* and *Alcaligenes faecalis* under Cd(II) stress: Changes in chemical components and adsorption performance. *Environ. Sci. Pollut. Res.* **2022**, *29*, 75883–75895. [CrossRef]
40. Abd-El-Haleem, D. Alpha-glucan: A novel bacterial polysaccharide and its application as a biosorbent for heavy metals. *J. Genet. Eng. Biotechnol.* **2023**, *21*, 133. [CrossRef]
41. Qu, C.; Duan, C.; Li, W.; Wu, X.; Liu, Z.; Feng, F.; Tang, X.; Chai, X.; Tang, C.-J. Understanding the slight inhibition of high As(III) stress on nitritation process: Insights from arsenic speciation and microbial community analyses. *J. Hazard. Mater.* **2022**, *435*, 128957. [CrossRef]
42. Pan, X.; Zhang, S.; Li, T.; Ouyang, J.; Gong, G.; Wang, G.; Xu, X.; Pu, Y.; Long, L.; Jia, Y. Response of microbiomes with different abundances to removal of metal fractions by soil washing. *Ecotoxicol. Environ. Saf.* **2022**, *242*, 113862. [CrossRef] [PubMed]
43. Shi, L.; Zhang, P.; He, Y.; Zeng, F.; Xu, J.; He, L. Enantioselective effects of cyflumetofen on microbial community and related nitrogen cycle gene function in acid-soil. *Sci. Total Environ.* **2021**, *771*, 144831. [CrossRef] [PubMed]
44. Wu, X.; Rensing, C.; Han, D.; Xiao, K.-Q.; Dai, Y.; Tang, Z.; Liesack, W.; Peng, J.; Cui, Z.; Zhang, F.; et al. Genome-Resolved Metagenomics Reveals Distinct Phosphorus Acquisition Strategies between Soil Microbiomes. *mSystems* **2022**, *7*, e0110721. [CrossRef] [PubMed]
45. Dedysh, S.N.; Beletsky, A.V.; Kulichevskaya, I.S.; Mardanov, A.V.; Ravin, N.V.; Cameron Thrash, J. Complete Genome Sequence of *Paludibaculum fermentans* P105T, a Facultatively Anaerobic Acidobacterium Capable of Dissimilatory Fe(III) Reduction. *Microbiol. Resour. Announc.* **2021**, *10*, e01313-20. [CrossRef]
46. Zhao, Z.; Liu, L.; Sun, Y.; Xie, L.; Liu, S.; Li, M.; Yu, Q. Combined microbe-plant remediation of cadmium in saline-alkali soil assisted by fungal mycelium-derived biochar. *Environ. Res.* **2024**, *240*, 117424. [CrossRef]
47. Shan, S.; Guo, Z.; Lei, P.; Li, Y.; Wang, Y.; Zhang, M.; Cheng, W.; Wu, S.; Wu, M.; Du, D. Increased biomass and reduced tissue cadmium accumulation in rice via indigenous *Citrobacter* sp. XT1-2-2 and its mechanisms. *Sci. Total Environ.* **2020**, *708*, 135224. [CrossRef] [PubMed]
48. Liu, X.; Chen, C.; Wang, J.; Zou, S.; Long, X. Phosphorus solubilizing bacteria *Bacillus thuringiensis* and *Pantoea ananatis* simultaneously promote soil inorganic phosphate dissolution and soil Pb immobilization. *Rhizosphere* **2021**, *20*, 100448. [CrossRef]
49. Huang, H.; Jia, Q.; Jing, W.; Dahms, H.-U.; Wang, L. Screening strains for microbial biosorption technology of cadmium. *Chemosphere* **2020**, *251*, 126428. [CrossRef]
50. Yu, H.; Li, C.; Yan, J.; Ma, Y.; Zhou, X.; Yu, W.; Kan, H.; Meng, Q.; Xie, R.; Dong, P. A review on adsorption characteristics and influencing mechanism of heavy metals in farmland soil. *RSC Adv.* **2023**, *13*, 3505–3519. [CrossRef]
51. Zhang, Y.-H.; Liu, C.-S.; Tian, Y.; Wang, J.; Xin, S.; Sheng, X. An eco-friendly photo-responsive hyaluronic acid-based supramolecular polysaccharide hybrid hydrogels for plant growth regulation and heavy metal ions adsorption. *Int. J. Biol. Macromol.* **2023**, *242*, 125194. [CrossRef]
52. Sun, J.; He, X.; Le, Y.; Al-Tohamy, R.; Ali, S.S. Potential applications of extremophilic bacteria in the bioremediation of extreme environments contaminated with heavy metals. *J. Environ. Manag.* **2024**, *352*, 120081. [CrossRef] [PubMed]
53. Karthik, C.; Oves, M.; Thangabalu, R.; Sharma, R.; Santhosh, S.B.; Indra Arulselvi, P. *Cellulosimicrobium funkei*-like enhances the growth of *Phaseolus vulgaris* by modulating oxidative damage under Chromium(VI) toxicity. *J. Adv. Res.* **2016**, *7*, 839–850. [CrossRef] [PubMed]
54. Zeng, W.; Li, F.; Wu, C.; Yu, R.; Wu, X.; Shen, L.; Liu, Y.; Qiu, G.; Li, J. Role of extracellular polymeric substance (EPS) in toxicity response of soil bacteria *Bacillus* sp. S3 to multiple heavy metals. *Bioprocess Biosyst. Eng.* **2019**, *43*, 153–167. [CrossRef] [PubMed]
55. K, K.R.; Sardar, U.R.; Bhargavi, E.; Devi, I.; Bhunia, B.; Tiwari, O.N. Advances in exopolysaccharides based bioremediation of heavy metals in soil and water: A critical review. *Carbohydr. Polym.* **2018**, *199*, 353–364. [CrossRef] [PubMed]
56. Musarurwa, H.; Tavengwa, N.T. Application of carboxymethyl polysaccharides as bio-sorbents for the sequestration of heavy metals in aquatic environments. *Carbohydr. Polym.* **2020**, *237*, 116142. [CrossRef] [PubMed]

57. Wei, L.; Li, Y.; Noguera, D.R.; Zhao, N.; Song, Y.; Ding, J.; Zhao, Q.; Cui, F. Adsorption of Cu^{2+} and Zn^{2+} by extracellular polymeric substances (EPS) in different sludges: Effect of EPS fractional polarity on binding mechanism. *J. Hazard. Mater.* **2017**, *321*, 473–483. [CrossRef]
58. Iyer, A.; Mody, K.; Jha, B. Accumulation of hexavalent chromium by an exopolysaccharide producing marine *Enterobacter cloaceae. Mar. Pollut. Bull.* **2004**, *49*, 974–977. [CrossRef]
59. Cui, D.; Tan, C.; Deng, H.; Gu, X.; Pi, S.; Chen, T.; Zhou, L.; Li, A. Biosorption Mechanism of Aqueous Pb^{2+}, Cd^{2+}, and Ni^{2+} Ions on Extracellular Polymeric Substances (EPS). *Archaea* **2020**, *2020*, 8891543. [CrossRef]
60. Vimalnath, S.; Subramanian, S. Studies on the biosorption of Pb(II) ions from aqueous solution using extracellular polymeric substances (EPS) of *Pseudomonas aeruginosa. Colloids Surf. B Biointerfaces* **2018**, *172*, 60–67. [CrossRef]
61. Zainab, N.; Amna; Din, B.U.; Javed, M.T.; Afridi, M.S.; Mukhtar, T.; Kamran, M.A.; Ul Ain, O.; Khan, A.A.; Ali, J.; et al. Deciphering metal toxicity responses of flax (*Linum usitatissimum* L.) with exopolysaccharide and ACC-deaminase producing bacteria in industrially contaminated soils. *Plant Physiol. Biochem.* **2020**, *152*, 90–99. [CrossRef]
62. Al-Dhabi, N.A.; Esmail, G.A.; Valan Arasu, M. Sustainable conversion of palm juice wastewater into extracellular polysaccharides for absorption of heavy metals from Saudi Arabian wastewater. *J. Clean. Prod.* **2020**, *277*, 124252. [CrossRef]
63. Morillo Pérez, J.A.; García-Ribera, R.; Quesada, T.; Aguilera, M.; Ramos-Cormenzana, A.; Monteoliva-Sánchez, M. Biosorption of heavy metals by the exopolysaccharide produced by *Paenibacillus jamilae. World J. Microbiol. Biotechnol.* **2008**, *24*, 2699–2704. [CrossRef]
64. Vandana; Das, S. Genetic regulation, biosynthesis and applications of extracellular polysaccharides of the biofilm matrix of bacteria. *Carbohydr. Polym.* **2022**, *291*, 119536. [CrossRef] [PubMed]
65. Gong, C.; Ma, L.; Cheng, H.; Liu, Y.; Xu, D.; Li, B.; Liu, F.; Ren, Y.; Liu, Z.; Zhao, C.; et al. Characterization of the particle size fraction associated heavy metals in tropical arable soils from Hainan Island, China. *J. Geochem. Explor.* **2014**, *139*, 109–114. [CrossRef]
66. Huang, B.; Yuan, Z.; Li, D.; Zheng, M.; Nie, X.; Liao, Y. Effects of soil particle size on the adsorption, distribution, and migration behaviors of heavy metal(loid)s in soil: A review. *Environ. Sci. Process. Impacts* **2020**, *22*, 1596–1615. [CrossRef]
67. Song, J.; Brookes, P.C.; Shan, S.; Xu, J.; Liu, X. Effects of remediation agents on microbial community structure and function in soil aggregates contaminated with heavy metals. *Geoderma* **2022**, *425*, 116030. [CrossRef]
68. Wu, N.; Pan, H.-X.; Qiu, D.; Zhang, Y.-M. Feasibility of EPS-producing bacterial inoculation to speed up the sand aggregation in the Gurbantunggut Desert, Northwestern China. *J. Basic Microbiol.* **2014**, *54*, 1378–1386. [CrossRef]
69. Zhao, R.; Kuzyakov, Y.; Zhang, H.; Wang, Z.; Li, T.; Shao, L.; Jiang, L.; Wang, R.; Li, M.; Jianxin Sun, O.; et al. Labile carbon inputs offset nitrogen-induced soil aggregate destabilization via enhanced growth of saprophytic fungi in a meadow steppe. *Geoderma* **2024**, *443*, 116841. [CrossRef]
70. Mohite, B.V.; Koli, S.H.; Patil, S.V. Heavy Metal Stress and Its Consequences on Exopolysaccharide (EPS)-Producing Pantoea agglomerans. *Appl. Biochem. Biotechnol.* **2018**, *186*, 199–216. [CrossRef]
71. Zhang, K.; Chen, S.; Ping, Y.; Song, F.; Fan, X.; Wang, L.; Zang, C. Coinoculation of arbuscular mycorrhizal fungi and rhizobia stimulates atrazine dissipation by changing the atrazine-degrading bacterial community at the soil aggregate scale. *Environ. Pollut.* **2024**, *345*, 123507. [CrossRef]
72. Fontana, M.B.; Novelli, L.E.; Sterren, M.A.; Uhrich, W.G.; Rondán, G.A.; Barbagelata, P.A.; Benintende, S.M. Cover crop benefit bacteria and increase aggregate-associate soil C and N storage. *Geoderma Reg.* **2024**, *36*, e00743. [CrossRef]
73. Wang, X.; Huang, P.; Ma, M.; Shan, K.; Wu, S. Effects of riparian pioneer plants on soil aggregate stability: Roles of root traits and rhizosphere microorganisms. *Sci. Total Environ.* **2024**, *940*, 173584. [CrossRef] [PubMed]
74. Pellegrino, E.; Piazza, G.; Helgason, T.; Ercoli, L. Microbiome structure and interconnection in soil aggregates across conservation and conventional agricultural practices allow to identify main prokaryotic and fungal taxa related to soil functioning. *Soil Biol. Biochem.* **2022**, *175*, 108833. [CrossRef]
75. Sarder, M.P.; Kamruzzaman, M.; Siddique, M.A.B.; Halder, M. Stability and heavy metals accumulation of soil aggregates under different land uses in the southwest coastal Bangladesh. *Heliyon* **2024**, *10*, e37806. [CrossRef]
76. Mann, C.; Lynch, D.; Fillmore, S.; Mills, A. Relationships between field management, soil health, and microbial community composition. *Appl. Soil Ecol.* **2019**, *144*, 12–21. [CrossRef]

soil systems

MDPI

Article

Volatile Organic Compounds from Candelilla-Associated PGPR Enhance *Arabidopsis thaliana* Seedling Growth Under Salinity Stress

María Teresa Salazar-Ramírez [1,2], Rubén Palacio-Rodríguez [3], Jesús Josafath Quezada-Rivera [3,*], Tania Elizabeth Velásquez-Chávez [3,4], Gisela Muro-Pérez [3], Hortencia Ivone Ortega-Reyes [2], Jorge Arnaldo Orozco-Vidal [2], Antonio Gerardo Yescas-Coronado [2], Gerardo Antonio Verástegui-Hernández [2] and Jorge Sáenz-Mata [3,*]

1 Faculty of Biological Sciences, Autonomous University of Coahuila, Carretera Torreón-Matamoros Km 7.5, Ciudad Universitaria C.P., Torreón 27276, México; salazar.teresa@uadec.edu.mx
2 Torreon Institute of Technology, National Technological Institute of Mexico, Carretera Torreón-San Pedro Km 7.5, Torreón 27170, México; ivoneortega1995@gmail.com (H.I.O.-R.); joorvi66@hotmail.com (J.A.O.-V.); antonio.yc@torreon.tecnm.mx (A.G.Y.-C.); antonioverastegui99@gmail.com (G.A.V.-H.)
3 Microbial Ecology Laboratory, Faculty of Biological Sciences, Juárez University of the State of Durango, Av. Universidad S/N, Col. Filadelfia, Gómez Palacio 35010, México; ruben.palacio@ujed.mx (R.P.-R.); tania.vc@itslerdo.edu.mx (T.E.V.-C.); giselamuro@ujed.mx (G.M.-P.)
4 Lerdo Institute of Technology, National Technological Institute of Mexico, Av. Tecnológico S/N, Col. Periférico C.P., Lerdo 35150, México
* Correspondence: josafath.quezada@ujed.mx (J.J.Q.-R.); jsaenz_mata@ujed.mx (J.S.-M.)

Abstract

Soil salinity imposes a critical constraint on plant productivity, highlighting the need for sustainable biological strategies to enhance stress tolerance. This study assessed the effects of volatile organic compounds (VOCs) emitted by ten plant-growth-promoting rhizobacteria (PGPR) isolated from the rhizosphere of *Euphorbia antisyphilitica* on the growth of *Arabidopsis thaliana* seedlings exposed to 0, 50, and 100 mM NaCl. A divided Petri dish system was used to quantify biomass, root architecture, proline accumulation, sodium content, and chlorophyll concentration. Three strains—*Siccibacter colletis* CASEcto12, *Enterobacter quasihormaechei* NFbEcto18, and *Bacillus wiedmannii* NFbEndo12—significantly enhanced seedling development under saline and non-saline conditions ($p \leq 0.05$). At 50 mM NaCl, *S. colletis* CASEcto12 increased primary root length from 40.25 to 64.81 mm and fresh weight from 45.05 to 133.33 mg, while *E. quasihormaechei* NFbEcto18 elevated lateral root number from 10 to 24, compared to the uninoculated control. Under 100 mM NaCl, *E. quasihormaechei* NFbEcto18 increased proline accumulation (0.564–1.378 mmol g^{-1} FW) and reduced Na^+ content (0.146–0.084 mmol g^{-1} FW), indicating improved osmotic and ionic regulation. VOC profiling using SPME-GC-MS revealed aldehydes, ketones, and alcohols as predominant classes. Overall, these findings demonstrate the potential of candelilla-associated PGPR VOCs as promising biostimulants for enhancing plant performance in salt-affected soils.

Keywords: PGPR; salt stress; VOCs; *Arabidopsis thaliana*; stress alleviation

1. Introduction

Soil salinization is characterized by the excessive accumulation of soluble salts in the soil profile, a critical issue that severely reduces crop yields, increases plant mortality, degrades soil quality, and often leads to the abandonment of agricultural land [1]. Globally, approximately 1 billion hectares are affected by salinity, with arid and semi-arid areas being the most impacted [2]. In irrigated lands, 20–30% are already affected by salinity, significantly limiting their productivity [1]. Salt accumulation can stem from both primary (natural or geological) and secondary (human activity) causes, but its effect on plants is uniform: it disrupts multiple physiological functions by creating osmotic imbalances, inhibiting root growth, and altering root architecture, thereby reducing water and nutrient uptake [3]. Furthermore, high salt levels trigger the overproduction of reactive oxygen species (ROS), compromising cell viability, photosynthetic pigment stability, membrane integrity, and hormonal balance [4,5].

To mitigate the impacts of salinity on soils and crops, researchers have explored the use of various strategies, including plant-growth-promoting rhizobacteria (PGPR). These beneficial microorganisms colonize the rhizosphere and promote plant development, enhancing tolerance to both abiotic and biotic stresses [6]. Their mechanisms include phytohormone production (indole-3-acetic acid, gibberellic acid, and cytokinins); improved nutrient availability through phosphate solubilization and siderophore production; ACC deaminase activity; and the release of secondary metabolites such as volatile organic compounds (VOCs) [7].

Growing interest in VOCs produced by microorganisms has led to several lines of research aimed at understanding their influence on plants and their potential applications in agriculture and biotechnology. VOCs—low-molecular-weight, lipophilic compounds with distinctive odors, high vapor pressures, and low boiling points—easily diffuse through soil pores. Nearly 2000 VOCs from about 1000 microbial species have been described [8,9]. Microbial VOCs include alkenes, alcohols, aldehydes, ketones, terpenes, benzenoids, pyrazines, aromatic and aliphatic hydrocarbons, acids, esters, and nitrogen- or sulfur-containing compounds [10–13]. These metabolites can induce plant resistance to abiotic and biotic stresses and promote growth.

Microbial volatile organic compounds (mVOCs) produced by plant-growth-promoting rhizobacteria (PGPR) have been recognized as chemical signals capable of modulating plant development and physiology, although they still receive less attention than other plant–microbe interaction mechanisms. Several reviews have highlighted that mVOCs can induce pathways associated with growth, defense, and abiotic stress tolerance through hormonal regulation, redox balance modulation, and ion homeostasis adjustment [14,15]. Experimental studies have shown that certain bacterial VOCs improve salt tolerance in model plants and crops, mainly through strains of widely studied genera such as Bacillus and Pseudomonas [16,17]. However, significant gaps in knowledge remain: there are a limited number of studies that systematically characterize the VOCs emitted by less studied PGPR, as well as their effects on *Arabidopsis thaliana* under controlled salinity conditions [18]. This gap hinders our understanding of the specific mechanisms involved and highlights the need for studies that integrate chemical analyses of VOCs with the physiological and molecular responses of the plant, such as the one proposed here.

Therefore, this study aimed to evaluate the effects of VOCs emitted by PGPR on *A. thaliana* seedlings' development under varying salinity conditions, using physiological and biochemical assays alongside volatile profile analysis.

2. Materials and Methods

2.1. Rhizobacterial Strains and Growth Conditions

The rhizobacterial strains used in this study were isolated from the rhizosphere of *Euphorbia antisyphilitica Zucc.* (candelilla), a plant endemic to the Chihuahuan Desert in northern Mexico. Rhizosphere samples were collected from the municipality of Viesca,

Coahuila, Mexico (25°10′41″ N, 102°39′12″ W). In previous research, these bacterial strains were shown to possess multiple traits associated with plant growth promotion in *A. thaliana* seedlings [19].

From the original 21 strains identified as plant-growth-promoting rhizobacteria (PGPR), 10 high-performing strains were selected for this study. The selection was based on a dual assessment: the most efficient plant growth promotion in preliminary *A. thaliana* assays and the highest tolerance to salinity stress (where the tolerance criterion was confirmed by their growth on LB solid medium supplemented with high concentrations of NaCl). These 10 selected strains were further assessed for their ability to emit volatile organic compounds (VOCs). Detailed information on the isolation and initial PGPR characterization of the 21 strains is available in the previous publication [19]. The strains were molecularly identified through 16S rRNA gene sequencing. Although some strains in the previous publication were reported with low identity percentages or short sequence lengths [19], molecular identification in the present study was improved and updated (Table 1).

Table 1. Molecular identification of bacteria isolated from candelilla rhizosphere.

ID	Taxon	bp	Identity (%)	Access Number NCBI
LbEcto8	*Mixta theicola*	1373	98	NR_137415.1
CASEcto2	*Siccibacter colletis*	1371	99	NR_134807.1
CASEcto4	*Staphylococcus epidermidis*	1402	99	NR_036904.1
CASEcto12	*Siccibacter colletis*	1411	99	NR_134807.1
CASEcto13	*Staphylococcus epidermidis*	1377	95	NR_113957.1
NFbEcto18	*Enterobacter quasihormaechei*	1369	99	NR_180451.1
LbEndo7	*Staphylococcus epidermidis*	1396	99	NR_113957.1
LbEndo13	*Bacillus subtilis*	1366	95	NR_104873.1
CASEndo10	*Bacillus cereus*	1398	99	NR_074540.1
NFbEndo12	*Bacillus wiedmannii*	1398	99	NR_152692.1

Molecular identification using 16SrRNA gene of rhizobacteria isolated from candelilla [14]. ID: identification, bp: base pairs.

2.2. Plant Material and In Vitro Evaluation of PGPR-Emitted VOCs Under Salinity Stress

A. thaliana seeds (ecotype Col-0) were surface-sterilized in Eppendorf tubes using 20% (*v*/*v*) solution of commercial sodium hypochlorite (which corresponds to an approximate concentration of 1.0–1.2% active NaOCl), with agitation on a vortex for 5 min [20]. The seeds were then rinsed four times with sterile distilled water. To synchronize germination, the seeds underwent vernalization at 4 °C for 48 h. A total of 45 seeds were sown on Petri dishes containing 0.2× Murashige and Skoog (MS) medium prepared using the standard Murashige and Skoog basal salt mixture and vitamin formulation (PhytoTech Labs, product number M519) and diluted to 20% of its final working concentration. The plates were incubated in a bioclimatic chamber under a 16 h/8 h light/dark photoperiod at 25 °C [21]. Germination began approximately four days after sowing.

To assess the effects of VOCs emitted by the 10 PGPR selected, a divided Petri dish system was employed (SYM brand). Each dish contained MS medium supplemented with NaCl at concentrations of 50 and 100 mM to induce salinity stress, alongside a control treatment without NaCl added. On one side of each divided dish were four *A. thaliana* seedlings that had germinated for four days. The opposite side was inoculated with 10 μL of a bacterial suspension at a concentration of 1×10^9 CFU mL^{-1}; bacteria were previously reactivated as individual strains in Luria–Bertani (LB) medium at 28–30 °C for 24 h, inoculated into LB liquid culture, and the concentration adjusted. All Petri dishes for the treatments were carefully sealed with sealing film. Dishes without bacterial inoculation were included as a control [21].

The plates were positioned vertically in a growth chamber under a 16 h/8 h light/dark cycle at 25 °C. Twelve days after inoculation with rhizobacteria, the following parameters were recorded: fresh weight of seedlings using an analytical balance, number of lateral roots, and length of primary root measured with digital caliper. Each treatment included 4 replicates with a total of 16 seedlings.

2.3. *Biochemical Analysis of A. thaliana*

2.3.1. Proline Quantification

Proline extraction was performed following the protocol described by Bates [22]. Plant samples previously exposed to VOCs emitted by rhizobacteria were used. A total of 100 mg of aerial plant tissue was weighed using an analytical balance (Ohaus Pioneer, Model PX224, manufactured in Shanghai, China) and transferred to an Eppendorf tube containing 140 mM sulfosalicylic acid. The tissue was homogenized with a micro pestle, and the resulting solution was filtered through a 0.20 μm membrane. The filtrate was mixed in a 1:1:1 ratio with 140 mM acid ninhydrin and glacial acetic acid, incubated at 100 °C for one hour, and then cooled on ice for 5 min. An equal volume of toluene was added, followed by vigorous vortexing. After standing for 10 min, the organic phase was separated and transferred to a quartz cuvette for spectrophotometric analysis at 520 nm (Thermo Fisher Scientific, Genesys 20, Model 4001/4, manufactured in Rochester, NY, USA) to quantify the ninhydrin–proline complex. A standard curve was constructed using commercial proline solutions at concentrations of 10, 20, 40, 50, 75, and 100 μg.

2.3.2. Sodium (Na^+) Content Determination

Sodium determination was carried out following the methodology of Levinsh [23]. Aerial tissue from *A. thaliana* seedlings previously exposed to rhizobacterial VOCs was used for sodium quantification. A total of 0.1 g of aerial tissue was weighed, ground into a green paste using a mortar and pestle, and diluted in 50 mL of Milli-Q water in Falcon tubes. Sodium content was measured using a sodium-ion-selective electrode via potential (Horiba, model 1512A, manufactured in Kyoto, Japan). A standard curve was prepared using NaCl and NaOH solutions at concentrations of 0, 3, 7, 30, 60, 120, and 240 mM.

2.3.3. Chlorophyll Content Determination

Chlorophyll content was assessed in four 20-day-old *A. thaliana* seedlings exposed to VOCs, following the protocol of Zhang and Huang [24]. Total chlorophyll was extracted from leaf tissue using N, N-dimethylformamide (DMF) and incubated for 24 h at 4 °C in the dark. Chlorophyll A and chlorophyll B concentrations were measured spectrophotometrically at 664 nm and 647 nm (Thermo Fisher Scientific, Genesys 20, Model 4001/4), respectively. The concentrations were calculated using the following equations:

$$\text{Chlorophyll (A)} = 12.7 \times \text{A664} - 2.79 \times \text{A647};$$

$$\text{Chlorophyll (B)} = 20.7 \times \text{A647} - 4.62 \times \text{A664};$$

$$\text{Total Chlorophyll (A + B)} = 17.90 \times \text{A647} + 8.08 \times \text{A664}.$$

2.4. *Identification of VOCs from Rhizobacteria*

VOCs released by rhizobacteria were collected from 2-day-old colonies grown on plastic Petri dishes (90 × 15 mm) containing solid MS medium, which were sealed with parafilm. Using solid-phase microextraction (SPME) followed by gas chromatography–mass spectrometry (GC–MS) analysis, volatile compounds were adsorbed using a blue SPME fiber (Polydimethylsiloxane/Divinylbenzene) and desorbed at 180 °C for 30 s into the injector port of a gas chromatograph (Agilent 7890B; Agilent Technologies, Foster City, CA, USA) equipped with a mass spectrometry detector (Agilent 5977A), and data acquisition were performed using the MassHunter Workstation software (2010 version) (Agilent Technologies, Santa Clara, CA, USA).

A capillary column with a fatty acid phase and a film thickness of 0.25 mm was used. Helium served as the carrier gas at a flow rate of 1 mL min^{-1}. The injector temperature was set to 250 °C. The column temperature was initially held at 40 °C for 5 min, then increased at a rate of 3 °C per minute to a final temperature of 220 °C, which was maintained for 5 min. Three independent measurements were performed for each sample. The ion source pressure was 7 Pa, the filament voltage was 70 eV, and the scan rate was 1.9 scans per second.

Volatile compounds were identified by comparing the obtained mass spectra with entries in the NIST/EPA/NIH Mass Spectral Library (2011 version).

2.5. Statistical Analysis

A completely randomized design was employed to evaluate the ten PGPR strains and a non-inoculated control under three NaCl concentrations (0, 50, and 100 mM), along with their respective controls without NaCl added to the MS medium. Each treatment included four replicates, resulting in a total of 132 experimental units.

Data from the measured variables were subjected to analysis of variance (ANOVA), and treatment means were compared using Tukey's test at a significance level of $p \leq 0.05$. All statistical analyses were performed using the SAS software package (Statistical Analysis System Institute, version 9.1). Graphs were generated using GraphPad Prism version 6.

3. Results

3.1. Rhizobacteria Promote *A. thaliana* Seedling Growth via VOCs

Of the ten rhizobacterial strains tested, three (*Siccibacter colletis* CASEcto12, *Enterobacter quasihormaechei* NFbEcto18, and *Bacillus wiedmannii* NFbEndo12) significantly increased biomass production in *A. thaliana* seedlings under non-saline conditions through exposure to rhizobacterial volatile organic compounds for 12 days.

Siccibacter colletis CASEcto12 exhibited the largest effect relative to the uninoculated control. Seedlings exposed to its VOCs had an average root length of 51.75 mm and a fresh weight of 68.08 mg, compared to control averages of 40.06 mm and 40.58 mg, respectively (Figures 1a and 2b). In terms of lateral root number, *Enterobacter quasihormaechei* NFbEcto18 induced an average of 14.63 lateral roots per seedling, significantly more than the uninoculated control's 8.31 roots (Figures 1a and 2d).

These results demonstrate the plant-growth-promoting potential of *S. colletis* CASEcto12, *E. quasihormaechei* NFbEcto18, and *B. wiedmannii* NFbEndo12 rhizobacteria isolated from the rhizosphere of candelilla (*Euphorbia antisyphilitica* Zucc.) and confirm that their VOCs can stimulate the development of *A. thaliana* seedlings.

3.2. Effect of Rhizobacterial VOCs on *A. thaliana* Under Salinity Stress

A. thaliana seedlings grown on MS medium supplemented with 50 mM NaCl and exposed to rhizobacterial VOCs for 12 days exhibited significantly greater growth than the uninoculated control. For root length and fresh weight, *S. colletis* CASEcto12 yielded the highest values, averaging 64.81 mm and 133.33 mg, respectively, compared to 40.25 mm and 45.05 mg in the uninoculated control (Figures 1b and 2b).

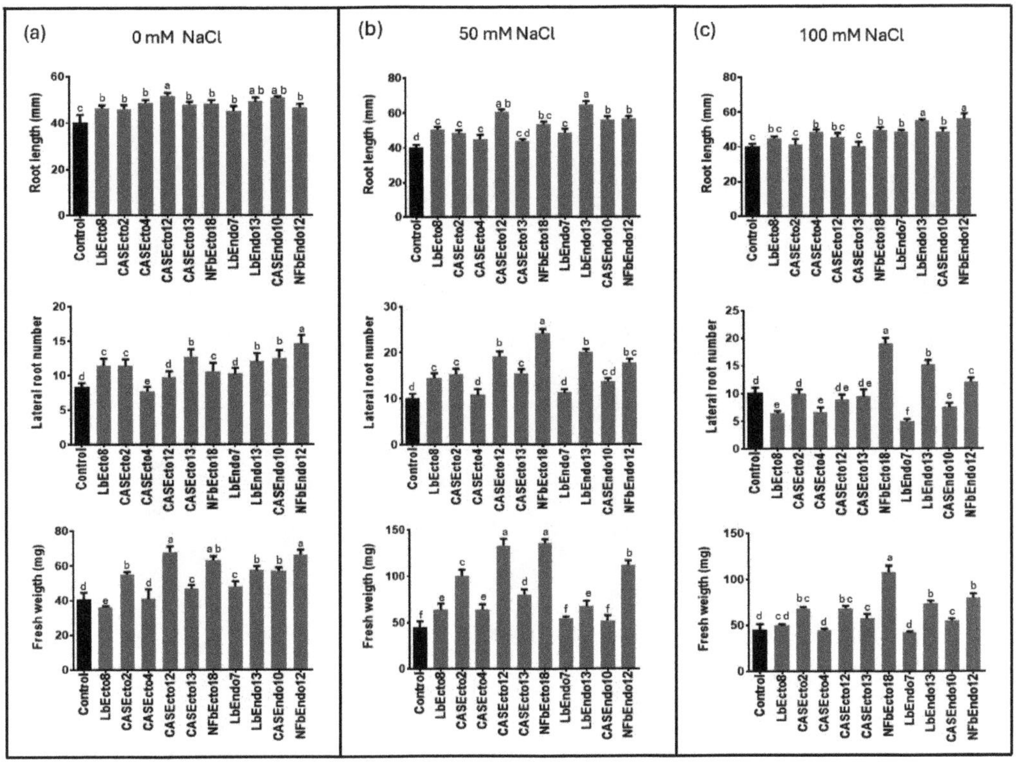

Figure 1. Effect of rhizobacterial inoculation on primary root length, lateral root number, and fresh weight of *A. thaliana* seedlings after 12 days of indirect exposure in divided Petri dishes under (**a**) no salinity, (**b**) 50 mM NaCl, and (**c**) 100 mM NaCl. Error bars represent the standard deviation of the mean ($n = 4$). Different letters indicate statistically significant differences between means ($p \leq 0.05$).

E. quasihormaechei NFbEcto18 induced the most lateral roots—24 per seedling—and the greatest total fresh weight, averaging 135.78 mg; both were significantly higher than in the uninoculated control with 10 roots per seedling and 45.05 mg of fresh weight (Figures 1b and 2c).

Under 100 mM NaCl, similar trends were observed. *Enterobacter quasihormaechei* NFbEcto18 achieved an average total fresh of 107.95 mg and 19.06 lateral roots on average, while *Bacillus wiedmannii* NFbEndo12 promoted the longest primary root length (56 mm). All measured parameters differed significantly from the uninoculated control: 45.05 mg, 10 lateral roots, and 40.25 mm, respectively (Figures 1c and 2c,d).

3.3. Extraction and Measurement of Proline in A. thaliana Plants Under Salinity Stress in Interaction with VOCs

Under non-saline conditions (MS medium without NaCl), no significant differences in proline content from aerial tissue of *A. thaliana* seedlings were observed among treatments exposed to rhizobacterial VOCs, with values ranging from 0.153 to 0.175 mmol g^{-1} fresh weight (FW).

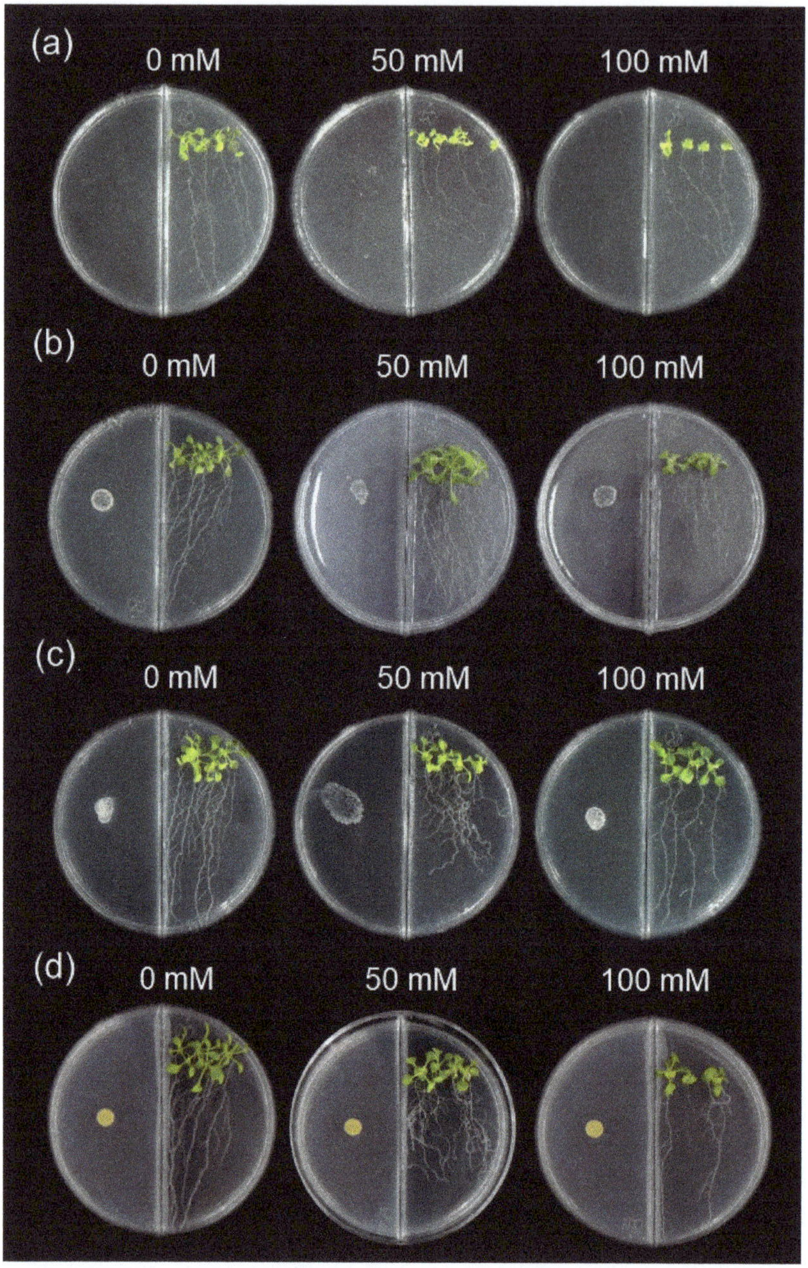

Figure 2. Effect of rhizobacterial inoculation on growth promotion in *A. thaliana* seedlings after 12 days of PGPR interactions. Seedlings were grown in divided Petri dishes on MS medium supplemented with 0, 50, or 100 mM NaCl to assess the impact of VOC emissions on growth: (**a**) uninoculated control; (**b**) *Siccibacter colletis* CASEcto12; (**c**) *Enterobacter quasihormaechei* NFbEcto18; (**d**) *Bacillus wiedmannii* NFbEndo12.

In contrast, under salt stress (50 mM and 100 mM NaCl), the aerial tissue of *A. thaliana* seedlings exposed to *Enterobacter quasihormaechei* NFbEcto18 VOCs at 12 days post-inoculation accumulated significantly more proline than the uninoculated control. At

50 mM NaCl, the proline content reached 0.657 mmol g^{-1} FW vs. 0.334 mmol g^{-1} FW in the control. At 100 mM NaCl, the proline content increased to 1.378 mmol g^{-1} FW in inoculated seedlings compared to 0.564 mmol g^{-1} FW in the control (Table 2).

Table 2. Proline accumulation in *A. thaliana* aerial tissues under salinity stress conditions of 0, 50, and 100 mM NaCl.

Rhizobacteria	Proline Concentration (mmol g^{-1} FW) Under Salinity Stress Conditions		
	0 mM	**50 mM**	**100 mM**
Control	0.166 ^{a}	0.334 ^{e}	0.564 ^{f}
Mixta theicola LBEcto8	0.166 ^{a}	0.464 ^{d}	1.150 ^{ef}
Siccibacter colletis CASEcto2	0.175 ^{a}	0.642 ^{a}	1.088 ^{e}
Staphylococcus epidermidis CASEcto4	0.155 ^{a}	0.644 ^{a}	1.179 ^{d}
Siccibacter colletis CASEcto12	0.168 ^{a}	0.566 ^{b}	1.296 ^{abc}
Staphylococcus epidermidis CASEcto13	0.158 ^{a}	0.479 ^{cd}	1.168 ^{ef}
Enterobacter quasihormaechei NFbEcto18	0.169 ^{a}	0.657 ^{a}	1.378 ^{a}
Staphylococcus epidermidis LbEndo7	0.153 ^{a}	0.644 ^{a}	1.209 ^{cd}
Bacillus subtilis LBEndo13	0.175 ^{a}	0.544 ^{b}	1.292 ^{abc}
Bacillus cereus CASEndo10	0.171 ^{a}	0.520 ^{bcd}	1.304 ^{ab}
Bacillus wiedmannii NFbEndo12	0.173 ^{a}	0.535 ^{bc}	1.227 ^{bcd}

Note: Means with different letters in each column represent significant statistical differences (Tukey, $p \leq 0.05$).

3.4. Na^{+} Concentration in *A. thaliana* Leaves Under Salinity Stress with VOC Exposure

Under non-saline conditions (MS medium without NaCl), no significant differences in Na^{+} content were observed among the PGPR treatments. In uninoculated seedlings, sodium accumulation was 0.062 mmol g^{-1} fresh weight (FW), similar to the values recorded in rhizobacteria-treated seedlings after 12 days (0.056–0.067 mmol g^{-1} FW).

Under salt stress (50 mM and 100 mM NaCl), *Enterobacter quasihormaechei* NFbEcto18 significantly reduced Na^{+} accumulation in aerial seedlings tissues, reaching only 0.074 mmol g^{-1} FW at 50 mM and 0.084 mmol g^{-1} FW at 100 mM, compared to the results in the uninoculated control with amounts of 0.102 and 0.146 mmol g^{-1} FW, respectively. These results indicate that inoculation with *Enterobacter quasihormaechei* NFbEcto18 effectively mitigated Na^{+} accumulation in *A. thaliana* under saline stress (Table 3).

Table 3. Sodium accumulation (Na^{+}) in aerial tissue *A. thaliana* seedlings under salinity stress conditions of 0, 50, and 100 mM NaCl.

Rhizobacteria	Na^{+} Concentration (mmol g^{-1} FW) Under Salinity Stress Conditions		
	0 mM	**50 mM**	**100 mM**
Control	0.062 ^{a}	0.102 ^{c}	0.146 ^{f}
Mixta theicola LBEcto8	0.067 ^{a}	0.078 ^{a}	0.110 ^{def}
Siccibacter colletis CASEcto2	0.061 ^{a}	0.110 ^{d}	0.091 ^{b}
Staphylococcus epidermidis CASEcto4	0.064 ^{a}	0.091 ^{ab}	0.086 ^{b}
Siccibacter colletis CASEcto12	0.076 ^{a}	0.076 ^{a}	0.106 ^{ef}
Staphylococcus epidermidis CASEcto13	0.062 ^{a}	0.091 ^{ab}	0.095 ^{c}
Enterobacter quasihormaechei NFbEcto18	0.059 ^{a}	0.074 ^{a}	0.084 ^{a}
Staphylococcus epidermidis LbEndo7	0.056 ^{a}	0.091 ^{ab}	0.096 ^{cd}
Bacillus subtilis LBEndo13	0.061 ^{a}	0.093 ^{ab}	0.111 ^{def}
Bacillus cereus CASEndo10	0.057 ^{a}	0.075 ^{a}	0.097 ^{cd}
Bacillus wiedmannii NFbEndo12	0.060 ^{a}	0.094 ^{bc}	0.125 ^{ef}

Note: Means with different letters in each column represent significant statistical differences (Tukey, $p \leq 0.05$).

3.5. Chlorophyll Concentration in A. thaliana Leaves Under Salinity Stress with VOC Exposure

Chlorophyll A, chlorophyll B, and total chlorophyll (A + B) were measured in *A. thaliana* leaves following exposure to rhizobacterial VOCs under 0, 50, and 100 mM NaCl.

Under non-saline conditions (0 mM NaCl), *Siccibacter colletis* CASEcto12 significantly increased chlorophyll A to 32.429 μg mL^{-1} compared with 30.214 μg mL^{-1} in the uninoculated control. *Enterobacter quasihormaechei* NFbEcto18-inoculated seedlings showed a higher chlorophyll B concentration (33.287 μg mL^{-1}) than the control (23.117 μg mL^{-1}), resulting in a greater total chlorophyll content (60.594 μg mL^{-1} vs. 53.313 μg mL^{-1}).

At 50 mM NaCl, chlorophyll A did not differ significantly among treatments; the uninoculated control measured 31.495 μg mL^{-1}. In contrast, *Staphylococcus epidermidis* CASEcto13 treatment increased chlorophyll B to 29.982 μg mL^{-1} vs. 20.574 μg mL^{-1} in the control, driving a higher total chlorophyll level (60.349 μg mL^{-1} vs. 51.069 μg mL^{-1}).

Under high salinity (100 mM NaCl), *S. epidermidis* CASEcto4 elevated the total chlorophyll to 56.481 μg mL^{-1} (control: 52.293 μg mL^{-1}). *E. quasihormaechei* NFbEcto18 again yielded the highest chlorophyll B, (33.112 μg mL^{-1}; control; 21.215 μg mL^{-1}) and total chlorophyll (62.612 μg mL^{-1}; control: 52.293 μg mL^{-1}). These data confirm the beneficial impact of VOC exposure on chlorophyll retention under salt stress (Table 4).

Table 4. Chlorophyll A, chlorophyll B, and total chlorophyll content (A + B) in *A. thaliana* tissues under salinity stress conditions of 0, 50, and 100 mM NaCl.

Rhizobacteria	Chlorophyll A			Chlorophyll B			Total Chlorophyll A + B		
	0 mM	50 Mm	100 mM	0 mM	50 mM	100 mM	0 mM	50 mM	100 mM
Control	30.214 ^{ab}	31.495 ^{a}	53.313 ^{b}	23.117 ^{b}	20.574 ^{b}	21.215 ^{bc}	53.313 ^{b}	51.069 ^{b}	52.293 ^{bc}
Mixta theicola LBEcto8	22.564 ^{d}	23.772 ^{e}	26.653 ^{e}	4.095 ^{ef}	9.507 ^{e}	3.298 ^{e}	26.653 ^{e}	32.714 ^{g}	22.925 ^{f}
Siccibacter colletis CASEcto2	22.979 ^{d}	27.434 ^{cd}	41.332 ^{d}	18.366 ^{bc}	11.235 ^{ef}	18.579 ^{cd}	41.332 ^{d}	38.007 ^{ef}	50.411 ^{cd}
Staphylococcus epidermidis CASEcto4	27.987 ^{bc}	27.230 ^{d}	29.211 ^{e}	1.230 ^{e}	15.761 ^{bcd}	23.862 ^{b}	29.211 ^{e}	42.191 ^{ef}	56.481 ^{b}
Siccibacter colletis CASEcto12	32.429 ^{a}	28.999 ^{abcd}	52.506 ^{b}	20.093 ^{bc}	12.689 ^{cde}	17.445 ^{cd}	52.506 ^{b}	40.963 ^{def}	47.740 ^{cd}
Staphylococcus epidermidis CASEcto13	24.835 ^{cd}	31.661 ^{a}	32.446 ^{e}	7.619 ^{d}	29.982 ^{a}	17.526 ^{cd}	32.446 ^{e}	60.349 ^{a}	47.167 ^{d}
Enterobacter quasihormaechei NFbEcto18	27.329 ^{bc}	31.088 ^{ab}	60.594 ^{a}	33.287 ^{a}	21.578 ^{b}	33.112 ^{a}	60.594 ^{a}	51.640 ^{b}	62.612 ^{a}
Staphylococcus epidermidis LbEndo7	31.827 ^{a}	30.460 ^{abc}	51.021 ^{b}	19.209 ^{bc}	20.473 ^{b}	20.984 ^{bc}	51.021 ^{b}	49.947 ^{bc}	51.147 ^{cd}
Bacillus subtilis LBEndo13	27.172 ^{bc}	28.297 ^{bcd}	43.573 ^{cd}	16.415 ^{c}	8.140 ^{e}	14.316 ^{d}	43.573 ^{cd}	35.862 ^{fg}	33.358 ^{e}
Bacillus cereus CASEndo10	29.517 ^{ab}	30.157 ^{abcd}	63.428 ^{a}	33.934 ^{a}	18.219 ^{bc}	19.286 ^{bc}	63.428 ^{a}	48.064 ^{bc}	49.408 ^{cd}
Bacillus wiedmannii NFbEndo12	30.008 ^{ab}	28.721 ^{abcd}	49.260 ^{bc}	19.264 ^{bc}	18.025 ^{bc}	20.241 ^{bc}	49.260 ^{bc}	44.833 ^{cd}	48.946 ^{cd}

Note: Means with different letters in each column represent significant statistical differences (Tukey, $p \leq 0.05$).

3.6. Identification of VOCs

VOC profiling using SPME-GC-MS was performed on two bacterial strains—*Enterobacter quasihormaechei* NFbEcto18 and *Siccibacter colletis* CASEcto12—both of which significantly influenced multiple physiological and biochemical parameters in *A. thaliana* seedlings, including root length, number of lateral roots, fresh weight, proline content, and sodium accumulation.

Aldehydes, ketones, and alcohols were among the predominant volatile compounds detected in both strains across all tested NaCl concentrations (0, 50, and 100 mM) (Tables 5 and 6).

Table 5. Volatile organic compounds identified in the *Enterobacter quasihormaechei* NFBEcto18 in 0, 50, and 100 mM NaCl conditions.

		Average (% Area)		
Compound	**RT (min)**	**0 mM**	**50 mM**	**100 mM**
Methanethiol	2.582	0	0.94	0.68
3-tetradecanone	5.263	7.73	2.30	0.39
2-pentanamine	6.260	0.00	1.23	0.91
2-pentadecanol	12.998	9.23	1.79	0.24
2,3-Butanedione	15.330	0.00	0.30	0.26
1-hexanol	17.213	29.66	5.09	1.26
Dimethyl disulfide	21.885	7.30	1.75	0.23
1-octanol	24.836	0.00	1.05	0.81
3-Methyl-2-butanol	25.209	0.00	0.91	0.40
Acetophenone	26.861	6.94	2.23	0.90
6-methyl-2-heptanol	29.316	3.52	1.15	0.94
2,3-Butanediol	32.261	0.00	0.92	0.42

Relative proportion of volatile compound = (peak area of volatile compound)/(total peak area of all volatile compounds). Values represent the average of three replicates.

Table 6. Volatile organic compounds identified in *Siccibacter colletis* CASEcto12 in 0, 50, and 100 mM NaCl conditions.

		Average (% Area)		
Compound	**RT (min)**	**0 mM**	**50 mM**	**100 mM**
Dimethyl disulfide	5.263	0.00	0.76	0.21
3-tetradecanone	6.260	11.32	5.30	1.30
9-octadecanona	12.998	13.08	8.05	2.17
3-Methyl-2-butanol	15.330	0.00	1.35	0.36
1-hexanol	18.213	10.31	6.68	1.75
1-tridecanol	23.850	2.74	0.82	0.30
Acetophenone	25.630	53.47	29.36	8.35
2,3-Butanedione	26.861	0.00	2.04	1.79
6-methyl-2-heptanol	28.210	6.25	1.05	0.84
2,3-Butanediol	32.261	0.00	0.94	0.21
1-butanol	34.261	1.14	0.74	0.22

Relative proportion of volatile compound = (peak area of volatile compound)/(total peak area of all volatile compounds). Values represent the average of three replicates.

3.6.1. *Enterobacter quasihormaechei* NFbEcto18

VOC emissions from *E. quasihormaechei* NFbEcto18 declined markedly with increasing salinity. In the absence of NaCl, the most abundant emitted compound was 1-hexanol, comprising 29.66% of the total VOCs profile. Its relative abundance dropped to 5.09% at 50 mM and to just 1.26% at 100 mM, representing a reduction of over 95% relative to the control. Other compounds that showed substantial decreases included 2-pentadecanol (from 9.23% to 0.24%), 3-tetradecanone (from 7.73% to 0.39%), dimethyl disulfide (from 7.30% to 0.23%), acetophenone (from 6.94% to 0.90%), and 6-methyl-2-heptanol (from 3.52% to 0.94%) (Table 5).

Despite the overall decline, a few compounds were newly detected or slightly increased under saline conditions, albeit in low proportions. For example, 2-pentanamine was absent in the control but appeared at 1.23% in 50 mM and 0.91% in 100 mM. Similarly, 1-octanol and 3-methyl-2-butanol emerged under saline conditions, though neither exceeded 1.05% (Table 5).

3.6.2. *Siccibacter colletis* CASEcto12

Similarly to *E. quasihormaechei* NFbEcto18, exposure to salinity led to a pronounced reduction in VOC emissions from *S. colletis* CASEcto12. Under non-saline conditions, acetophenone was the predominant compound (53.47%), but its abundance declined to 29.36% at 50 mM and to 8.35% at 100 mM NaCl—an overall reduction of more than 80%. Other compounds with notable decreases included 9-octadecanone (from 13.08% to 2.17%), 3-tetradecanone (from 11.32% to 1.30%), 1-hexanol (from 10.31% to 1.75%), and 6-methyl-2-heptanol (from 6.25% to 0.84%) (Table 6).

Despite the general decline in emissions, several compounds were induced by salinity: 2,3-butanedione, absent in the control, appeared at 2.04% (50 mM) and 1.79% (100 mM). Similarly, 3-methyl-2-butanol emerged at 1.35% at 50 mM, and dimethyl disulfide reached 0.76% under this condition (Table 6).

4. Discussion

The current study advances our understanding of PGPR-mediated salinity stress alleviation by identifying two potent strains—*Siccibacter colletis* CASEcto12 and *Enterobacter quasihormaechei* NFbEcto18—isolated from the rhizosphere of Candelilla (*Euphorbia antisyphilitica* Zucc.) and characterized as PGPR in the published work of Salazar-Ramirez [19]. Candelilla is an endemic plant of the Chihuahuan desert, adapted to drought, salinity, and high temperatures. The ecological origin of these isolates suggests that they are preadapted to harsh environments, representing a unique resource for developing resilient biostimulants. Our primary finding is that the VOC blends emitted by both strains elicited significant growth promotion in *A. thaliana* seedlings under both optimal and saline conditions (50 and 100 mM NaCl). Although *S. colletis* and *E. quasihormaechei* have been previously reported as PGPR [25–27], their VOCs had not been characterized nor had their protective effects against stresses such as salinity been evaluated. This study provides the first report of these properties, positioning these PGPR as promising bioinoculants for mitigating the increasing challenges of soil salinization in agriculture.

The emission of VOCs by rhizobacteria is widely recognized as a key mechanism for promoting plant growth, acting as short- or long-distance signal molecules [16,24,28–30]. This study aligns with pioneering evidence from Ryu [28,30], who demonstrated that exposing *A. thaliana* to VOCs from *Bacillus amyloliquefaciens* GB03 resulted in an increase of up to fivefold in leaf area. Similarly, *Paenibacillus polymyxa* E681 has been reported to double leaf biomass in *A. thaliana* [31]. In our work, the strains *S. colletis* CASEcto12 and *E. quasihormaechei* NFbEcto18 markedly enhanced the primary root length, lateral root number, and fresh weight of *A. thaliana* seedlings under non-saline conditions (Figure 1a). These results support the hypothesis that bacterial VOCs modulate plant development, potentially through molecular pathways that modulate the expressions of cell division and auxin-related genes.

Beyond growth promotion, VOCs are known to induce tolerance to various abiotic stresses [32]. PGPR can activate diverse physiological and biochemical mechanisms in *Arabidopsis* seedlings to cope with saline stress. In this study, we observed that *A. thaliana* seedlings exposed to VOCs from *S. colletis* CASEcto12, *E. quasihormaechei* NFbEcto18, and *Bacillus wiedmannii* NFbEndo12 showed significantly enhanced development and growth compared with seedlings not exposed to VOCs under salt stress (50 and 100 mM NaCl) (Figure 2). These findings align with Bhattacharyya [33] and Luo [18], who observed that VOCs emitted by *Alcaligenes faecalis* JBCS1294 and *Burkholderia pyrrocinia* CNUC9, respectively, improved fresh weight, primary root length, and lateral root number in *A. thaliana* exposed to 100 mM NaCl.

The root system can uptake water and nutrients from the surroundings; thus, more developed radical systems could improve resistance to various stresses [34,35]. Root development in *A. thaliana* seedlings was markedly impaired by salinity, as evidenced by reduced primary and lateral root growth in the controls. However, these effects were substantially mitigated by exposure to VOCs from PGPR. VOCs from *S. colletis* CASEcto12 notably restored primary root elongation and biomass at 50 mM NaCl, while *E. quasihormaechei* NFbEcto18 promoted a significant increase in lateral root formation, highlighting strain-specific impacts on root system architecture (Figure 2). Although the precise mechanisms by which these PGPR alter root hormonal networks remain to be elucidated, these findings are consistent with reports that PGPR volatiles influence auxin- and cytokinin-regulated developmental pathways. This suggests that the distinct VOC profiles of these strains differentially modulate hormonal networks to enhance seedling resilience under salt stress [28,36].

Under saline stress, plants undergo physiological and biochemical changes. Proline accumulation is a key indicator of the stress response. One major adaptation to salinity is the accumulation of compatible solutes such as proline, which act as osmoprotectants and antioxidant defense molecules, helping to maintain osmotic balance under salinity stress [18,37–39]. PGPR inoculation has been shown to enhance proline accumulation in plants under salt stress [40]. In our study, *E. quasihormaechei* NFbEcto18 significantly increased proline levels in *A. thaliana* seedlings under both 50 and 100 mM NaCl, suggesting VOC-mediated activation of osmotic defense pathways (Table 3). Similar results were reported for *Kocuria rhizophila* Y1, which increased proline content in maize under salinity stress [41]. Nawaz [3] also demonstrated that consortium inoculation with *Pseudomonas fluorescens*, *Bacillus pumilus*, and *Exiguobacterium aurantiacum* significantly increased free proline content in wheat (*Triticum aestivum* L.) under saline stress conditions.

Chlorophyll—the vital green pigment that enables plants to absorb light energy for growth and development—is also affected by salinity. Under saline conditions, chlorophyll content decreases which impairs photosynthesis [42]. Ledger [43] showed that VOCs from *Bacillus* sp. GB03 and *Paraburkholderia phytofirmans* PsJN modulate chlorophyll content in *A. thaliana*, while Taïbi [44] documented progressive chlorophyll and carotenoid loss in *Phaseolus vulgaris* with increasing salinity. Our results demonstrated that exposure to VOCs from *E. quasihormaechei* NFbEcto18 under 100 mM NaCl yielded the highest chlorophyll B and total chlorophyll levels (Table 4). These findings are consistent with Liu [39] and Luo [18], who reported that VOCs from *Bacillus amyloliquefaciens* FZB42 and *Burkholderia pyrrocinia* CNUC9 increased the total chlorophyll in salinity-stressed *A. thaliana*.

Plants exposed to salt stress experience ion toxicity due to the increased concentration of Na^+. Beneficial microbes such as PGPR can alleviate Na^+ toxicity in plants through various mechanisms, including restricting uptake, extruding ions from tissues, sequestering them in vacuoles, redirecting Na^+ from shoots to roots, and increasing the K^+/Na^+ ratio [45,46]. For example, in *A. thaliana*, seedlings exposed to VOCs from *Bacillus subtilis* GB03 accumulated less Na^+ in both roots and shoots [28]. Similarly, inoculation of maize plants with *Bacillus amyloliquefaciens* SQR9 lowered Na^+ content compared with non-inoculated plants [47]. Liu [39] reported that VOCs from *Bacillus amyloliquefaciens* FZB42 significantly decreased Na^+ content in *A. thaliana* and induced the expression of genes such as *NHX1* (Na^+/H^+ exchanger 1) and *HKT1* (high-affinity K^+ transporter 1) which function to alleviate Na^+ toxicity. In the present study, we also found that VOCs emitted by *E. quasihormaechei* NFbEcto18 reduced Na^+ accumulation in the shoots of *A. thaliana* seedlings compared with control seedlings under salinity stress that were not exposed to NFbEcto18

VOCs (Table 4). Although this study did not explore the transcriptional regulation of *NHX1* and *HKT1* in *Arabidopsis*, these transcripts are likely involved in the observed decrease in Na^+ in the aerial parts of plants exposed to PGPR VOCs from NFbEcto18.

SPME-GC-MS profiling of VOCs emitted by *E. quasihormaechei* NFbEcto18 and *S. colletis* CASEcto12 revealed that aldehydes, ketones, and alcohols dominated their spectra (Tables 5 and 6). Notably, 1-hexanol and acetophenone were the most abundant compounds in the non-saline controls for *E. quasihormaechei* NFbEcto18 and *S. colletis* CASEcto12, respectively. Importantly, the abundance of these VOCs decreased markedly when 50 mM NaCl was applied, and this reduction was further accentuated at 100 mM NaCl. However, some VOCs were detected only after NaCl application (50 mM) and even at higher concentrations under greater stress (100 mM). Salinity shifted the VOC profiles, inducing compounds such as methanethiol, 2-pentanamine, 2,3-butanedione, and 2,3-butanediol, which may reflect stress-triggered alterations in bacterial secondary metabolism. VOC production varies greatly with bacterial genotype, growth phase, nutrient availability, oxygen concentration, and inoculum density [48]. Shestivska [49] documented up to 100-fold differences in VOC profiles among 36 *Pseudomonas aeruginosa* strains grown under similar conditions, underscoring strain-specific emissions. Specific VOCs have been linked to salt tolerance; e.g., *Raoultella aquatilis* JZ-GX1 emits 2,3-butanediol which reduces Na^+ accumulation and boosts antioxidant activity in *Robinia pseudoacacia* under 100 mM NaCl [17]. Ayuso-Calles [50] observed that salinity triggers the loss of some volatiles and the emergence of others (e.g., acetoin, 3-heptanone) in *Rhizobium* sp., supporting adaptive modulation of VOC profiles. Reviews of halotolerant PGPR also highlight compounds such as dimethyl disulfide, 2,3-butanediol, acetoin, and tridecane as differentially regulated under saline conditions [51]. The current data regarding VOC biosynthesis pathways are insufficient to elucidate these phenomena. Consequently, a more in-depth investigation into the metabolic processes and genetic regulation governing VOC biosynthesis is imperative, along with the determination of their mechanisms of action on diverse plant species.

5. Conclusions

Our results demonstrate that VOCs emitted by rhizobacteria play a pivotal role in promoting the growth of *A. thaliana* and enhancing its tolerance to salt stress. Specifically, *Siccibacter colletis* CASEcto12, *Enterobacter quasihormaechei* NFbEcto18, and *Bacillus wiedmannii* NFbEndo12 each significantly increased seedling biomass, root development, and proline accumulation, underscoring their potential as agricultural biostimulants.

Moreover, exposure to these bacterial VOCs significantly reduced sodium accumulation in *A. thaliana* tissues under saline conditions, suggesting a protective mechanism linked to the modulation of proline metabolism and regulation of ion transport. VOC profiling identified aldehydes, ketones, and alcohols as dominant compound classes, with 1-hexanol, acetophenone, 2,3-Butanedione, and 2-pentanamine emerging as promising candidate bioactives. Although the beneficial effects of VOC-producing PGPR are evident, further research is needed to identify the specific compounds responsible for growth promotion and salinity tolerance and to validate their individual bioactivity through pure-compound assays. Additional steps include elucidating the molecular signaling pathways activated in plants, evaluating VOC stability and diffusion in soil environments, and testing these strains under greenhouse and field conditions, where environmental variability may influence performance. A key challenge is scaling VOC-based technologies to agricultural applications while ensuring consistent activity across different crop species and soil types. Importantly, VOC-emitting PGPR hold strong potential for crops impacted by salinity, such as sugarcane, where salt stress reduces root vigor, sucrose accumulation, and overall yield. Implementing these strains—or their most active VOCs—could enhance root architecture, improve nutrient and water uptake, and strengthen physiological resilience, contributing to more sustainable and productive sugarcane cultivation in saline or marginal lands. Taken together, these findings confirm the dual function of rhizobacterial VOCs as growth-promoting agents and stress-mitigating factors. They position bacterial volatiles as promising biotechnological tools for the development of sustainable agricultural practices in saline and otherwise challenging environments.

Author Contributions: Conceptualization, M.T.S.-R. and J.S.-M.; methodology, R.P.-R.; software, J.J.Q.-R.; validation, T.E.V.-C. and G.M.-P.; formal analysis, G.M.-P.; investigation, M.T.S.-R.; resources, H.I.O.-R.; data curation, A.G.Y.-C.; writing—original draft preparation, M.T.S.-R., J.J.Q.-R., and J.S.-M.; writing—review and editing, J.S.-M., J.J.Q.-R., and R.P.-R.; visualization, G.A.V.-H.; supervision, J.S.-M.; project administration, J.A.O.-V. All authors have read and agreed to the published version of the manuscript.

Funding: This research was funded by SECIHTI (Secretaría de Ciencia, Humanidades, Tecnología e Innovación) through a doctoral scholarship awarded to María Teresa Salazar Ramírez (number 355051).

Institutional Review Board Statement: Not applicable.

Informed Consent Statement: Not applicable.

Data Availability Statement: The original contributions presented in this study are included in the article. Further inquiries can be directed to the corresponding authors.

Acknowledgments: During the preparation of this manuscript, the authors used Microsoft Copilot (version as of August 2025) for the purposes of language assistance during manuscript preparation, specifically for improving syntax, punctuation, and grammar. The authors have reviewed and edited the output and take full responsibility for the content of this publication.

Conflicts of Interest: The authors declare no conflicts of interest.

References

1. Zaman, M.; Shahid, S.A.; Heng, L. *Guideline for Salinity Assessment, Mitigation and Adaptation Using Nuclear and Related Techniques*; Springer International Publishing: Cham, Switzerland, 2018. [CrossRef]
2. Vargas, R. *Mapping of Salt-Affected Soils—Technical Manual*; FAO: Rome, Italy, 2020. [CrossRef]
3. Nawaz, A.; Shahbaz, M.; Asadullah; Imran, A.; Marghoob, M.U.; Imtiaz, M.; Mubeen, F. Potential of Salt Tolerant PGPR in Growth and Yield Augmentation of Wheat (*Triticum aestivum* L.) Under Saline Conditions. *Front. Microbiol.* **2020**, *11*, 2019. [CrossRef]
4. Ali, B.; Wang, X.; Saleem, M.H.; Sumaira; Hafeez, A.; Afridi, M.S.; Khan, S.; Zaib-Un-Nisa; Ullah, I.; Amaral Júnior, A.T.d.; et al. PGPR-Mediated Salt Tolerance in Maize by Modulating Plant Physiology, Antioxidant Defense, Compatible Solutes Accumulation and Bio-Surfactant Producing Genes. *Plants* **2022**, *11*, 345. [CrossRef]
5. Cappellari, L.d.R.; Chiappero, J.; Palermo, T.B.; Giordano, W.; Banchio, E. Volatile Organic Compounds from Rhizobacteria Increase the Biosynthesis of Secondary Metabolites and Improve the Antioxidant Status in *Mentha piperita* L. Grown under Salt Stress. *Agronomy* **2020**, *10*, 1094. [CrossRef]
6. Ha-Tran, D.M.; Nguyen, T.T.M.; Hung, S.-H.; Huang, E.; Huang, C.-C. Roles of Plant Growth-promoting Rhizobacteria (PGPR) in Stimulating Salinity Stress Defense in Plants: A Review. *Int. J. Mol. Sci.* **2021**, *22*, 3154. [CrossRef]
7. Basu, A.; Prasad, P.; Das, S.N.; Kalam, S.; Sayyed, R.Z.; Reddy, M.S.; El Enshasy, H. Plant Growth Promoting Rhizobacteria (PGPR) as Green Bioinoculants: Recent Developments, Constraints, and Prospects. *Sustainability* **2021**, *13*, 1140. [CrossRef]
8. Kottb, M.; Gigolashvili, T.; Großkinsky, D.K.; Piechulla, B. *Trichoderma* Volatiles Effecting *Arabidopsis*: From Inhibition to Protection against Phytopathogenic Fungi. *Front. Microbiol.* **2015**, *6*, 995. [CrossRef]
9. Lemfack, M.C.; Gohlke, B.-O.; Toguem, S.M.T.; Preissner, S.; Piechulla, B.; Preissner, R. MVOC 2.0: A Database of Microbial Volatiles. *Nucleic Acids Res.* **2018**, *46*, D1261–D1265. [CrossRef]
10. Almeida, O.A.C.; de Araujo, N.O.; Dias, B.H.S.; de Sant'Anna Freitas, C.; Coerini, L.F.; Ryu, C.-M.; de Castro Oliveira, J.V. The Power of the Smallest: The Inhibitory Activity of Microbial Volatile Organic Compounds against Phytopathogens. *Front. Microbiol.* **2023**, *13*, 951130. [CrossRef]
11. Macías-Rubalcava, M.L.; Sánchez-Fernández, R.E.; Roque-Flores, G.; Lappe-Oliveras, P.; Medina-Romero, Y.M. Volatile Organic Compounds from *Hypoxylon anthochroum* Endophytic Strains as Postharvest Mycofumigation Alternative for Cherry Tomatoes. *Food Microbiol.* **2018**, *76*, 363–373. [CrossRef]

12. Morath, S.U.; Hung, R.; Bennett, J.W. Fungal Volatile Organic Compounds: A Review with Emphasis on Their Biotechnological Potential. *Fungal Biol. Rev.* **2012**, *26*, 73–83. [CrossRef]
13. Zhao, X.; Zhou, J.; Tian, R.; Liu, Y. Microbial Volatile Organic Compounds: Antifungal Mechanisms, Applications, and Challenges. *Front. Microbiol.* **2022**, *13*, 922450. [CrossRef]
14. Garbeva, P.; Weisskopf, L. Airborne medicine—Bacterial volatiles and their influence on plant health: Transley review. *New Phytol.* **2020**, *226*, 32–43. [CrossRef]
15. Montejano-Ramírez, V.; Ávila-Oviedo, J.L.; Campos-Mendoza, F.J.; Valencia-Cantero, E. Microbial volatile organic compounds: Insights into plant defense. *Plants* **2024**, *13*, 2013. [CrossRef]
16. Liu, X.-M.; Zhang, H. The effects of bacterial volatile emissions on plant abiotic stress tolerance. *Front. Plant Sci.* **2015**, *6*, 774. [CrossRef]
17. Li, P.-S.; Kong, W.-L.; Wu, X.-Q.; Zhang, Y. Volatile organic compounds of the plant growth-promoting rhizobacteria JZ-GX1 enhanced the tolerance of *Robinia pseudoacacia* to salt stress. *Front. Plant Sci.* **2021**, *12*, 753332. [CrossRef]
18. Luo, H.; Riu, M.; Ryu, C.-M.; Yu, J.M. Volatile organic compounds emitted by *Burkholderia pyrrocinia* CNUC9 trigger induced systemic salt tolerance in *Arabidopsis thaliana*. *Front. Microbiol.* **2022**, *13*, 1050901. [CrossRef]
19. Salazar-Ramírez, M.T.; Sáenz-Mata, J.; Preciado-Rangel, P.; Fortis-Hernández, M.; Rueda-Puente, E.O.; Yescas-Coronado, P.; Orozco-Vidal, J.A. Plant Growth-Promoting Rhizobacteria associated to candelilla rhizosphere (*Euphorbia antisyphilitica*) and its effects on *Arabidopsis thaliana* seedlings. *Not. Bot. Horti Agrobot. Cluj-Napoca* **2021**, *49*, 12294. [CrossRef]
20. Yang, Y.; Meier, F.; Ann Lo, J.; Yuan, W.; Lee Pei Sze, V.; Chung, H.J.; Yuk, H.-G. Overview of recent events in the microbiological safety of sprouts and new intervention technologies: Sprout safety and control. *Compr. Rev. Food Sci. Food Saf.* **2013**, *12*, 265–280. [CrossRef]
21. López-Bucio, J.; Campos-Cuevas, J.C.; Hernández-Calderón, E.; Velásquez-Becerra, C.; Farías-Rodríguez, R.; Macías-Rodríguez, L.I.; Valencia-Cantero, E. Bacillus megaterium rhizobacteria promote growth and alter root-system architecture through an auxin-and ethylene-independent signaling mechanism in Arabidopsis thaliana. *Mol. Plant-Microbe Interact.* **2007**, *20*, 207–217. [CrossRef]
22. Bates, L.; Waldren, R.P.; Teare, I.D. Rapid determination of free proline for water stress studies. *Plant Soil.* **1973**, *39*, 205–207. [CrossRef]
23. Ievinsh, G.; Meža, S.; Karlsons, A.; Osvalde, A. Validation of ion-selective electrodes for measurement of Na^+ and K^+ concentration in halophyte tissues. In Proceedings of the 76th Scientific Conference of the University of Latvia, Riga, Latvia, January–April 2018; p. 75.
24. Zhang, Z.; Huang, R. Analysis of Malondialdehyde, Chlorophyll Proline, Soluble Sugar, and Glutathione Content in *Arabidopsis* seedling. *Bio-Protocol* **2013**, *3*, e817. [CrossRef]
25. Chamkhi, I.; Zwanzig, J.; Ibnyasser, A.; Cheto, S.; Geistlinger, J.; Saidi, R.; Ghoulam, C. *Siccibacter colletis* as a member of the plant growth-promoting rhizobacteria consortium to improve faba-bean growth and alleviate phosphorus deficiency stress. *Front. Sustain. Food Syst.* **2023**, *7*, 1134809. [CrossRef]
26. Goulart, J.T.d.S.S.; Quintanilha-Peixoto, G.; Esteves, B.d.S.; de Souza, S.A.; Lopes, P.S.; da Silva, N.D.; Soares, J.R.; Barroso, L.M.; Suzuki, M.S.; Intorne, A.C. Isolation and Characterization of Plant-Growth-Promoting Bacteria Associated with *Salvinia auriculata* Aublet. *Microorganisms* **2024**, *12*, 1842. [CrossRef]
27. Misra, S.; Semwal, P.; Pandey, D.D.; Mishra, S.K.; Chauhan, P.S. Siderophore-producing *Spinacia oleracea* bacterial endophytes enhance nutrient status and vegetative growth under iron-deficit conditions. *J. Plant Growth Regul.* **2024**, *43*, 1317–1330. [CrossRef]
28. Ryu, C.M.; Farag, M.A.; Hu, C.H.; Reddy, M.S.; Wei, H.X.; Paré, P.W.; Kloepper, J.W. Bacterial volatiles promote growth in *Arabidopsis*. *Proc. Natl. Acad. Sci. USA* **2003**, *100*, 4927–4932. [CrossRef]
29. Park, Y.S.; Dutta, S.; Ann, M.; Raaijmakers, J.M.; Park, K. Promotion of plant growth by *Pseudomonas fluorescens* strain SS101 via novel volatile organic compounds. *Biochem. Biophys. Res. Commun.* **2015**, *461*, 361–365. [CrossRef]
30. Ryu, C.-M.; Farag, M.A.; Hu, C.-H.; Reddy, M.S.; Kloepper, J.W.; Paré, P.W. Bacterial Volatiles Induce Systemic Resistance in *Arabidopsis*. *Plant Physiol.* **2004**, *134*, 1017–1026. [CrossRef]
31. Lee, B.; Farag, M.A.; Park, H.B.; Kloepper, J.W.; Lee, S.H.; Ryu, C.M. Induced Resistance by a Long-Chain Bacterial Volatile: Elicitation of Plant Systemic Defense by a C13 Volatile Produced by *Paenibacillus polymyxa*. *PLoS ONE* **2012**, *7*, e48744. [CrossRef]
32. Orozco-Mosqueda, M.C.; Velázquez-Becerra, C.; Macías-Rodríguez, L.I.; Santoyo, G.; Flores-Cortez, I.; Alfaro-Cuevas, R.; Valencia-Cantero, E. *Arthrobacter agilis* UMCV2 Induces Iron Acquisition in *Medicago truncatula* (Strategy I Plant) in Vitro via Dimethylhexadecylamine Emission. *Plant Soil.* **2013**, *362*, 51–66. [CrossRef]
33. Bhattacharyya, D.; Yu, S.-M.; Lee, Y.H. Volatile Compounds from *Alcaligenes faecalis* JBCS1294 Confer Salt Tolerance in *Arabidopsis thaliana* through the Auxin and Gibberellin Pathways and Differential Modulation of Gene Expression in Root and Shoot Tissues. *Plant Growth Regul.* **2015**, *75*, 297–306. [CrossRef]

34. Wang, S.; Na, X.; Yang, L.; Liang, C.; He, L.; Jin, J.; Liu, Z.; Qin, J.; Li, J.; Wang, X.; et al. Bacillus megaterium strain WW1211 promotes plant growth and lateral root initiation via regulation of auxin biosynthesis and redistribution. *Plant Soil.* **2021**, *466*, 491–504. [CrossRef]
35. Li, Y.; Shao, J.; Xie, Y.; Jia, L.; Fu, Y.; Xu, Z.; Zhang, N.; Feng, H.; Xun, W.; Liu, Y.; et al. Volatile compounds from beneficial rhizobacteria Bacillus spp. promote periodic lateral root development in Arabidopsis. *Plant Cell Environ.* **2021**, *44*, 1663–1678. [CrossRef]
36. Cho, S.M.; Kim, H.S.; Kim, Y.H.; Ryu, C.M.; Kim, M.K. Bacterial volatile compound 2,3-butanediol enhances salt stress tolerance by regulating auxin and cytokinin balance in *Arabidopsis thaliana*. *Front. Plant Sci.* **2017**, *8*, 1056. [CrossRef]
37. Verslues, P.E.; Sharma, S. Proline Metabolism and Its Implications for Plant- Environment Interaction. *Arab. Book/Am. Soc. Plant Biol.* **2010**, *8*, e0140. [CrossRef]
38. Ashraf, M.; Harris, P.J. Potential biochemical indicators of salinity tolerance in plants. *Plant Sci.* **2004**, *166*, 3–16. [CrossRef]
39. Liu, S.; Tian, Y.; Jia, M.; Lu, X.; Yue, L.; Zhao, X.; Jin, W.; Wang, Y.; Zhang, Y.; Xie, Z.; et al. Induction of Salt Tolerance in *Arabidopsis thaliana* by Volatiles From *Bacillus amyloliquefaciens* FZB42 via the Jasmonic Acid Signaling Pathway. *Front. Microbiol.* **2020**, *11*, 562934. [CrossRef]
40. Abbas, R.; Rasul, S.; Aslam, K.; Baber, M.; Shahid, M.; Mubeen, F.; Naqqash, T. Halotolerant PGPR: A hope for cultivation of saline soils. *J. King Saud Univ. Sci.* **2019**, *31*, 1195–1201. [CrossRef]
41. Li, X.; Sun, P.; Zhang, Y.; Jin, C.; Guan, C. A novel PGPR strain *Kocuria rhizophila* Y1 enhances salt stress tolerance in maize by regulating phytohormone levels, nutrient acquisition, redox potential, ion homeostasis, photosynthetic capacity and stress-responsive genes expression. *Environ. Exp. Bot.* **2020**, *174*, 104023. [CrossRef]
42. Saghafi, K.; Ahmadi, J.; Asgharzadeh, A.; Bakhtiari, S. The effect of microbial inoculants on physiological responses of two wheat cultivars under salt stress. *Int. J. Adv. Biol. Biomed. Res.* **2013**, *1*, 421–431.
43. Ledger, T.; Rojas, S.; Timmermann, T.; Pinedo, I.; Poupin, M.J.; Garrido, T.; Richter, P.; Tamayo, J.; Donoso, R. Volatile-Mediated Effects Predominate in *Paraburkholderia phytofirmans* Growth Promotion and Salt Stress Tolerance of *Arabidopsis thaliana*. *Front. Microbiol.* **2016**, *7*, 1838. [CrossRef]
44. Taïbi, K.; Ait Abderrahim, L.; Boussaid, M.; Bissoli, G.; Taïbi, F.; Achir, M.; Souana, K.; Mulet, J.M. Salt-Tolerance of *Phaseolus vulgaris* L. Is a Function of the Potentiation Extent of Antioxidant Enzymes and the Expression Profiles of Polyamine Encoding Genes. *S. Afr. J. Bot.* **2021**, *140*, 114–122. [CrossRef]
45. Chen, X.; Zhang, X.; Jia, A.; Xu, G.; Hu, H.; Hu, X.; Hu, L. Jasmonate mediates salt-induced nicotine biosynthesis in tobacco (*Nicotiana tabacum* L.). *Plant Diver.* **2016**, *38*, 118–123. [CrossRef]
46. Upadhyay, S.K.; Singh, J.S.; Singh, D.P. Exopolysaccharide-producing plant growth-promoting rhizobacteria under salinity condition. *Pedosphere* **2011**, *21*, 214–222. [CrossRef]
47. Chen, L.; Liu, Y.; Wu, G.; Veronican Njeri, K.; Shen, Q.; Zhang, N.; Zhang, R. Induced maize salt tolerance by rhizosphere inoculation of *Bacillus amyloliquefaciens* SQR9. *Physiol. Plant.* **2016**, *158*, 34–44. [CrossRef]
48. Sánchez Aguilar, M.C.; Espuny Gómez, M.d.R. Compuestos Orgánicos Volátiles (COV) En Las Interacciones Planta-Microorganismo. Bachelor's Thesis, Universidad de Sevilla, Seville, Spain, 2021.
49. Shestivska, V.; Španěl, P.; Dryahina, K.; Sovová, K.; Smith, D.; Musílek, M.; Nemec, A. Variability in the Concentrations of Volatile Metabolites Emitted by Genotypically Different Strains of *Pseudomonas aeruginosa*. *J. Appl. Microbiol.* **2012**, *113*, 701–713. [CrossRef] [PubMed]
50. Ayuso-Calles, M.; Flores-Félix, J.D.; Amaro, F.; García-Estévez, I.; Jiménez-Gómez, A.; de Pinho, P.G.; Escribano-Bailón, M.T.; Rivas, R. Effect of *Rhizobium* Mechanisms in Improving Tolerance to Saline Stress in Lettuce Plants. *Chem. Biol. Technol. Agric.* **2023**, *10*, 89. [CrossRef]
51. Sunita, K.; Mishra, I.; Mishra, J.; Prakash, J.; Arora, N.K. Secondary Metabolites From Halotolerant Plant Growth Promoting Rhizobacteria for Ameliorating Salinity Stress in Plants. *Front. Microbiol.* **2020**, *11*, 567768. [CrossRef]

agriculture

MDPI

Article

Rhizobacteria's Effects on the Growth and Competitiveness of *Solidago canadensis* Under Nutrient Limitation

Zhi-Yun Huang [1], Ying Li [1], Hu-Anhe Xiong [1], Misbah Naz [2], Meng-Ting Yan [1], Rui-Ke Zhang [1], Jun-Zhen Liu [1], Xi-Tong Ren [1,3], Guang-Qian Ren [1], Zhi-Cong Dai [1,3,4,*] and Dao-Lin Du [5,*]

1 Institute of Environment and Ecology, School of the Environment and Safety Engineering, Jiangsu University, Zhenjiang 212013, China; huangzhiyun2024@163.com (Z.-Y.H.); liying24117@163.com (Y.L.); xhah0518@163.com (H.-A.X.); yanmengting0412@163.com (M.-T.Y.); zhangruike0601@163.com (R.-K.Z.); liujunzhen1111@163.com (J.-Z.L.); r1010828705@163.com (X.-T.R.); rgq@ujs.edu.cn (G.-Q.R.)

2 Key Laboratory of Green Pesticide and Agricultural Bioengineering, Ministry of Education, Guizhou University, Guiyang 550025, China; misbahnaz.ray@yahoo.com

3 School of Agricultural Engineering, Jiangsu University, Zhenjiang 212013, China

4 Jiangsu Collaborative Innovation Center of Technology and Material of Water Treatment, Suzhou University of Science and Technology, Suzhou 215009, China

5 Jingjiang College, Jiangsu University, Zhenjiang 212013, China

* Correspondence: daizhicong@163.com (Z.-C.D.); ddl@ujs.edu.cn (D.-L.D.)

Abstract

The role of rhizosphere bacteria in facilitating plant invasion is increasingly acknowledged, yet the influence of specific microbial functional traits remains insufficiently understood. This study addresses this gap by isolating two bacterial strains, *Bacillus* sp. ScRB44 and *Pseudomonas* sp. ScRB22, from the rhizosphere of the invasive weed *Solidago canadensis*. We assessed their nitrogen utilization capacity and indoleacetic acid (IAA) production capabilities to evaluate their ecological functions. Our three-stage experimental design encompassed strain promotion, nutrient stress, and competition phases. *Bacillus* sp. ScRB44 demonstrated robust IAA production and significantly improved the nitrogen utilization efficiency, significantly enhancing *S. canadensis* growth, especially under nutrient-poor conditions, and promoting a shift in biomass allocation toward the roots, thereby conferring a competitive advantage over native species. Conversely, *Pseudomonas* sp. ScRB22 exhibited limited functional activity and a negligible impact on plant performance. These findings underscore that the ecological impact of rhizosphere bacteria on invasive weeds is closely linked to their specific growth-promoting functions. By enhancing stress adaptation and optimizing resource allocation, certain microorganisms may facilitate the establishment of invasive weeds in adverse environments. This study highlights the significance of microbial functional traits in invasion ecology and suggests novel approaches for microbiome-based invasive weed management, with potential applications in agricultural soil health improvement and ecological restoration.

Keywords: microbial functional traits; competitive advantage; plant growth promotion; plant–microbe interactions; nitrogen use efficiency

1. Introduction

Biological invasion is a phenomenon in which alien species successfully establish and spread within new ecosystems due to natural or anthropogenic factors [1]. These invasions not only alter ecological communities but also significantly affect the global distribution of organisms, posing a serious threat to global biodiversity [2]. With the deepening of globalization, invasive alien species, e.g., invasive weeds, have caused substantial ecological and economic damage worldwide [3]. This disruption can severely alter ecosystem structure and function [4], impacting nutrient cycling and energy flow. For example, invasive plants can influence the carbon cycle by increasing community productivity and contributing more biomass to the soil [5].

Invasive weeds adapt well to new environments and frequently form beneficial microbial interactions [6,7] and enhance their competition for resources like nutrients, light, and water [8,9]. Studies emphasize rhizospheric microorganisms' role in plant performance, as plants selectively recruit them via root exudates, where these microbes influence plant development and stress responses [10].

The rhizosphere harbors diverse microorganisms that promote plant health by enhancing nutrient uptake and pathogen resistance [11], though their efficacy depends on specific environmental conditions [12]. Root-associated microbes can enhance plant performance by modulating the bacterial community for beneficial traits like hormone production and nutrient solubilization [13], while plants employ innate immunity against pathogens [14], a system these microbes can influence to regulate plant growth [15].

Nitrogen-fixing bacteria are among the most extensively studied plant-beneficial microbes due to their ability to convert atmospheric nitrogen into a form usable by plants [16]. Beyond nitrogen fixation, these bacteria possess multiple functional traits that enhance plant growth and soil health, including (1) symbiotic associations with host plants [17], (2) production of growth-promoting hormones (e.g., auxins, cytokinins), (3) phosphate solubilization, (4) soil structure modification, and (5) organic matter decomposition [18–20]. Other growth-promoting bacteria, such as species of *Herbaspirillum*, also contribute to nitrogen acquisition through nitrogen fixation, nitrification, and phytohormone production [21]. These hormones can influence root architecture, increasing the surface area for nutrient absorption [22,23].

Soil microbial interactions influence not only plant growth and nutrient acquisition but also resistance to stress and competition with other plant species [24]. These microbes benefit from root exudates as carbon sources [25], and in turn, plants can shape microbial communities by promoting or suppressing specific populations [26]. Invasive weeds often manipulate rhizosphere microbial structures to enhance their growth and competitive ability, creating soil conditions that support their continued spread [27]. Increased microbial diversity in the rhizosphere reinforces mutual feedback between plants and microbes, driving ecosystem-level changes [28]. Therefore, understanding plant microbe interactions in the rhizosphere is essential for uncovering the mechanisms of plant invasion.

Among rhizospheric microbes, plant growth-promoting rhizobacteria (PGPR) play a particularly important role. While nitrogen-fixing bacteria (e.g., *Herbaspirillum*) are a subset of PGPR, other PGPR strains enhance plant growth through distinct mechanisms such as indole-3-acetic acid (IAA) production, which stimulates root branching and nutrient uptake [29]. These changes in root structure expand the surface area for water and nutrient absorption, enabling plants to more effectively access critical soil nutrients like phosphorus and potassium. Furthermore, certain PGPR strains possess nitrogen-fixing capabilities, converting atmospheric nitrogen (N_2) into ammonia (NH_3), a form of nitrogen that plants can readily absorb [21]. By boosting both hormone production and nutrient uptake, these microbes significantly enhance plant growth, development, and overall productivity. They give plants a competitive advantage in resource-limited environments, supporting their establishment and survival under various nutrient-deficient conditions [30]. Beyond agriculture, their role in shaping plant microbe dynamics during biological invasions is gaining interest [31].

Microorganisms, particularly PGPRs, play a key role in the invasion process by enhancing the growth and competitiveness of invasive weeds [32]. PGPR can interact with root exudates and chemosensory signals released by invasive weeds, thereby altering the composition and structure of the soil microbial community. These beneficial microbes not only enhance the growth potential of invasive weeds by improving nutrient uptake and stimulating hormone production but also suppress certain soil pathogens and native plant species, ultimately reshaping belowground biodiversity. Through these mechanisms, PGPRs provide invasive weeds with substantial growth and a competitive advantage, enabling them to better adapt to and spread in new environments [33–35].

Solidago canadensis L. originates from North America [36], and it has successfully invaded Asia, Europe, Oceania, etc. In particular, *S. canadensis* has become a notorious weed in diverse habitats in East China, including on roadsides, in abandoned and agricultural fields, and even in open barren areas [37]. This is a highly invasive species due to its rapid reproduction, fast spread, strong growth advantages, and ability to adapt to various environments [38]. It aggressively displaces native plants, causing severe damage to local vegetation and crops, threatening natural and agricultural ecosystems, and reducing biodiversity. Notably, its remarkable adaptability has made it a focal species in plant invasion research [39]. While numerous studies have examined *S. canadensis'* growth and invasive characteristics [40–42], the specific role of rhizospheric bacteria in facilitating its invasion success remains poorly understood. Recent work has begun to explore microbial contributions to invasion success [43], but the mechanistic understanding remains limited.

Therefore, this study aims to explore the ecological effects and underlying mechanisms of rhizobacteria on the growth and competitive advantage of *S. canadensis*, addressing two key questions: (1) Do rhizobacteria contribute to the positive growth responses of *S. canadensis* under nutrient-limited conditions? (2) Do rhizobacteria enhance the competitiveness of *S. canadensis*? The results will contribute to generating a comprehensive understanding of the role of rhizospheric bacteria in alien plant/weed invasions and further enrich the field of invasion ecology.

2. Materials and Methods

To explore the ecological effects of rhizospheric bacteria on invasive weed *Solidago canadensis*, three experiments were conducted: (1) isolation and identification of rhizobacteria (Figure 1A), (2) plant growth-promoting experiment under low nutrient stress (Figure 1B), and (3) competition with a native plant (Figure 1C).

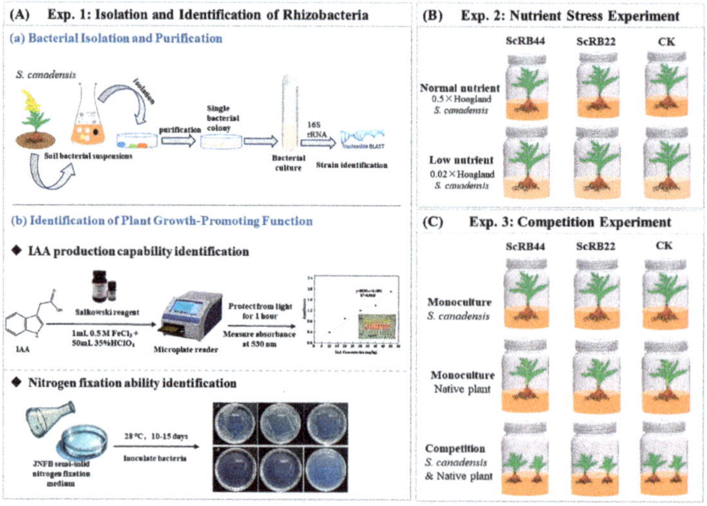

Figure 1. Experimental design and flowchart.

2.1. Isolation and Identification of Rhizospheric Bacteria of *S. canadensis*

Rhizosphere soil samples of *S. canadensis* were collected from Zhenjiang, Jiangsu Province (32°10′45.92″ N, 119°30′34.04″ E), during the maturation stage (late November). After the bulk soil was shaken off the plant root, the remaining rhizosphere soil, which was intimately attached to the fine roots, was collected using a paintbrush, placed in envelopes, and temporarily stored in 4 °C iceboxes. The samples were immediately transported to the laboratory where they were homogenized and passed through a 2 mm sieve. Soil bacterial suspensions were prepared using sterile distilled water. Then, the isolation [44], purification, and identification of rhizospheric bacteria were conducted according to standard microbiological methods [45], including 16S rRNA sequencing. The experimental procedure was conducted as follows: Root samples of S. canadensis were sequentially washed and oscillated in sterile water to obtain bacterial suspensions, which were then gradient-diluted and spread on LB agar plates for bacterial isolation. Pure cultures were acquired through single-colony picking and streak plating methods, followed by expansion in LB liquid medium and long-term preservation in 50% glycerol under dual conditions of −80 °C and −20 °C. All operations were performed under strictly aseptic conditions in a laminar flow cabinet to ensure the reliability of the strain isolation and purification.

From the numerous isolated strains, two bacterial strains were selected for further experimentation: *Bacillus* sp. ScRB44 and *Pseudomonas* sp. ScRB22. Single colonies of these two strains were selected and picked out into tubes containing LB (Luria–Bertani) culture solution and further incubated and activated in a shaker (200 rpm, 30 °C) for further use.

2.2. Bacterial IAA Production Capacity Assay

The IAA production ability of the bacterial strains was measured using a microplate reader. The specific steps are as follows: To prepare a 50 mg/L IAA standard solution, begin by pouring 500 mL of ultrapure water into a 1000 mL beaker. Then, measure 50 mg of solid IAA using an analytical balance and gently add it to the water. Place the beaker on a magnetic stirrer and stir continuously until the IAA powder is mostly dissolved. Then, slowly add an aqueous sodium hydroxide solution to ensure complete dissolution. Once fully dissolved, transfer the solution to a 1000 mL volumetric flask and carefully adjust the volume to the calibration mark. This stock solution is reserved for later use. Similarly, IAA solutions with concentrations of 40 mg/L, 30 mg/L, 20 mg/L, and 10 mg/L were prepared. The absorbance of each concentration was measured using the microplate reader to generate a standard curve. For the quantification of IAA produced by rhizosphere bacteria, bacteria isolated from the root surface of *S. canadensis* (stored at −20 °C) were activated and then cultured on R_2A medium using a single colony isolation method. The cultures were incubated in a shaker at 200 rpm and 30 °C for three days. After incubation, the bacterial suspensions were centrifuged at 8000 rpm for 15 min, and the supernatant was carefully collected using a pipette. To determine the IAA concentration, Salkowski's chromogenic solution was prepared by mixing 1 mL of 0.5 M $FeCl_3$ and 50 mL of 35% $HClO_4$ (maintaining a supernatant-to-reagent ratio of 2:3). The reaction was allowed to proceed for 1 h in the dark. Finally, the absorbance of each sample was measured at a wavelength of 530 nm using a spectrophotometer, enabling the quantification of IAA production capacity in bacteria strains.

2.3. Determination of Bacterial Nitrogen Utilization Capacity

The activated bacterial strains were individually inoculated into nitrogen-free Ashby solid nitrogen fixation medium, following the method described by reference [46]. The cultures were incubated at 28 °C for 10 to 15 days to assess their nitrogen utilization ability.

plants inoculated with ScRB44 and ScRB22 exhibited significantly greater underground fresh weights compared to the control group. The root-to-shoot ratio also increased significantly in these groups. However, only ScRB44-inoculated plants showed significant increases in aboveground and root length. While plant height and leaf number slightly increased with the addition of ScRB44, these differences were not statistically significant. In contrast, plants inoculated with ScRB22 displayed significant decreases in aboveground fresh weight, plant height, and root length compared to the control group.

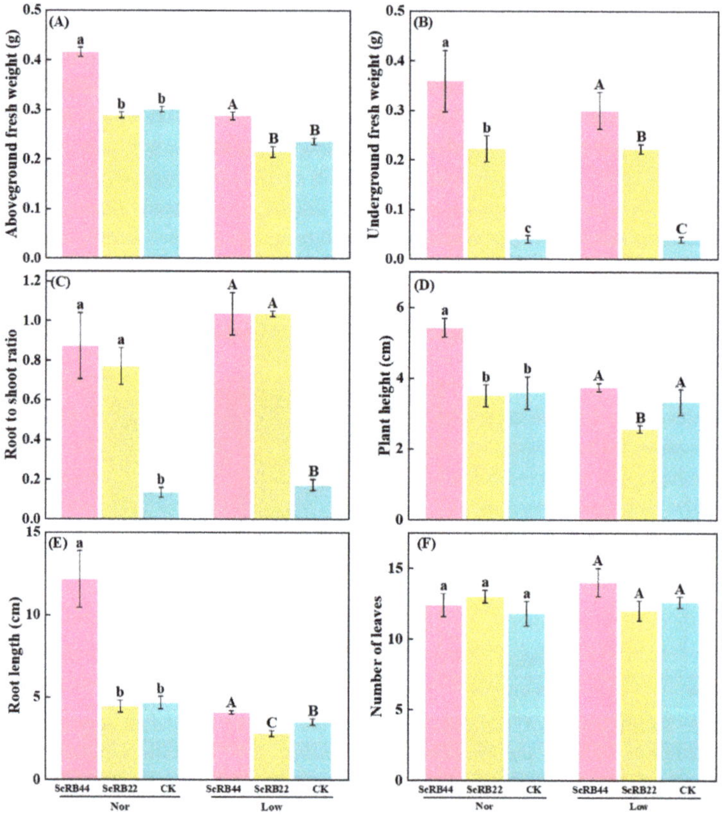

Figure 3. Effects of rhizospheric *Bacillus* sp. ScRB44 and *Pseudomonas* sp. ScRB22 on biomass (**A**,**B**), root-to-shoot radio (**C**), number of leaves (**D**), plant height (**E**), and root length (**F**) of *S. canadensis* under different nutrient conditions. Nor—normal nutrient treatment; Low—low nutrient treatment. Lowercase letters indicate statistically significant differences under normal nutrient conditions, while uppercase letters denote significant differences under low-nutrient conditions. Different letters indicate significant differences at $p < 0.05$.

3.3. Effects of Rhizospheric Bacteria on the Competitiveness of *S. canadensis*

After inoculation with ScRB44, both *S. canadensis* and native plants showed increased biomass under monoculture conditions. This increase in biomass was more pronounced in *S. canadensis* compared to the native plant, with significant differences observed in both the aboveground and underground portions (Figure 4A,B). Furthermore, the root-to-shoot ratios also increased for both plant types under monoculture conditions, with ScRB44 having a more significant effect on *S. canadensis*. In contrast, inoculation with ScRB22 promoted underground biomass in *S. canadensis* under monoculture conditions but significantly inhibited both aboveground and underground biomass in native plants. Under mixed planting conditions, the aboveground biomass of *S. canadensis* increased significantly after inoculation with ScRB44, while no effect was observed on the aboveground biomass of native plants. However, both *S. canadensis* and native plant showed significant increases in underground biomass after inoculation, with a more significant promotion effect observed for native plants (Figure 4C,D). Comparing their root-to-shoot ratios under mixed planting conditions, both *S. canadensis* and the native plant showed significant increases, with a more pronounced promotion effect observed for native plants (Figure 4G,H). In contrast, ScRB22 inoculation inhibited both plant types, significantly reducing the underground biomass of *S. canadensis* and the aboveground biomass of native plants.

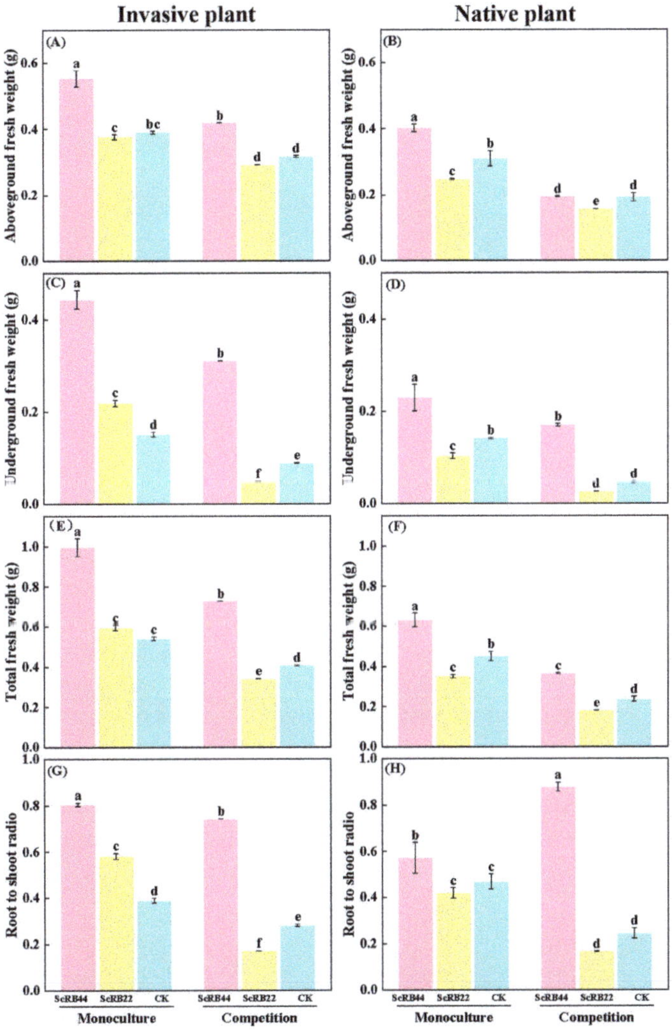

Figure 4. Effects of rhizospheric *Bacillus* sp. ScRB44 and *Pseudomonas* sp. ScRB22 on aboveground fresh weight (**A**,**B**), underground fresh weight (**C**,**D**), total fresh weight (**E**,**F**), and root-to-shoot radio (**G**,**H**) of *S. canadensis* and native plant under competition. Different lowercase letters indicate significant differences at $p < 0.05$.

From the graph, it can be seen that under monoculture conditions, inoculation with ScRB44 significantly increased plant height and root length in both *S. canadensis* and native plants. Specifically, the promotion of root length was more pronounced in *S. canadensis* compared to native plants. In contrast, ScRB22 inoculation exhibited no significant effect on *S. canadensis* but significantly reduced both plant height and root length in native plants. Under mixed planting conditions, inoculation with ScRB44 did not have a significant effect on plant height for either *S. canadensis* or native plants (Figure 5A,B). However, there was a significant increase in root length for both species, with a more pronounced effect observed in *S. canadensis* (Figure 5C,D). Notably, ScRB22 inoculation had no change on either plant height or root length in both plant types under mixed planting conditions.

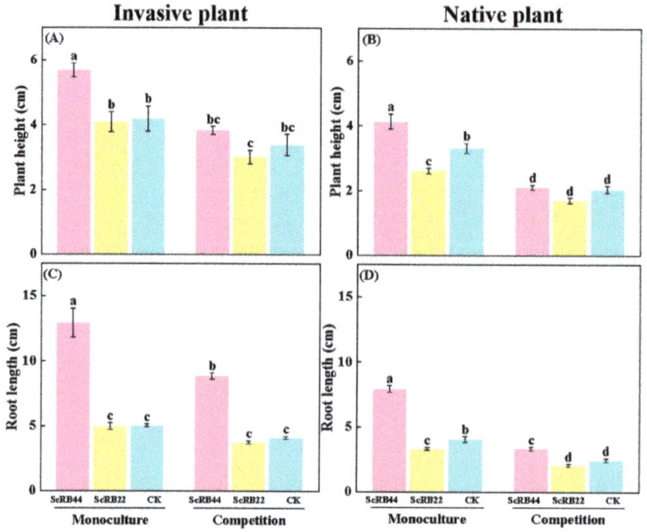

Figure 5. Effects of rhizospheric *Bacillus* sp. ScRB44 and *Pseudomonas* sp. ScRB22 on plant height (**A**,**B**) and root length (**C**,**D**) of *S. canadensis* and native plant under competition. Different lowercase letters indicate significant differences at $p < 0.05$.

After inoculating ScRB44, the aboveground biomass RCI and total biomass RCI of *S. canadensis* were significantly higher than those of the native plants (Figure 6A,C), while the underground biomass RCI showed a more pronounced promotion in native plants(Figure 6B). This indicates that after inoculation with ScRB44, *S. canadensis* exhibits a higher relative competitive strength compared to native plants. In contrast, ScRB22 inoculation increased the aboveground biomass RCI of S. canadensis but had no significant effect on its underground biomass RCI. Moreover, both plant types showed significantly reduced underground biomass RCI compared to the CK.

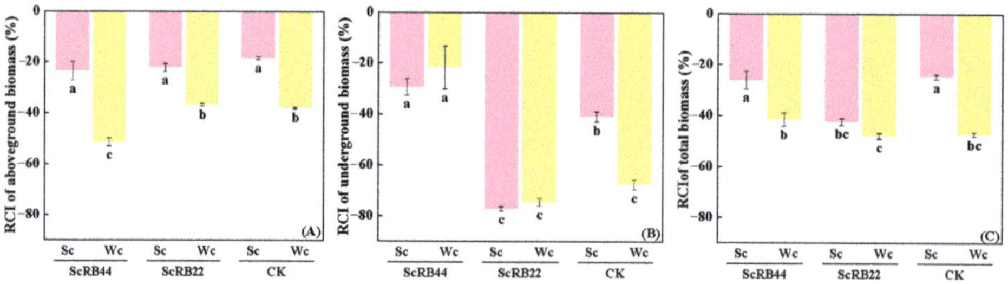

Figure 6. Effects of rhizospheric *Bacillus* sp. ScRB44 and *Pseudomonas* sp. ScRB22 on RCI of aboveground biomass (**A**), RCI of belowground biomass (**B**), and RCI of total biomass (**C**) of *S. canadensis* (Sc) and native plant *W. trilobata* (Wc). Different lowercase letters indicate significant differences at $p < 0.05$.

4. Discussion

4.1. Biofunctional Differentiation of Rhizobacteria Endows Their Distinct Growth-Promoting Ability on Invasive Weed S. canadensis

This study illustrates that the invasive potential of *S. canadensis* is strongly influenced by the functional characteristics of its association with rhizobacteria. Specifically, *Bacillus* sp. ScRB44 significantly enhanced root development and biomass accumulation by producing elevated levels of indole-3-acetic acid (IAA), thereby boosting plant growth under both nutrient-rich and nutrient-poor conditions. These findings align with previous research, such as the ability of *Bacillus megaterium* to promote soybean growth through phytohormone regulation and increased root surface area [51]. Furthermore, *Bacillus* species are well-known for their ability to stimulate plant growth by synthesizing IAA, a key plant hormone involved in growth regulation [52]. These findings have important implications for agriculture, particularly in nutrient-deficient soils, as *Bacillus* sp. ScRB44 has the potential to enhance soil fertility and improve crop resilience, especially under low-input agricultural practices [53]. For example, studies have shown that similar plant growth-promoting rhizobacteria (PGPR) can enhance maize growth in saline soils [54]. Furthermore, a recent analysis confirmed that *Bacillus* spp. plays a crucial role in improving crop yield and plant health [55].

In contrast, *Pseudomonas* sp. ScRB22 produced only minimal levels of IAA (1.7 mg/L) and displayed negligible nitrogen utilization activity, leading to a substantially weaker growth-promoting effect. This stark contrast between the two strains underscores the ecological advantage that multifunctional bacteria like *Bacillus* sp. ScRB44 have, a pattern consistent with studies showing that invasive plants often associate with microbes exhibiting complementary growth-promoting traits [56,57]. While our data focus on IAA and nitrogen utilization efficiency, these results align with broader evidence that microbial synergies contribute to plant invasiveness (e.g., enhanced nutrient acquisition and stress tolerance) [58,59]. These results suggest that *S. canadensis* may actively recruit rhizobacteria with complementary and synergistic growth-enhancing traits, enabling it to overcome environmental challenges and establish dominance in new habitats.

The enrichment of *Bacillus* sp. ScRB44 in the rhizosphere of *S. canadensis* likely contributes to its enhanced resource acquisition and competitive performance, thereby facilitating its rapid establishment and ecological dominance. This microbial association provides a biological explanation for the adaptive success of *S. canadensis* as it spreads invasively across diverse environments. For instance, *Bacillus*-like strains can enhance the growth of invasive plants by improving nutrient utilization efficiency and strengthening stress tolerance [60], thereby outcompeting native microorganisms. Furthermore, such rhizobacteria have been demonstrated to modulate soil microbial communities to benefit host plants, which aligns with our observations of ScRB44′s growth-promoting effects on *S. canadensis* [61,62].

4.2. Rhizospheric Bacteria Can Effectively Resist Nutrient Stress for S. canadensis

Under nutrient-limited conditions, *Bacillus* sp. ScRB44 provided dual adaptive benefits to *S. canadensis*: enhanced nutrient uptake through IAA-induced root proliferation and improved nitrogen availability via biological nitrogen fixation. These findings are closely related to agricultural practices, suggesting that employing PGPR can potentially reduce dependence on chemical fertilizers, thus promoting sustainable agriculture [63]. By increasing nutrient use efficiency, *Bacillus* sp. ScRB44 emerges as a promising candidate for enhancing crop productivity, particularly in nutrient-deficient agricultural soils. Previous studies have demonstrated that plant hormones such as IAA play a crucial role in modulating plant responses to environmental stress [64]. PGPRs are also known to help plants acquire phosphorus under limiting conditions by excreting organic acids or solubilizing inorganic phosphorus. For example, certain *Pseudomonas* and *Bacillus* strains can release phosphorus from phytate through phytase activity, thereby increasing phosphorus bioavailability [65]. In contrast, *Pseudomonas* sp. ScRB22, lacking both IAA synthesis and nitrogen-fixing ability, showed limited capacity to support plant growth under nitrogen-deficient conditions and may even exert suppressive effects. Thus, the dual growth-promoting mechanism root growth stimulation via IAA and improved nitrogen availability contribute to reshaping the competitive landscape in favor of the invader. The capacity of *S. canadensis* to selectively recruit functionally beneficial microbes highlights a key microbial-mediated pathway by which invasive species strengthen their ecological dominance.

Ecologically, the recruitment of ScRB44 equips *S. canadensis* with a robust microbial strategy to resist nutrient stress and outcompete neighboring native species. This microbial facilitation reshapes the competitive landscape by providing the invader with physiological advantages that are not equally available to its competitors, thereby contributing to its invasive success.

4.3. Rhizobacteria Restructure Resource Competition Dynamics in S. canadensis

The contrasting functional traits of the two rhizobacterial strains led to differing competitive outcomes in both monoculture and coculture systems. In monoculture, treatment with ScRB44 resulted in a 65% increase in belowground biomass allocation in *S. canadensis*, compared to a 51% increase in *W. chinensis*. The root-to-shoot ratio also shifted more significantly in *S. canadensis*, reflecting a more efficient resource allocation strategy. In co-culture systems, ScRB44-treated *S. canadensis* displayed enhanced competitiveness, driven by the synergistic effects of IAA-induced root growth and improved nitrogen utilization efficiency. In contrast, ScRB22 failed to provide similar benefits due to its limited functional capabilities, offering no significant advantage to *S. canadensis*.

Overall, our findings suggest that a critical factor in the invasive success of *S. canadensis* is its ability to shape the rhizosphere microbial community in its favor. By selectively recruiting multifunctional rhizobacteria, *S. canadensis* not only enhances soil fertility but also restructures microbial community composition, creating an environment conducive to its dominance. This study underscores the pivotal role of microbial partners in facilitating plant invasion and provides new insights for agricultural soil management, particularly in improving crop productivity through the manipulation of microbial communities.

5. Conclusions

Our research elucidates the critical role of plant–microbe symbiosis in the invasive success of *S. canadensis*. Specifically, the association with plant growth-promoting rhizobacteria, such as *Bacillus* sp. ScRB44, enhances the plant's nutrient acquisition, stress resilience, and competitive capacity against *W. chinensis*. This mutualistic interaction not only supports the establishment and spread of *S. canadensis* in invaded ecosystems where *W. chinensis* is present but also underscores the significance of rhizosphere microbiota in shaping specific plant community dynamics. Given the reciprocal effects between *S. canadensis*, *W. chinensis*, and their associated soil microbial communities, targeting these microbial associations presents a promising avenue for developing sustainable and ecologically sound invasive species management strategies in terms of both invasive species control and agricultural soil improvement. Future research should focus on manipulating these specific plant microbe interactions to mitigate the impacts of invasive weeds, enhance sustainable agricultural productivity, and promote the restoration of native biodiversity and ecosystem health.

Author Contributions: Conceptualization, Z.-C.D. and D.-L.D.; methodology, Z.-Y.H., Y.L., H.-A.X. and M.-T.Y.; software, Z.-Y.H. and R.-K.Z.; data curation, Z.-Y.H., X.-T.R. and J.-Z.L.; writing—original draft preparation, Z.-Y.H. and M.N.; writing—review and editing, G.-Q.R. and Z.-C.D.; supervision, Z.-C.D.; funding acquisition, Z.-C.D. and D.-L.D. All authors have read and agreed to the published version of the manuscript.

Funding: This research was funded by the National Natural Science Foundation of China (32271587, 32401311), Carbon peak and carbon neutrality technology innovation foundation of Jiangsu Province (BK20220030), and the National College Students Innovation Training Program.

Institutional Review Board Statement: Not applicable.

Data Availability Statement: The data presented in this study are available on request from the corresponding author.

Conflicts of Interest: The authors declare no conflicts of interest.

References

1. Yang, Y.; Bian, Z.; Ren, W.; Wu, J.; Liu, J.; Shrestha, N. Spatial patterns and hotspots of plant invasion in China. *Glob. Ecol. Conserv.* **2023**, *43*, e02424. [CrossRef]
2. Latombe, G.; Pysek, P.; Jeschke, J.M.; Blackburn, T.M.; Bacher, S.; Capinha, C.; Costello, M.J.; Fernández, M.; Gregory, R.D.; Hobern, D.; et al. A vision for global monitoring of biological invasions. *Biol. Conserv.* **2017**, *213*, 295–308. [CrossRef]
3. Liu, Y.; Oduor, A.M.O.; Zhang, Z.; Manea, A.; Tooth, I.M.; Leishman, M.R.; Xu, X.; van Kleunen, M. Do invasive alien plants benefit more from global environmental change than native plants? *Glob. Change Biol.* **2017**, *23*, 3363–3370. [CrossRef] [PubMed]
4. Hulme, P.E. Thematic mapping of biosecurity highlights divergent conceptual foundations in human, animal, plant and ecosystem health. *NeoBiota* **2024**, *95*, 221–239. [CrossRef]
5. He, Y.; Li, J.; Siemann, E.; Li, B.; Xu, Y.; Wang, Y. Plant Invasion Increases Soil Microbial Biomass Carbon: Meta-Analysis and Empirical Tests. *Glob. Change Biol.* **2025**, *31*, e70109. [CrossRef]
6. Zhang, H.-Y.; Goncalves, P.; Copeland, E.; Qi, S.-S.; Dai, Z.-C.; Li, G.-L.; Wang, C.-Y.; Du, D.-L.; Thomas, T. Invasion by the weed *Conyza canadensis* alters soil nutrient supply and shifts microbiota structure. *Soil Biol. Biochem.* **2020**, *143*, 107739. [CrossRef]
7. Gong, T.Y.; Xin, X.F. Phyllosphere microbiota: Community dynamics and its interaction with plant hosts. *J. Integr. Plant Biol.* **2021**, *63*, 297–304. [CrossRef]
8. Wang, W.J.; Zhu, Q.Y.; Dai, S.Y.; Meng, L.; He, M.Q.; Chen, S.D.; Zhao, C.; Dan, X.Q.; Cai, Z.C.; Zhang, J.B.; et al. Effects of *Solidago canadensis* L. on mineralization-immobilization turnover enhance its nitrogen competitiveness and invasiveness. *Sci. Total Environ.* **2023**, *882*, 163641. [CrossRef]
9. Zhou, J.; Xu, Z.; Zhong, S.; Yu, Y.; Xu, Z.; Du, D.; Wang, C. Nitrogen Influence to the Independent Invasion and the Co-Invasion of *Solidago canadensis* and *Conyza canadensis* via Intensified Allelopathy. *Sustainability* **2022**, *14*, 11970. [CrossRef]
10. Hahl, T.; Moorsel, S.J.; Schmid, M.W.; Zuppinger-Dingley, D.; Schmid, B.; Wagg, C. Plant responses to diversity-driven selection and associated rhizosphere microbial communities. *Funct. Ecol.* **2020**, *34*, 707–722. [CrossRef]
11. Leach, J.E.; Triplett, L.R.; Argueso, C.T.; Trivedi, P. Communication in the Phytobiome. *Cell* **2017**, *169*, 587–596. [CrossRef]
12. Diagne, N.; Ngom, M.; Djighaly, P.I.; Fall, D.; Hocher, V.; Svistoonoff, S. Roles of Arbuscular Mycorrhizal Fungi on Plant Growth and Performance: Importance in Biotic and Abiotic Stressed Regulation. *Diversity* **2020**, *12*, 370. [CrossRef]
13. Howard, M.M.; Muñoz, C.A.; Kao-Kniffin, J.; Kessler, A. Soil Microbiomes from Fallow Fields Have Species-Specific Effects on Crop Growth and Pest Resistance. *Front. Plant Sci.* **2020**, *11*, 1171. [CrossRef] [PubMed]
14. Ma, K.W.; Niu, Y.L.; Jia, Y.; Ordon, J.; Copeland, C.; Emonet, A.; Geldner, N.; Guan, R.; Stolze, S.C.; Nakagami, H.; et al. Coordination of microbe-host homeostasis by crosstalk with plant innate immunity. *Nat. Plants* **2021**, *7*, 814–826. [CrossRef] [PubMed]
15. Sun, W.; Shahrajabian, M.H.; Soleymani, A. The Roles of Plant-Growth-Promoting Rhizobacteria (PGPR)-Based Biostimulants for Agricultural Production Systems. *Plants* **2024**, *13*, 613. [CrossRef]
16. Chen, Y.X.; Fu, W.J.; Xiao, H.; Zhai, Y.K.; Luo, Y.; Wang, Y.Z.; Liu, Z.H.; Li, Q.; Huang, J.A. A Review on Rhizosphere Microbiota of Tea Plant (*Camellia sinensis* L): Recent Insights and Future Perspectives. *J. Agric. Food Chem.* **2023**, *71*, 19165–19188. [CrossRef]
17. Walker, L.; Lagunas, B.; Gifford, M.L. Determinants of Host Range Specificity in Legume-Rhizobia Symbiosis. *Front. Microbiol.* **2020**, *11*, 585749. [CrossRef]
18. Ribeiro, N.V.d.S.; Vidal, M.S.; Barrios, S.C.L.; Baldani, V.L.D.; Baldani, J.I. Genetic diversity and growth promoting characteristics of diazotrophic bacteria isolated from 20 genotypes of *Brachiaria* spp. *Plant Soil* **2019**, *451*, 187–205. [CrossRef]

19. Jiang, S.; Li, J.; Wang, Q.; Yin, C.; Zhan, Y.; Yan, Y.; Lin, M.; Ke, X. Maize Growth Promotion by Inoculation with an Engineered Ammonium-Excreting Strain of Nitrogen-Fixing Pseudomonas stutzeri. *Microorganisms* **2022**, *10*, 1986. [CrossRef]
20. Aasfar, A.; Bargaz, A.; Yaakoubi, K.; Hilali, A.; Bennis, I.; Zeroual, Y.; Meftah Kadmiri, I. Nitrogen Fixing Azotobacter Species as Potential Soil Biological Enhancers for Crop Nutrition and Yield Stability. *Front. Microbiol.* **2021**, *12*, 628379. [CrossRef]
21. Santhoshkumar, R.; Parvathy, A.H.; Soniya, E.V. Modifications in Metabolomic Profile of Andrographis paniculata by Arsenic Tolerance *Herbaspirillum* sp.: Perception into Plant–Microbe Interactions. *J. Plant Growth Regul.* **2025**, *44*, 1–17. [CrossRef]
22. Ding, Z.; Bai, Y. The current and future studies on plant root development and root microbiota. *Sci. Sin. Vitae* **2021**, *51*, 1447–1456. [CrossRef]
23. Han, S.H.; Kim, S.; Chang, H.; Li, G.; Son, Y. Increased soil temperature stimulates changes in carbon, nitrogen, and mass loss in the fine roots of *Pinus koraiensis* under experimental warming and drought. *Turk. J. Agric. For.* **2019**, *43*, 80–87. [CrossRef]
24. Jones, P.; Garcia, B.J.; Furches, A.; Tuskan, G.A.; Jacobson, D. Plant Host-Associated Mechanisms for Microbial Selection. *Front. Plant Sci.* **2019**, *10*, 862. [CrossRef] [PubMed]
25. Fields, B.; Friman, V.P. Microbial eco-evolutionary dynamics in the plant rhizosphere. *Curr. Opin. Microbiol.* **2022**, *68*, 102153. [CrossRef] [PubMed]
26. Fitzpatrick, C.R.; Copeland, J.; Wang, P.W.; Guttman, D.S.; Kotanen, P.M.; Johnson, M.T.J. Assembly and ecological function of the root microbiome across angiosperm plant species. *Proc. Natl. Acad. Sci. USA* **2018**, *115*, E1157–E1165. [CrossRef] [PubMed]
27. Bi, B.Y.; Wang, K.; Zhang, H.; Wang, Y.; Fei, H.Y.; Pan, R.P.; Han, F.P. Plants use rhizosphere metabolites to regulate soil microbial diversity. *Land Degrad. Dev.* **2021**, *32*, 5267–5280. [CrossRef]
28. Kong, Z.; Liu, H. Modification of Rhizosphere Microbial Communities: A Possible Mechanism of Plant Growth Promoting Rhizobacteria Enhancing Plant Growth and Fitness. *Front. Plant Sci.* **2022**, *13*, 920813. [CrossRef]
29. Jung, B.K.; Khan, A.R.; Hong, S.-J.; Park, G.-S.; Park, Y.-J.; Kim, H.-J.; Jeon, H.-J.; Khan, M.A.; Waqas, M.; Lee, I.-J.; et al. Quorum sensing activity of the plant growth-promoting rhizobacterium *Serratia glossinae* GS2 isolated from the sesame (*Sesamum indicum* L.) rhizosphere. *Ann. Microbiol.* **2017**, *67*, 623–632. [CrossRef]
30. Slimani, A.; Oufdou, K.; Meddich, A. Intercropping with alfalfa and co-inoculation of AMF and PGPR improve growth, yield, grain bioactive quality, and soil fertility of barley. *Arch. Agron. Soil Sci.* **2023**, *69*, 3469–3483. [CrossRef]
31. Berlow, M.; Mesa, M.; Creek, M.; Duarte, J.G.; Carpenter, E.; Phinizy, B.; Andonian, K.; Dlugosch, K.M. Plant G × Microbial E: Plant Genotype Interaction with Soil Bacterial Community Shapes Rhizosphere Composition During Invasion. *Microb. Ecol.* **2024**, *87*, 133. [CrossRef]
32. Kalske, A.; Blande, J.D.; Ramula, S. Soil microbiota explain differences in herbivore resistance between native and invasive populations of a perennial herb. *J. Ecol.* **2022**, *110*, 2649–2660. [CrossRef]
33. Elsheikh, E.A.E.; El-Keblawy, A.; Mosa, K.A.; Okoh, A.I.; Saadoun, I. Role of Endophytes and Rhizosphere Microbes in Promoting the Invasion of Exotic Plants in Arid and Semi-Arid Areas: A Review. *Sustainability* **2021**, *13*, 13081. [CrossRef]
34. Li, X.; Zhang, Y.; Kong, F.-L.; Naz, M.; Zhou, J.-Y.; Qi, S.-S.; Dai, Z.-C.; Du, D.-L. Invasive Plant *Alternanthera philoxeroides* Benefits More Competition Advantage from Rhizosphere Bacteria Regardless of the Host Source. *Plants* **2023**, *12*, 2085. [CrossRef] [PubMed]
35. Wang, Z.-J.; Li, X.; Wang, J.-H.; Qi, S.-S.; Dai, Z.-C.; Du, D.-L. Effect of nitrogen-fixing bacteria on resource investment of the root system in an invasive clonal plant under low nutritional environment. *Flora* **2022**, *297*, 152166. [CrossRef]
36. Judžentienė, A.; Būdienė, J.; Labanauskas, L.; Stancelytė, D.; Nedveckytė, I. Allelopathic Activity of Canadian Goldenrod (*Solidago canadensis* L.) Extracts on Seed Germination and Growth of Lettuce (*Lactuca sativa* L.) and Garden Pepper Cress (*Lepidium sativum* L.). *Plants* **2023**, *12*, 1421. [CrossRef]
37. Czortek, P.; Królak, E.; Borkowska, L.; Bielecka, A. Impacts of soil properties and functional diversity on the performance of invasive plant species *Solidago canadensis* L. on post-agricultural wastelands. *Sci. Total Environ.* **2020**, *729*, 139077. [CrossRef]
38. Qi, S.; Wang, J.; Wan, L.; Dai, Z.; da Silva Matos, D.M.; Du, D.; Egan, S.; Bonser, S.P.; Thomas, T.; Moles, A.T. Arbuscular Mycorrhizal Fungi Contribute to Phosphorous Uptake and Allocation Strategies of *Solidago canadensis* in a Phosphorous-Deficient Environment. *Front. Plant Sci.* **2022**, *13*, 831654. [CrossRef]
39. Gala-Czekaj, D.; Dziurka, M.; Bocianowski, J.; Synowiec, A. Autoallelopathic potential of aqueous extracts from Canadian goldenrod (*Solidago canadensis* L.) and giant goldenrod (*S. gigantea* Aiton). *Acta Physiol. Plant.* **2021**, *44*, 1. [CrossRef]
40. Wang, S.; Wei, M.; Wu, B.; Cheng, H.; Wang, C. Combined nitrogen deposition and Cd stress antagonistically affect the allelopathy of invasive alien species Canada goldenrod on the cultivated crop lettuce. *Sci. Hortic.* **2020**, *261*, 108955. [CrossRef]
41. Adomako, M.O.; Ning, L.; Tang, M.; Du, D.-L.; van Kleunen, M.; Yu, F.-H. Diversity- and density-mediated allelopathic effects of resident plant communities on invasion by an exotic plant. *Plant Soil* **2019**, *440*, 581–592. [CrossRef]
42. Pan, L.; He, F.; Liang, Q.; Bo, Y.; Lin, X.; Javed, Q.; Ullah, M.S.; Sun, J. Allelopathic Effects of Caffeic Acid and Its Derivatives on Seed Germination and Growth Competitiveness of Native Plants (*Lantana indica*) and Invasive Plants (*Solidago canadensis*). *Agriculture* **2023**, *13*, 1719. [CrossRef]
43. Fahey, C.; Flory, S.L. Soil microbes alter competition between native and invasive plants. *J. Ecol.* **2021**, *110*, 404–414. [CrossRef]

44. Shang, L.; Bai, X.; Chen, C.; Liu, L.; Li, M.; Xia, X.; Wang, Y. Isolation and identification of a *Bacillus megaterium* strain with ochratoxin A removal ability and antifungal activity. *Food Control* **2019**, *106*, 106743. [CrossRef]
45. Panchami, P.S.; Geetha Thanuja, K.; Karthikeyan, S. Isolation and Characterization of Indigenous Plant Growth-Promoting Rhizobacteria (PGPR) from Cardamom Rhizosphere. *Curr. Microbiol.* **2020**, *77*, 2963–2981. [CrossRef] [PubMed]
46. Yi-qing, X.; Hai-lang, L.; Na, L.; Fu-rong, Z.; Mao-yao, P.; Yun-tong, M. Isolation, identification and functional verification of bacteria from seeds of *Coptis chinensis*. *Nat. Prod. Res. Dev.* **2023**, *35*, 191–199. [CrossRef]
47. Dai, Z.-C.; Fu, W.; Wan, L.-Y.; Cai, H.-H.; Wang, N.; Qi, S.-S.; Du, D.-L. Different Growth Promoting Effects of Endophytic Bacteria on Invasive and Native Clonal Plants. *Front. Plant Sci.* **2016**, *7*, 706. [CrossRef]
48. Asiloglu, R.; Shiroishi, K.; Suzuki, K.; Turgay, O.C.; Murase, J.; Harada, N. Protist-enhanced survival of a plant growth promoting rhizobacteria, *Azospirillum* sp. B510, and the growth of rice (*Oryza sativa* L.) plants. *Appl. Soil Ecol.* **2020**, *154*, 103599. [CrossRef]
49. Song, J.; Fan, Y.; Li, X.; Li, Y.; Mao, H.; Zuo, Z.; Zou, Z. Effects of daily light integral on tomato (*Solanum lycopersicon* L.) grafting and quality in a controlled environment. *Int. J. Agric. Biol. Eng.* **2022**, *15*, 44–50. [CrossRef]
50. Tunio, M.H.; Gao, J.; Qureshi, W.A.; Sheikh, S.A.; Chen, J.; Chandio, F.A.; Lakhiar, I.A.; Solangi, K.A. Effects of droplet size and spray interval on root-to-shoot ratio, photosynthesis efficiency, and nutritional quality of aeroponically grown butterhead lettuce. *Int. J. Agric. Biol. Eng.* **2022**, *15*, 79–88. [CrossRef]
51. Park, Y.G.; Mun, B.G.; Kang, S.M.; Hussain, A.; Shahzad, R.; Seo, C.W.; Kim, A.Y.; Lee, S.U.; Oh, K.Y.; Lee, D.Y.; et al. *Bacillus aryabhattai* SRB02 tolerates oxidative and nitrosative stress and promotes the growth of soybean by modulating the production of phytohormones. *PLoS ONE* **2017**, *12*, e0173203. [CrossRef] [PubMed]
52. Sharifi, M.; Ghorbanli, M.; Ebrahimzadeh, H. Improved growth of salinity-stressed soybean after inoculation with salt pre-treated mycorrhizal fungi. *J. Plant Physiol.* **2007**, *164*, 1144–1151. [CrossRef]
53. Luisa, L.-X.; Gonzalez-Hernandez, A.I.; Camanes, G.; Vicedo, B.; Scalschi, L.; Llorens, E. Harnessing Green Helpers: Nitrogen-Fixing Bacteria and Other Beneficial Microorganisms in Plant-Microbe Interactions for Sustainable Agriculture. *Horticulturae* **2024**, *10*, 621. [CrossRef]
54. Zhu, Q.; Zhou, J.; Sun, M.; Li, H.; Han, Y.; Lv, J.; Li, Y.; Zhang, X.; George, T.S.; Liu, W.; et al. A newly isolated Bacillus megaterium OQ560352 promotes maize growth in saline soils by altering rhizosphere microbial communities and organic phosphorus utilization. *Rhizosphere* **2023**, *27*, 100746. [CrossRef]
55. Zhang, Y.; Zhang, N.; Bi, X.; Bi, T.; Baloch, F.B.; Miao, J.; Zeng, N.; Li, B.; An, Y. Growth promotion on maize and whole-genome sequence analysis of *Bacillus velezensis* D103. *Microbiol. Spectr.* **2024**, *12*, e01147-24. [CrossRef]
56. Meng, Y.; Geng, X.; Zhu, P.; Bai, X.; Zhang, P.; Ni, G.; Hou, Y. Enhanced mutualism: A promotional effect driven by bacteria during the early invasion of *Phytolacca americana*. *Ecol. Appl.* **2022**, *34*, e2742. [CrossRef]
57. Wu, T.; Duan, Q.; Zhang, H.; Xiu, W.; Jiang, Z. Differences in soil nutrient and microbial characteristics between invasive *Ageratina adenophora* and native plant communities. *PLoS ONE* **2025**, *20*, e0325193. [CrossRef]
58. Li, X.; Sun, P.; Zhang, Y.; Jin, C.; Guan, C. A novel PGPR strain *Kocuria rhizophila* Y1 enhances salt stress tolerance in maize by regulating phytohormone levels, nutrient acquisition, redox potential, ion homeostasis, photosynthetic capacity and stress-responsive genes expression. *Environ. Exp. Bot.* **2020**, *174*, 104023. [CrossRef]
59. Wen, Z.; Yang, M.; Han, H.; Fazal, A.; Liao, Y.; Ren, R.; Yin, T.; Qi, J.; Sun, S.; Lu, G.; et al. Mycorrhizae Enhance Soybean Plant Growth and Aluminum Stress Tolerance by Shaping the Microbiome Assembly in an Acidic Soil. *Microbiol. Spectr.* **2023**, *11*, e03310. [CrossRef]
60. Vasques, N.C.; Nogueira, M.A.; Hungria, M. Increasing Application of Multifunctional Bacillus for Biocontrol of Pests and Diseases and Plant Growth Promotion: Lessons from Brazil. *Agronomy* **2024**, *14*, 1654. [CrossRef]
61. Chen, L.; Fang, K.; Zhou, J.; Yang, Z.-P.; Dong, X.-F.; Dai, G.-H.; Zhang, H.-B. Enrichment of soil rare bacteria in root by an invasive plant *Ageratina adenophora*. *Sci. Total Environ.* **2019**, *683*, 202–209. [CrossRef]
62. Sardar, M.F.; Chen, Z.; Tang, C.; Zhang, S.; Fang, L.; Miao, D.; Li, Y.; Zhang, Q.; Li, Y. Seasonal linkages between soil nitrogen mineralization and the microbial community in broadleaf forests with Moso bamboo (*Phyllostachys edulis*) invasion. *Sci. Total Environ.* **2023**, *899*, 165557. [CrossRef]
63. Agbodjato, N.A.; Babalola, O.O. Promoting sustainable agriculture by exploiting plant growth-promoting rhizobacteria (PGPR) to improve maize and cowpea crops. *PeerJ* **2024**, *12*, e16836. [CrossRef]
64. Shaffique, S.; Hussain, S.; Kang, S.M.; Imran, M.; Injamum-Ul-Hoque, M.; Khan, M.A.; Lee, I.J. Phytohormonal modulation of the drought stress in soybean: Outlook, research progress, and cross-talk. *Front. Plant Sci.* **2023**, *14*, 1237295. [CrossRef]
65. Bach, E.; Seger, G.D.D.; Fernandes, G.D.; Lisboa, B.B.; Passaglia, L.M.P. Evaluation of biological control and rhizosphere competence of plant growth promoting bacteria. *Appl. Soil Ecol.* **2016**, *99*, 141–149. [CrossRef]

microorganisms

MDPI

Review

Mechanisms of Microorganisms Alleviating Drought and Salt Stresses in Plants

Di Feng [1,*,†], Wenxiang Li [1,†], Pengfei Huang [2], Meiying Gu [1], Guangmu Tang [1], Yanhong Ding [1], Gang Cao [1] and Wanli Xu [1,*]

[1] Key Laboratory of Saline–Alkali Soil Improvement and Utilization (Saline–Alkali Land in Arid and Semi-Arid Regions), Ministry of Agriculture and Rural Affairs of the People's Republic of China, Xinjiang Academy of Agricultural Sciences, Urumchi 830091, China

[2] Farmland Irrigation Research Institute, Chinese Academy of Agricultural Sciences, Xinxiang 453003, China

* Correspondence: fengdi2008sunny@163.com (D.F.); wlxu2005@163.com (W.X.)

† These authors contributed equally to this work.

Abstract

Drought and salt stresses are critical environmental constraints affecting plant growth and development, and microorganisms can enhance plant tolerance to these abiotic stresses through complex mechanisms. This review systematically synthesizes the core mechanisms by which microorganisms regulate plant physiological and biochemical processes under such stresses, specifically including the following: (1) regulating the perception and transduction of abiotic stress signals to enhance plant adaptive responses; (2) boosting gene expression and protein synthesis for overall plant metabolic regulation; (3) activating the antioxidant system to strengthen plant tolerance; (4) modulating plant hormone levels to stimulate growth in response to adversity; (5) enhancing plant nutrition and absorption to improve resilience; (6) optimizing the photosynthesis system to promote the synthesis of essential substances, safeguarding plant growth and development amidst adversity. Finally, the application of microbial inoculants in saline–alkali soil improvement and crop cultivation in arid areas and prospective research directions are discussed.

Keywords: abiotic stress; drought stress; salt stress; endophytes; rhizosphere microorganisms; plant–microbe interactions

1. Introduction

Within the heterogeneous landscape of abiotic stresses, drought and salt stresses are preeminent constraints that severely impair plant vitality. They trigger physiological and biochemical disruptions, such as enzyme dysfunction, metabolic imbalance and altered hormone signaling [1]. These disruptions further disrupt plant growth and development, leading to reduced germination, diminished vigor, yield losses or even death. Beyond biological harm, these stresses also cause substantial agricultural economic losses and undermine ecological balance [1,2]. To maintain dynamic homeostasis under stress, plants have evolved intrinsic regulatory mechanisms: they alleviate or counteract stress-induced damage by activating antioxidant systems, initiating osmotic protection mechanisms, and inducing the expression of stress-tolerant proteins [3,4]. However, abiotic stress forces plants into a growth–defense trade-off—they sacrifice growth rate, yield, or quality to prioritize stress defense [5]. As stress intensity escalates, the plant's regulatory system fails to sustain the balance between growth and defense, ultimately leading to severe or irreversible impairment of growth and development processes.

https://doi.org/10.3390/microorganisms13112565

Beneficial microorganisms can enhance plant abiotic stress resistance by establishing interactions with plants—either endophytically (colonizing internal plant tissues) or rhizospherically (associating with root surfaces) [6]. Based on their colonization niche, plant-associated microorganisms are broadly categorized into endophytes and rhizosphere microorganisms [7,8]. Endophytes exhibit localized or systemic colonization of plant organs (e.g., roots, stems, leaves, fruits, rhizomes); for instance, endophytic fungi extend hyphae through plant intercellular spaces to acquire nutrients and carbohydrates from the apoplast [9]. Beneficial endophytes enhance plant growth and abiotic stress tolerance via multiple pathways, including improving nutrient and water uptake, increasing water use efficiency, regulating endogenous hormone levels, and boosting viability competitiveness [10–12].

Rhizosphere microorganisms, by contrast, are tightly attached to rhizosphere soil particles. This niche is defined by a mutualistic "inter-utilization" relationship: plant roots and their exudates provide nutrients and energy for rhizosphere microbes, while microbes reciprocate by producing plant-growth-promoting substances or degrading phytotoxic compounds [13].

Prior reviews have explored microbial roles in plant stress tolerance, but critical gaps remain. Feng et al. summarized mechanisms by which exogenous substances (including chemicals and microorganisms) alleviate plant salt and drought stress, proposing six common pathways—inducing osmolyte synthesis, activating antioxidant enzymes, regulating hormone production, mediating gene expression and signal transduction, improving photochemical systems, and microbial regulation [14,15]. However, their coverage of microbial regulatory roles was limited and unsystematic. Kumar et al. (2020) focused on plant-growth-promoting bacteria (PGPB) and the phytomicrobiome, reviewing their roles in mitigating plant salt stress via phytohormone regulation, extracellular polymeric substance (EPS) production, and osmolyte synthesis with antioxidant activity induction [16]. For rhizosphere microorganisms, Braud et al. (2009) and Hayat et al. (2010) classified plant-growth-promoting rhizobacteria (PGPR)-mediated abiotic stress tolerance into direct pathways (e.g., promoting nutrient absorption) and indirect pathways (e.g., inducing systemic tolerance via microbial metabolites) [17,18]. Meanwhile, Gupta et al. (2022) reviewed plant associations with PGPR in mediating plant drought and salt tolerance mechanisms, including the core defense network of reactive oxygen species (ROS) balance, regulation pathways of osmotic and ion homeostasis, and expression [19].

In summary, existing studies focus on either drought or salt stress, or on one microbial group at a time, with limited integration of microbial regulatory pathways across both stress types.

Given the rapid advancement of this field, a comprehensive, timely synthesis is needed to foster a systematic understanding of plant–microbe interactions, as well as the functional relationships between endophytes and rhizosphere microorganisms. This review aims to fill this gap by synthesizing current research on microbial involvement in regulating plant stress tolerance mechanisms in response to both drought and salt stress.

2. Plant Regulatory Pathways in Response to Drought and Salt Stresses

Drought and salt stresses disrupt plant cellular homeostasis via osmotic imbalance, ionic toxicity, and oxidative damage, driving plants to activate intrinsic regulatory pathways for survival—these pathways form the core targets for microbial intervention, mainly including three interconnected modules [3,4].

2.1. Stress Signal Perception and Transduction

Plants sense stress via membrane-localized sensors. For drought, plasma membrane aquaporins and osmosensors (e.g., *Arabidopsis* AtHK1) detect changes in turgor pressure/osmolarity, triggering the mitogen-activated protein kinase/calcium-dependent protein kinase (MAPK/CDPK) phosphorylation cascade [20,21]. For salt, Salt Overly Sensitive 1 (SOS1) transporters (Na^+ extruders) interact with SOS3 calcium sensors to activate the SOS pathway [21]. Abscisic acid (ABA) amplifies signals: stress-induced ABA binds PYR/PYL/RCAR receptors, inhibiting protein phosphatases 2C (PP2Cs) and activating sucrose non-fermenting-1-related protein kinases 2 (SnRK2s), which in turn regulate stomatal closure and ion homeostasis [20].

2.2. Stress-Responsive Gene Expression Regulation

Plants reprogram transcriptomes via transcription factors (TFs). DREB/CBF family (e.g., DREB2A) binds DRE/CRT elements to upregulate LEA protein/osmoprotectant synthase genes [21]. NAC TFs (e.g., ANAC019) induced by ABA/stress regulate cell wall modification and ROS scavenging genes [20]. Epigenetic modifications (e.g., histone H3K9 acetylation) also fine-tune expression—e.g., salt stress boosts SOS1 transcription via this modification [4].

2.3. Antioxidant System Activation

Stress-induced ROS (O_2^-, H_2O_2) cause cellular damage, so plants activate both enzymatic and non-enzymatic antioxidant systems. The enzymatic antioxidant system includes superoxide dismutase (SOD), ascorbate peroxidase (APX), Catalase (CAT) and glutathione reductase (GR), etc. [22–25]. The non-enzymatic antioxidant system includes ascorbic acid (AsA), glutathione (GSH), and carotenoids [22–25]. For example, drought-stressed wheat upregulates wheat superoxide dismutase (TaSOD) and wheat ascorbate peroxidase (TaAPX) to maintain ROS balance [3]; the ABA-SnRK2 signaling pathway further coordinates this system by upregulating antioxidant enzyme genes [20].

These pathways form a robust network, but high stress (e.g., severe salt overwhelming SOS pathway) limits their efficacy [21]—creating opportunities for microorganisms to intervene by targeting key nodes, as discussed later.

3. Advances in the Study of Microbial Strategies Against Drought and Salt Stresses in Plants

Research on microbial-mediated mitigation of plant drought and salt stress emerged in the late 20th century [26] and advanced rapidly in the 21st century [27–29]. This field covers diverse microbial functions, enabling systematic insights into underlying mechanisms. We summarized ~50 microorganisms involved in regulating plant stress resistance under drought and salt stress.

To ensure comprehensiveness of the literature collection, we conducted searches on core academic databases including Web of Science and ScienceDirect. The search was performed using keywords related to adversity types (e.g., "drought stress," "salt stress") and microbial species (e.g., "plant-growth-promoting rhizobacteria," "arbuscular mycorrhizal fungi"), with no restrictions on authors, publication years, or journal types. Articles with content irrelevant to the research theme were excluded during the screening process. The final search primarily covered studies published in the past 30 years, resulting in the identification of approximately 1200 articles focusing on the interactions between microbes and plant stress responses. These studies mainly address key research areas such as plant oxidative stress regulatory mechanisms, plant–microbe interaction dynamics, maintenance of plant hormone homeostasis, enhancement of plant stress tolerance, microbial colonization strategies, and modulation of plant growth processes under stress conditions.

3.1. Advances in Microorganisms Involved in Plant Drought Stress Response

Global warming and frequent droughts increase drought stress incidence, causing plant physiological water deficit, even death, and global crop yield declines [30]. Plants can adapt to aridity by enhancing soil microbial ecosystems via morphological, molecular, physiological, and biochemical adjustments [6,31].

Under drought stress, microorganisms mitigate harm mainly through the following mechanisms: (1) Synthesizing growth-promoting substances—drought-tolerant microorganisms thicken plant cell walls, induce dormancy, accumulate osmotic regulators, and produce EPS to meet plant nutritional needs and improve survival [32]; *Bacillus cereus* L90 secretes cytokinins (CTKs) to promote stomatal opening, enhance photosynthesis, and reduce chlorophyll loss in walnuts [33]; nitrogen-fixing bacteria produce indole-3-acetic acid (IAA) to boost wheat root development and nutrient/water absorption [13]; *Bacillus megaterium* BOFC15 excretes spermidine (SPD) to increase ABA content and drought tolerance in *Arabidopsis thaliana* [34]; 1-amino-cyclopropane-1-carboxylate (ACC) deaminase-producing microorganisms can moderately reduce ethylene, balance stomatal opening and closing, while also slowing leaf senescence and extending the plant's photosynthetic cycle [35]. (2) Enhancing nutrient/water uptake—inoculation with AZP2 strains elongates pistachio root hairs and induces lateral roots, expanding uptake area for water and minerals [36]; long-term changes in microbial communities (driven by root exudate shifts) promote nutrient mineralization, aiding plant drought recovery [37]. (3) Improving antioxidant ability—soybeans inoculated with *Penicillium anisopliae* show reduced lipid peroxidation and linolenic acid (LNA) accumulation, plus increased APX, peroxidase (POX), CAT, and SOD activities [32]. Table 1 provides a comprehensive overview of microbial effects on plants under drought stress.

Table 1. Effects of microorganisms on plants under drought stress.

Name of Microorganism	Test Crop	Level of Stress	Microorganism Dosage Usage	Response of Microbial Inoculated Plants to Drought Stress as Compared to Controls	Mechanism of Action	Reference
Arbuscular mycorrhizal fungi (*AMF*)	Maize (*Zea mays* L.)	35%, 55%, 80% field water holding capacity	10 mL 10^8 CFU/mL	Enhances root colonization, water utilization and root hydraulic conductivity, thereby improving nutrient uptake in the corn root system and aboveground	(3) (4) (5) (6)	[38]
Chaetomium globosum ND35	Wheat (*Triticum aestivum* L.)	30% of maximum water holding capacity	10^6 CFU/mL	Promote root and plant development during the seedling stage of winter wheat, allowing wheat to enter the three-leaf stage earlier, enhance drought avoidance, and at the same time improve root activity and increase drought resistance	(2) (3) (5)	[39]
Glomous mosseae	Chinese wildrye (*Leymus chinensis*)	10% PEG	5 g *AMF* strain	Inhibition of Na^+, Cl^- uptake, enhancement of Ka^+ uptake, elevated proline content, elevated antioxidant defense enzyme content	(1) (3) (5)	[35]

Table 1. *Cont.*

Name of Microorganism	Test Crop	Level of Stress	Microorganism Dosage Usage	Response of Microbial Inoculated Plants to Drought Stress as Compared to Controls	Mechanism of Action	Reference
Azospirillum lipoferum AZ1, *Azospirillum lipoferum* AZ45, *Azospirillum lipoferum* AZ9	Wheat (*Triticum aestivum* L.)	80, 50, 25% field water holding capacity	3.2×10^9 CFU/mL	Indole-3-acetic acid (IAA), and proteins, polyamines, nitrogen fixation, root growth promotion	(1) (4) (5)	[40]
Bacillus cereus L90	Walnut (*Juglans regia* L.)	Water content 34.64%	2×10^8 CFU/mL	Promotes secretion of cytokinin (CTK), which increases net photosynthetic rate, stomatal conductance, intercellular CO_2 concentration and chlorophyll content	(1) (3) (4) (6)	[33]
Bacillus megatherium BOFC15	Arabidopsis thaliana (*Arabidopsis thaliana* L. Heynh.)	200 min of dehydration	1 mL bacterial diluent	Increase plant biomass, improved root structure, and enhanced photosynthetic capacity.	(1) (2) (3) (4) (6)	[34]
Sarthrobacter protophormiae SA3 *Dietzia natronolimnaea* STR1 *Bacillus subtilis* LDR2	Wheat (*Triticum aestivum* L.)	10% PEG	25 mL 10^5 CFU/mL	Increase IAA content, photosynthetic efficiency, reduced abscisic acid and Enzyme 1-amino-cyclopropane-1-carboxylate (ACC) content	(1) (2) (4) (6)	[41]
AMF	Spring wheat (*Triticum aestivum* L.)	40% soil moisture content	Inoculums (1600 propagules/g) were mixed with wheat seeds at 10 mL/kg pre-wetting rate.	Increase N and P concentrations in stems and grains resulted in a significant increase in the plant's water use efficiency	(1) (5) (6)	[42]
Pseudomonas fluorescens strains FB-49	Acacia martius (*Acacia farnesiana* L. Willd.)	Keep 20% water content	15 mL 10^8 CFU/mL	Increase root length, aboveground node length and dry biomass of plants	(1) (5)	[43]
Azotobacter brazilensis *Bacillus* sp.	Tropical trees (*Pinus tropicalis* Morelet)	14, 30% humidity	50 mL 10^6 CFU/mL	Induce greater accumulation of secondary compounds and increased leaf area.	(1) (3)	[44]

Note: (1) regulating the perception and transduction of abiotic stress signals to enhance plant adaptive responses; (2) boosting gene expression and protein synthesis for overall plant metabolic regulation; (3) activating the antioxidant system to strengthen plant tolerance; (4) modulating plant hormone levels to stimulate growth in response to adversity; (5) enhancing plant nutrition and absorption to improve resilience; (6) optimizing the photosynthesis system to promote the synthesis of essential substances, safeguarding plant growth and development amidst adversity.

3.2. *Advancements in Microorganisms Involved in Plant Salt Stress Response*

Salt stress causes high osmotic pressure, leading to plant water loss, cell death, and growth/yield inhibition [45]; however, plant–microbe symbiosis can promote plant growth in saline soils [46]. Key microbial mitigation strategies under salt stress include the following: (1) Regulating ion balance—*Bacillus velezensis* JC-K3 reduces Na^+ uptake in wheat shoots, increases soluble sugars and chlorophyll, and alters endophytic/rhizosphere microbial diversity [14]; *Pseudomonas geniculate* MF-84 decreases maize Na^+ uptake while increasing root K^+/Ca^{2+} absorption, alleviating ionic toxicity [47]. (2) Enhancing nutrient absorption—arbuscular mycorrhizal fungi (*AMF*) form symbioses with plants to absorb diverse minerals, compensating for nutrient deficiencies [48]; *Trichoderma harzianum* (TH) increases S, Mn, Mg, Ca, and K contents in salt-stressed Indian mustard (by 7.0–36.3%) and reduces salt uptake [28]. (3) Modulating phytohormones and metabolites—bacteria with ACC deaminase reduce salt-induced ethylene (ETH) synthesis to enhance salt tolerance [49,50]; PGPR strains promote proline accumulation and POX activity in soybeans [51]; *Avicennia marina* bacteria secrete EPS to alleviate salt stress, solubilize inorganic phosphate, and improve nutrient uptake in rice [52]. (4) Activating plant signaling—endophytic bacteria in high-salinity soils induce plant synthesis of H_2O_2, salicylic acid (SA), and jasmonic acid (JA), upregulating sesquiterpene biosynthesis genes [53,54]; bacterial exocrines (IAA, ACCD deaminase) trigger plant regulatory pathways [16]. Table 2 provides a comprehensive overview of microbial effects on plants under salt stress. These microorganisms assist plants in coping with salt stress via multiple pathways.

Table 2. Effects of microorganisms on plants under salt stress.

Name of Microorganism	Test Crop	Level of Stress	Microorganism Dosage Usage	Response of Microbial Inoculated Plants to High Salt Stress as Compared to Controls	Mechanism of Action	Reference
Bacillus subtilis NCD-2	Tomato (*Solanum lycopersicum* L.)	100 mmol/L NaCl	1.0×10^9 CFU/mL	Enhanced resistance enzyme activity, increased ABA content, and enriched rhizosphere beneficial microbes	(1) (3) (4)	[55]
Brevibacterium sediminis Strain IBGE3C	Rice (*Oryza sativa* L.)	0.2–1.2% NaCl	Seed soaking	Improve rice varieties with different levels of salt tolerance	(1) (3) (4)	[56]
Burkholderia phytofirmans PsJN *Enterobacter* sp. FD17	Maize (*Zea mays* L.)	25 mmol/L NaCl	Mix 20 mL of bacterial suspension with 100 g of sterilized peat	Reduce xylem Na^+ concentration uptake, thereby maintaining nutrient balance and promoting plant growth	(1) (4) (5)	[57]
Pseudomonas fluorescens YsS6 *Pseudomonas migulae* 8R6	Tomato (*Solanum lycopersicum* L.)	165, 185 mmol/L NaCl	1.75×10^8–1.97×10^8 CFU/ml	Higher fresh and dried biomass, higher chlorophyll content and more flowers and buds reduce salt stress	(1) (4) (6)	[58]
Klebsiella pseudomonas Agrobacterium ochrobactrum	Peanut (*Arachis hypogaea* L.)	4, 8% NaCl	10^8 cells/mL	Increase solubilization of phosphorus; promotes stem length, root length, shoot and root growth in peanut plants	(1) (2) (3) (4) (5)	[59]
Bacillus subtilis 10-4	Babury Wolfberry Fruit (*Lycium chinense* Miller)	2% NaCl	10^5 CFU/mL	Inhibition of Salicylic acid (SA) accumulation, increase in water storage capacity in leaf tissues	(1) (3) (4) (5)	[60]
Bacillus amyloliquefaciens RWL-1	Rice (*Oryza sativa* L.)	40 g/L NaCl	10^8 CFU/mL	Increase essential amino acids and SA, decreased ABA levels	(1) (4) (5)	[61]
Bacillus cereous Pb25	Mung bean (*Vigna radiata* (L.) R. Wilczek)	Electric conductivity 9 dS/m	10^7–10^8 CFU/mL	Increase plant antioxidant enzyme activity, proline, potassium, nitrogen, and phosphorus accumulation; decreased sodium accumulation	(1) (3) (4) (5)	[62]
Staphylococcus equorum strain EN21	Tomato (*Solanum lycopersicum* L.)	30% NaCl	10^9 CFU/mL	Increase seed vigor index, branch length and root dry weight of plants	(1) (3) (4) (5) (6)	[63]
Pseudomonas strains AK-1 *Bacillus* strains SJ-5	Soybean (*Glycine max* (L.) Merr.)	100 mmol/L NaCl	10^8 CFU/mL	Increase plant biomass, leaf water content, photosynthetic activity; increased proline accumulation and peroxidase (POX) activity	(1) (2) (3) (4) (5) (6)	[51]
Avicennia marina	Rice (*Oryza sativa* L.)	0.5–22.5% NaCl	10^8 CFU/mL	Promotes solubilization of inorganic phosphate and enhances nutrient uptake	(1)	[52]
Piriformospora indica	Barley (*Hordeum vulgare* L.)	100, 300 mmol/L NaCl	The mycelial colonization rate is 50–60%	Enhancement of Ascorbate peroxidase (APX) activity in barley roots	(1) (3) (4) (6)	[64]

Table 2. *Cont.*

Name of Microorganism	Test Crop	Level of Stress	Microorganism Dosage Usage	Response of Microbial Inoculated Plants to High Salt Stress as Compared to Controls	Mechanism of Action	Reference
Pseudomonas strains PF1/TDK1	Rice (*Oryza sativa* L.)	100 mmol/L NaCl	0.2 g/L	Plant height, root length, aboveground and root dry weight were significantly increased	(1) (3) (4) (6)	[65]
Trichoderma harzianum	Indian mustard (*Brassica juncea* L.)	100, 200 mmol/L NaCl	2×10^9 CFU/mL	Increase oil content improves absorption of essential nutrients, enhances antioxidant and osmotic agent accumulation, and reduces salt absorption	(1) (3) (4) (5) (6)	[28]

Note: (1) regulating the perception and transduction of abiotic stress signals to enhance plant adaptive responses; (2) boosting gene expression and protein synthesis for overall plant metabolic regulation; (3) activating the antioxidant system to strengthen plant tolerance; (4) modulating plant hormone levels to stimulate growth in response to adversity; (5) enhancing plant nutrition and absorption to improve resilience; (6) optimizing the photosynthesis system to promote the synthesis of essential substances, safeguarding plant growth and development amidst adversity.

In terms of microbial response mechanisms, the two stresses share commonalities. Both can regulate the osmotic pressure of plant cells to cope with osmotic stress by synthesizing osmotic regulators such as EPS. Additionally, both enhance plants' nutrient uptake to support growth. However, differences exist: under drought stress, microorganisms focus on producing growth-promoting substances (e.g., secreting CTK and IAA) to regulate plant physiological processes; under salt stress, microorganisms prioritize ion balance regulation, reducing plants' Na^+ uptake while increasing the absorption of beneficial ions like K^+ and Ca^{2+}.

4. Mechanisms Behind Microbial Regulation of Plant Stress

After comprehensively summarizing the advancements in the field of microorganisms aimed at enhancing plant stress tolerance under the aforementioned drought stress and salt stress conditions, six distinct mechanisms of action have been proposed. These mechanisms include the following: (1) regulating signal perception and transduction, (2) enhancing gene expression and protein synthesis, (3) activating the antioxidant system, (4) modulating phytohormone concentrations, (5) enhancing plant nutrition and absorption capabilities, and (6) optimizing the photosynthetic system. These mechanisms are schematically depicted in Figure 1.

4.1. Regulating the Perception and Transduction of Abiotic Stress Signals to Enhance Plant Adaptive Responses

Plants possess sophisticated sensing, signaling, and response mechanisms to cope with external stressors [20,21]. Chemical signaling between plants and microorganisms is a collaborative outcome of their interactions [66]. Notably, most key regulators involved in plant defense responses operate within a complex network of modulators rather than following a linear pathway [67]. Prolonged exposure to abiotic stress can induce adaptive changes in specific microbial subgroups, enabling them to maintain efficient routine metabolic functions even under stressful conditions [68]. This adaptation leads to the emergence of microorganisms with enhanced stress tolerance, which in turn promotes crop growth. During plant responses to external stress, root exudates serve as a critical communication bridge between plants and microorganisms: they provide essential energy for microbial growth and facilitate direct interactions via specific signaling molecules [37]. Importantly, the composition and quantity of root exudates dynamically change throughout plant development [69], and stress factors further alter their properties and production [70]. Microorganisms can recognize root exudates, perceive stress signals, and recruit additional beneficial microbial populations to perform functional roles that benefit the host plant—fostering a symbiotic relationship that supports the mutual survival and development of both parties [71]. For example, a recent study demonstrated that under salt stress, wild soybeans secrete purine analogs as signaling molecules; upon recognition by *Pseudomonas aeruginosa*, these bacteria secrete secondary signaling substances to inversely regulate the plant's stress signal transduction pathway, ultimately enhancing plant salt tolerance [72]. Specifically, the secondary signaling substances secreted by *Pseudomonas aeruginosa* bind to plant root cell membrane receptors, activating the plant's MAPK signaling pathway and further regulating the expression of stress-responsive genes (e.g., *RD29A*), thereby linking microbial signals to plant stress regulatory pathways.

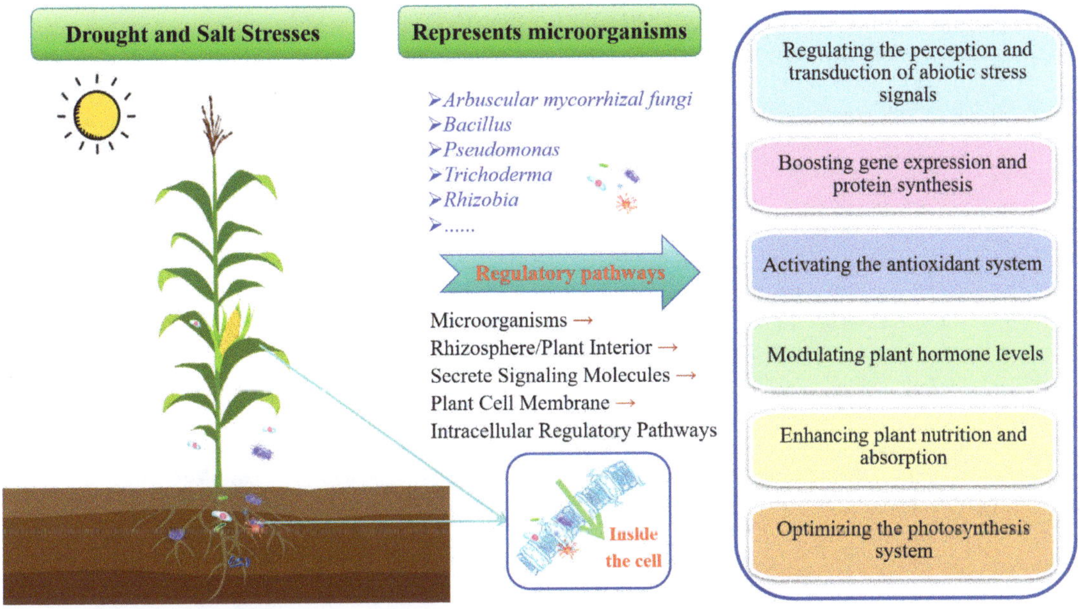

Figure 1. Schematic diagram illustrating the mechanisms of microbial regulation on plant abiotic stress responses. Note: Microorganisms secrete signaling molecules in the rhizosphere or inside plants, which act on plant intracellular regulatory pathways to alleviate drought and salt stresses. Red arrows indicate a process chain.

4.2. Boosting Gene Expression and Protein Synthesis for Overall Plant Metabolic Regulation

Gene regulatory strategies offer innovative methodologies for cultivating crops that demonstrate tolerance to abiotic stress conditions [73]. In response to abiotic stress environments, plants have naturally developed diverse mechanisms and regulatory networks that modulate the expression patterns of their genes, enabling them to adapt and thrive. Additionally, these regulatory systems facilitate interactions between plants and microorganisms, further enhancing their survival capabilities [74]. Firstly, the expression of endogenous genes can be modulated under stressful conditions, resulting in either upregulation or repression. For instance, *Pseudomonas fluorescens* enhances the expression of the ACC deaminase gene acdS, thereby reducing the concentration of ETH and improving resistance to abiotic stress [75]. Similarly, tomato plants inoculated with TH exhibit significant elevations in transcript levels of *NAC1*, dehydrin *TAS14*, and *P5CS* genes [76]. Additionally, abscisic acid-induced *myb1* (SlAIM1) [77] and stress-inducible genes such as *CaACCO* and *CaLTPI* are regulated in a comparable manner [78]. Additionally, endophytes harbor genes crucial for biological nitrogen fixation, a process that facilitates the conversion of atmospheric nitrogen (N_2) into ammonium and nitrate within the host plant [79]. Finally, it promotes the expression of the drought-induced gene wzy2 in wheat, thereby upregulating the expression of drought-related proteins and enhancing the dehydration tolerance and drought stress adaptability of two winter wheat varieties [39].

4.3. Activating the Antioxidant System to Strengthen Plant Tolerance

In plants, ROS play a pivotal role in regulating a broad spectrum of biological processes, such as cell signaling, growth, and development [22,23]. Nevertheless, when ROS concentrations exceed a critical threshold—often triggered by abiotic stresses like drought and salt—they induce oxidative damage to cellular components, including lipids, proteins, and nucleic acids [22]. Notably, abiotic stress typically leads to a marked surge in plant ROS levels, which in turn inhibits growth and development by disrupting metabolic homeostasis [24]. Beneficial microorganisms sustain plant antioxidant system functionality to mitigate stress-induced oxidative damage via two mechanisms: (1) directly secreting extracellular antioxidants (e.g., superoxide dismutase, SOD) to scavenge rhizospheric or apoplastic ROS; (2) indirectly enhancing plant endogenous antioxidant synthesis by activating stress-responsive signaling pathways (e.g., salicylic acid-mediated pathways) or upregulating antioxidant-related genes [25]. This dual regulation boosts the accumulation of both enzymatic antioxidants (e.g., SOD, CAT, peroxidase (POD)) and non-enzymatic antioxidants (e.g., glutathione (GSH), AsA, tocopherols, and carotenoids), effectively shielding plants from ROS-induced cellular damage [80]. For instance, inoculation with Penicillium anisopliae has been shown to reduce lipid peroxidation and enhance SOD, CAT, and ascorbate peroxidase (APX) activities in soybean under drought stress, mirroring this protective mechanism [32]. PGPB further reinforces antioxidant defense. Specifically, PGPB enhance both enzymatic and non-enzymatic antioxidant pathways, which collectively preserve cellular redox homeostasis, elevate ROS scavenging capacity, and optimize biosignature traits (e.g., nutrient uptake efficiency, photosynthetic stability). These effects ultimately translate to improved tolerance of plants to drought and salt stresses [44].

4.4. Modulating Plant Hormone Levels to Stimulate Growth in Response to Adversity

Phytohormones act as pivotal signaling molecules in plant responses to environmental stresses [81]. During interactions with microorganisms, plants can mitigate stress effects through hormone-mediated pathways, involving the regulation of IAA, ABA, ETH, SA, and JA [82]. This regulatory process operates through three primary mechanisms: First, symbiotic microorganisms either produce phytohormones or their analogs or induce plants to secrete these signaling molecules [83]. For example, several bacterial genera including *Pseudomonas*, *Pantoea*, *Hymenobacter*, and *Rhizobium* synthesize IAA analogs [84]. Similarly, TH produces metabolites such as hazylactone and 6-pentyl-α-pyrone that mimic IAA activity, effectively stimulating plant growth [85]. Conversely, *Variovorax* species modulate root development by regulating IAA metabolism—specifically targeting key enzymes in the IAA pathway through unique degradation sites to adjust hormone levels, thereby influencing root cell division, elongation, and differentiation [66]. Second, microorganisms significantly impact plant hormone metabolism and synthesis. A well-documented example is the reduction in ethylene levels by PGPR through ACC deaminase activity. This enzyme catalyzes the conversion of ACC—the immediate precursor of ethylene—into ammonia and α-ketobutyric acid [35,75,86]. Similar ethylene-modulating effects have been observed in *Arthrobacter* and *Bacillus* species [78]. Fungal inoculation, meanwhile, typically reduces ABA levels while altering SA and JA concentrations [87]. Additionally,

Bacillus licheniformis SA03 activates ABA-dependent nitric oxide (NO) synthesis, thereby enhancing the stress resistance of chrysanthemums. As a secondary messenger, NO further improves chrysanthemums' stress tolerance by regulating downstream stress-responsive pathways [88]. Specifically, NO functions as a signaling hub by mediating S-nitrosylation of target proteins, regulating components of signaling pathways (e.g., MAPK, cGMP), and maintaining ROS homeostasis. Through these mechanisms, it coordinates with plant hormones in a synergistic or antagonistic manner to regulate plant growth, development, and stress responses, with its regulatory effects exhibiting concentration dependence. Notably, NO also serves as a pivotal signaling molecule in plant–microbe interactions: it mediates the regulatory role of microorganisms in plant stress tolerance and participates in the crosstalk between microorganisms and plant hormones [89]. Third, microorganisms influence hormone transport and functional mechanisms within plants. *Pseudomonas malodorata*, for instance, modifies endogenous hormone accumulation in roots and shoots, increasing levels of gibberellins (GA), IAA, CTK, and ABA to mitigate stress impacts [90]. PGPR enhance wheat tolerance to abiotic stresses through a coordinated hormonal strategy: increasing IAA levels, reducing ABA and ACC concentrations, and regulating the expression of ethylene signaling pathway components (e.g., CTR1) and DREB2 transcription factors [41].

4.5. Enhancing Plant Nutrition and Absorption to Improve Tolerance

The rhizosphere is a crucial habitat for microorganisms, where root exudates attract microbial colonization and provide nutrients essential for microbial proliferation [68,69,91]. *Mycorrhizal fungi* (MF) play a pivotal role in facilitating host water acquisition under stress and stimulating lateral root formation, thereby increasing root surface area and promoting root uptake [92]. Additionally, radial mycelium extension induced by *AMF* inoculation effectively enhances nutrient uptake [93]. Mycelium forms a filamentous network via apical extension and subapical branching; its rapid apical elongation enables insertion into the plant cell wall [94], which facilitates the regulation of osmotic balance and composition. Mycorrhizal symbiosis significantly improves water retention, thereby alleviating dehydration and metabolic disorders [95]. Furthermore, microbial fertilizers can modulate the concentration of IAA, effectively reducing root sensitivity and promoting root elongation and growth [96].

Roots release chemical signals and nutrients that influence the rhizosphere microbial community [97]. *Pseudomonas, Rhizobium, Bacillus, Penicillium*, and *Aspergillus* are well-recognized efficient phosphate-solubilizing microorganisms (PSMs) [59]. These microbes secrete organic acids to lower the surrounding soil pH, thereby solubilizing phosphorus from mineral phosphates for subsequent plant absorption [98]. Additionally, soil phosphorus-solubilizing bacteria (PSB) not only prevent phosphorus fixation but also enrich the soil with micronutrients [99]. Mycorrhizal colonization further enhances plant P uptake [42], and phosphatases secreted by *AMF* and plant root colonization collectively improve plant P absorption efficiency [27]. Elevated ACC deaminase activity is associated with increased accumulation of nutrients such as N, P, K, and other essential elements, thereby promoting root growth and colonization [38]. Nitrogen-fixing bacteria (NFB) assist plants in amino acid synthesis by converting atmospheric nitrogen and regulating its utilization and absorption, thereby enhancing plant nutritional capacity [12]. A variety of metabolic compounds produced by plant-associated microorganisms—including growth-promoting substances—play key roles in rhizosphere function and indirectly enhance plant stress resistance. Among these compounds, ACC deaminase, IAA, siderophores, and organic acids are particularly pivotal [38].

4.6. Optimizing the Photosynthesis System to Promote the Synthesis of Essential Substances, Safeguarding Plant Growth and Development Amidst Adversity

Exogenous microorganisms can effectively regulate plant photosynthetic and alleviate stress-induced adverse effects. The most intuitive manifestation is increased photosynthetic pigment content in leaves. By enhancing water uptake efficiency and scavenging ROS, exogenous microorganisms improve leaf physiological activity, thereby significantly increasing plant photosynthetic rates [38]. Additionally, microorganisms maintain the dynamic balance between plant water status and photosynthesis by regulating stomatal conductance, reducing negative impacts from excessive transpiration—an effect *Trichoderma* exerts on various crops [6].

Acinetobacter calcoaceticus strain X128 secretes CTK, a hormone that promotes stomatal opening and inhibits chlorophyll degradation. This keeps photosynthetic limitations at the reversible stomatal level, delaying irreversible mesophyll cell damage from non-stomatal limitations and enhancing post-rehydration recovery of photosynthetic function [100]. *AMF* alleviate drought-induced inhibition of photosynthetic apparatus in wheat by maintaining PSII stability, improving plant water status (measured via RWC), and optimizing nutrient supply—ultimately increasing biomass [101]. For instance, *Acinetobacter calcoaceticus* X128-secreted CTK not only maintains stomatal opening but also upregulates the expression of chloroplast-related genes (e.g., *rbcL*, which encodes Rubisco large subunit), thereby protecting chloroplast structure and function. However, the direct mechanisms by which microorganisms regulate chloroplast function require further in-depth research.

Under salt stress, tomato plants inoculated with wild-type endophytes (*Pseudomonas fluorescens* YS56, *Pseudomonas migulae* 8R6) containing ACC deaminase exhibit significantly reduced endogenous ethylene levels, delayed chlorophyll decomposition, and maintained photosynthetic capacity [58]. Notably, *AMF* reduces Na^+ toxicity through selective K^+ uptake, while PGPR binds Na^+ via EPS secretion—reducing Na^+ transport to leaves. These actions preserve the activity of key enzymes (e.g., Rubisco) and protein synthesis processes [27,51].

5. Conclusions and Outlook

This review comprehensively summarizes the latest research progress (1994–2025) on the role of core microbial groups—including PGPR, *AMF*, and endophytes—in enhancing plant tolerance to drought and salt stresses. It further synthesizes and proposes six core mechanisms by which microorganisms improve plant stress resistance: (1) modulation of stress signal perception and transduction; (2) enhancement of gene expression and protein biosynthesis; (3) activation of antioxidant defense systems; (4) regulation of phytohormone homeostasis; (5) promotion of plant nutrient status and absorption efficiency; and (6) optimization of photosynthetic processes.

While this review outlines the six core mechanisms of microbial-mediated stress alleviation, critical gaps and inconsistencies in existing research must be acknowledged. First, microbial treatments do not uniformly enhance plant drought or salt tolerance. For example, certain PGPR strains (e.g., *Bacillus subtilis* strain GB03) significantly improve maize drought tolerance but show no obvious effect on wheat under the same stress intensity [102,103]; similarly, *AMF* colonization (e.g., *Rhizophagus irregularis*) enhances salt tolerance in tomato cultivar 'Moneymaker', yet fails to alleviate salt damage in 'Ailsa Craig' [104,105]. This variability is linked to four key factors: plant genotype (differences in root exudate composition and stress response pathways among species/cultivars affect microbial colonization and function), soil properties (acidic soils with low organic matter inhibit PGPR growth, while high bulk density impedes *AMF* hyphal extension), stress severity (mild salt stress stimulates microbial metabolite secretion, but extreme drought/high-salt conditions impair microbial viability), and microbial strain specificity (different strains of the same genus have distinct capacities for hormone production or osmolyte synthesis).

Second, the direct interaction mechanisms between microorganisms and plant chloroplast function remain unclear—such as how microbial signaling molecules regulate chloroplast-related gene expression or maintain photosystem II stability under stress. Additionally, lab-scale results lack systematic field validation, as most laboratory experiments rely on controlled conditions inconsistent with real agricultural scenarios. Notably, the lab-field gap is pronounced: laboratory tests use homogenized soil substrates, constant temperature/humidity, and single-stress treatments, which fail to replicate natural environmental complexity, including soil variability (e.g., heterogeneous texture, uneven nutrient distribution, fluctuating pH in field soils) and unstable microbial durability (microbial colonization ability and functional activity are reduced by extreme field conditions like sudden temperature changes or soil compaction). These limitations undermine the reliability of extrapolating lab findings to real-world agricultural production.

To address these gaps and inconsistencies, future research should prioritize the following: (1) Strengthening field validation under complex stress scenarios (e.g., "drought-salt combined stress" or "salt stress with soil degradation") by quantifying agronomic traits (e.g., yield, water use efficiency) and microbial functional stability (e.g., enzyme activity, metabolite secretion) to bridge the lab-field gap. (2) Deciphering tripartite interactions (microbes–plants–environment/agronomic factors) to optimize microbial colonization—for example, exploring how adjusting nitrogen fertilizer rates enhances *AMF* colonization in wheat roots, or developing microbial consortia (combining stress-tolerant PGPR and *AMF*) to reduce strain-specific variability. (3) Tailoring rhizosphere environments (via optimized irrigation, nutrient supply, or organic amendments like biochar) to promote beneficial microbial recruitment and proliferation, mitigating adverse soil-induced inhibition. (4) Expanding research on microbial metabolites (e.g., exopolysaccharides, cyclic peptides) to evaluate their roles in plant growth promotion and stress alleviation, plus synergistic application with low-dose pesticides/fertilizers. (5) Advancing interdisciplinary collaboration (integrating agriculture, microbiology, soil science, and molecular biology) to exploit microbial resources—such as using metagenomics to identify key stress-tolerant taxa in saline–alkali soils or engineering microbes with enhanced stress adaptability—providing practical support for sustainable agriculture and ecological restoration.

Author Contributions: D.F. and W.X. planned and designed the research. W.L. conducted literature retrieval, data analysis, and sorting. D.F., W.L. and P.H. prepared the first draft of the manuscript. G.T., M.G., Y.D., G.C. and W.X. revised the manuscript. D.F. and W.L. contributed equally. All authors have read and agreed to the published version of the manuscript.

Funding: This study was supported by the Major Science and Technology Special Projects of Xinjiang Uygur Autonomous Region (No. 2024A03008-3), the Chinese Academy of Sciences, and the Project of Fund for Stable Support to Agricultural Sci-Tech Renovation (No. xjnkywdzc-2025001-30).

Institutional Review Board Statement: Not applicable.

Informed Consent Statement: Not applicable.

Data Availability Statement: No new data were created or analyzed in this study. Data sharing is not applicable to this article.

Conflicts of Interest: The authors declare no conflict of interest.

References

1. Wang, W.; Vinocur, B.; Altman, A. Plant responses to drought, salinity and extreme temperatures: Towards genetic engineering for stress tolerance. *Planta* **2003**, *218*, 1–14. [CrossRef] [PubMed]

2. Lee, C.; Song, T. Green finance and food production: Evidence from cities in China. *J. Clean. Prod.* **2024**, *458*, 142423. [CrossRef]
3. Ahmad, A.; Aslam, Z.; BelliTürk, K.; Hussain, S.; Bibi, I. Soil application of cellulolytic Microbe–Enriched vermicompost modulated the morpho-physiological and biochemical responses of wheat cultivars under different moisture regimes. *J. Soil Sci. Plant Nutr.* **2022**, *22*, 4153–4167. [CrossRef]
4. Chang, Y.; Chen, Z.; Jiang, J.; Zhang, H.; Zhu, J.; Duan, C. Epigenetic regulation in plant abiotic stress responses. *J. Integr. Plant Biol.* **2020**, *62*, 563–580. [CrossRef]
5. Huot, B.; Yao, J.; Montgomery, B.L.; He, S.Y. Growth–Defense Tradeoffs in Plants: A balancing act to optimize fitness. *Mol. Plant* **2014**, *7*, 1267–1287. [CrossRef]
6. Cruz, C.C.; Cardoso, P.; Santos, J.; Matos, D.; Figueira, E. Bioprospecting soil bacteria from srid zones to increase plant tolerance to drought: Growth and biochemical status of maize inoculated with plant growth-promoting bacteria isolated from sal island, cape verde. *Plants* **2022**, *11*, 2912. [CrossRef]
7. Liu, H.; Carvalhais, L.C.; Crawford, M.; Singh, E.; Dennis, P.G.; Pieterse, C.M.J. Inner plant values: Diversity, colonization and benefits from endophytic bacteria. *Front. Microbiol.* **2017**, *8*, 2552. [CrossRef]
8. Eichmann, R.; Richards, L.; Schäfer, P. Hormones as go-betweens in plant microbiome assembly. *Plant J.* **2021**, *105*, 518–541. [CrossRef]
9. Christensen, M.J.; Bennett, R.J.; Ansari, H.A.; Koga, H.; Johnson, R.D.; Bryan, G.T.; Simpson, W.R.; Koolaard, J.; Nickless, E.; Voisey, C.R. Epichloë endophytes grow by intercalary hyphal extension in elongating grass leaves. *Fungal Genet. Biol.* **2008**, *45*, 84–93. [CrossRef]
10. Khan, A.L.; Hamayun, M.; Radhakrishnan, R.; Waqas, M.; Kang, S.M.; Kim, Y.H.; Shin, J.H.; Choo, Y.S.; Kim, J.G.; Lee, I.J. Mutualistic association of *Paecilomyces formosus* LHL10 offers thermotolerance to *Cucumis sativus*. *Antonie Van Leeuwenhoek* **2012**, *101*, 267–279. [CrossRef] [PubMed]
11. Kumar, A.; Verma, J.P. Does plant—Microbe interaction confer stress tolerance in plants: A review? *Microbiol. Res.* **2018**, *207*, 41–52. [CrossRef]
12. Lata, R.; Chowdhury, S.; Gond, S.K.; White, J.F. Induction of abiotic stress tolerance in plants by endophytic microbes. *Lett. Appl. Microbiol.* **2018**, *66*, 268–276. [CrossRef]
13. Lynch, J.M.; Moffat, A. Bioremediation—Prospects for the future application of innovative applied biological research. *Ann. Appl. Biol.* **2005**, *146*, 217–221. [CrossRef]
14. Feng, D.; Gao, Q.; Liu, J.; Tang, J.; Hua, Z.; Sun, X. Categories of exogenous substances and their effect on alleviation of plant salt stress. *Eur. J. Agron.* **2023**, *142*, 126656. [CrossRef]
15. Feng, D.; Liu, W.; Chen, K.; Ning, S.; Gao, Q.; Chen, J.; Liu, J.; Sun, X.; Xu, W. Exogenous substances used to relieve plants from drought stress and their associated underlying mechanisms. *Int. J. Mol. Sci.* **2024**, *25*, 9249. [CrossRef] [PubMed]
16. Kumar, A.; Singh, S.; Gaurav, A.K.; Srivastava, S.; Verma, J.P. Plant growth-promoting bacteria: Biological tools for the mitigation of salinity stress in plants. *Front. Microbiol.* **2020**, *11*, 1216. [CrossRef] [PubMed]
17. Braud, A.; Jézéquel, K.; Bazot, S.; Lebeau, T. Enhanced phytoextraction of an agricultural Cr- and Pb-contaminated soil by bioaugmentation with siderophore-producing bacteria. *Chemosphere* **2009**, *74*, 280–286. [CrossRef] [PubMed]
18. Hayat, R.; Ali, S.; Amara, U.; Khalid, R.; Ahmed, I. Soil beneficial bacteria and their role in plant growth promotion: A review. *Ann. Microbiol.* **2010**, *60*, 579–598. [CrossRef]
19. Gupta, A.; Mishra, R.; Rai, S.; Bano, A.; Pathak, N.; Fujita, M.; Sharma, M.; Hasanuzzaman, M. Mechanistic insights of plant growth promoting bacteria mediated drought and salt stress tolerance in plants for sustainable agriculture. *Int. J. Mol. Sci.* **2022**, *23*, 3741. [CrossRef]
20. Sabagh, A.E.; Islam, M.S.; Hossain, A.; Iqbal, M.A.; Mubeen, M.; Waleed, M.; Reginato, M.; Battaglia, M.L.; Ahmed, S.; Rehman, A.; et al. Phytohormones as growth regulators during abiotic stress tolerance in plants. *Front. Agron.* **2022**, *4*, 765068. [CrossRef]
21. Chen, X.; Ding, Y.; Yang, Y.; Song, C.; Wang, B.; Yang, S.; Guo, Y.; Gong, Z. Protein kinases in plant responses to drought, salt, and cold stress. *J. Integr. Plant Biol.* **2021**, *63*, 53–78. [CrossRef] [PubMed]
22. Castro, B.; Citterico, M.; Kimura, S.; Stevens, D.M.; Wrzaczek, M.; Coaker, G. Stress-induced reactive oxygen species compartmentalization, perception and signalling. *Nat. Plants* **2021**, *7*, 403–412. [CrossRef] [PubMed]
23. Farooq, M.; Niazi, A.; Akhtar, J.; Saifullah; Farooq, M.; Souri, Z.; Karimi, N.; Rengel, Z. Acquiring control: The evolution of ROS-Induced oxidative stress and redox signaling pathways in plant stress responses. *Plant Physiol. Biochem.* **2019**, *141*, 353–369. [CrossRef]
24. Mittler, R. Oxidative stress, antioxidants and stress tolerance. *Trends Plant Sci.* **2002**, *7*, 405–410. [CrossRef] [PubMed]
25. Gul, N.; Wani, I.A.; Mir, R.A.; Nowshehri, J.A.; Aslam, S.; Gupta, R.; Verma, S.; Aslam, S. Plant growth promoting microorganisms mediated abiotic stress tolerance in crop plants: A critical appraisal. *Plant Growth Regul.* **2023**, *100*, 7–24. [CrossRef]
26. Issa, A.A.; Abd-Alla, M.H.; Mahmoud, A.M. Effect of biological treatments on growth and some metabolic activities of barley plants grown in saline soil. *Microbiol. Res.* **1994**, *149*, 317–320. [CrossRef]

27. Aggarwal, A.; Kadian, N.; Karishma, K.; Neetu, N.; Tanwar, A.; Gupta, K. Arbuscular mycorrhizal symbiosis and alleviation of salinity stress. *J. Appl. Nat. Sci.* **2012**, *4*, 144–155. [CrossRef]
28. Ahmad, P.; Hashem, A.; Abd_Allah, E.F.; John, R.; Egamberdieva, D.; Gücel, S. Role of *Trichoderma harzianum* in mitigating NaCl stress in Indian mustard (*Brassica juncea* L.) through antioxidative defense system. *Front. Plant Sci.* **2015**, *6*, 868. [CrossRef]
29. Tiwari, S.; Lata, C. Heavy metal stress, signaling, and tolerance due to plant-associated microbes: An overview. *Front. Plant Sci.* **2018**, *9*, 452. [CrossRef]
30. Lesk, C.; Rowhani, P.; Ramankutty, N. Influence of extreme weather disasters on global crop production. *Nature* **2016**, *529*, 84–87. [CrossRef] [PubMed]
31. Iqbal, M.S.; Singh, A.K.; Ansari, M.I. Effect of drought stress on crop production. In *New Frontiers in Stress Management for Durable Agriculture*; Rakshit, A., Singh, H., Singh, A., Singh, U., Fraceto, L., Eds.; Springer: Singapore, 2020; pp. 35–47. [CrossRef]
32. Iqbal, S.; Iqbal, M.; Li, C.; Iqbal, A.; Abbas, R.Z. Overviewing drought and heat stress amelioration—From plant responses to microbe-mediated mitigation. *Sustainability* **2023**, *15*, 1671. [CrossRef]
33. Liu, F.; Ma, H.; Ma, B.; Du, Z.; Jing, D.; Xing, S. Effect of plant growth-promoting rhizobacteria on photosynthetic characteristics in walnut seedlings under drought stress. *Sci. Silvae Sin.* **2015**, *51*, 84–90.
34. Zhou, C.; Ma, Z.; Zhu, L.; Xiao, X.; Xie, Y.; Zhu, J.; Wang, J. Rhizobacterial strain *bacillus megaterium* BOFC15 induces cellular polyamine changes that improve plant growth and drought resistance. *Int. J. Mol. Sci.* **2016**, *17*, 976. [CrossRef]
35. Ravanbakhsh, M.; Sasidharan, R.; Voesenek, L.A.C.J.; Kowalchuk, G.A.; Jousset, A. Microbial modulation of plant ethylene signaling: Ecological and evolutionary consequences. *Microbiome* **2018**, *6*, 52. [CrossRef] [PubMed]
36. Timmusk, S.; El-Daim, I.A.A.; Copolovici, L.; Tanilas, T.; Kännaste, A.; Behers, L.; Nevo, E.; Seisenbaeva, G.; Stenström, E.; Niinemets, Ü. Drought-tolerance of wheat improved by rhizosphere bacteria from harsh environments: Enhanced biomass production and reduced emissions of stress volatiles. *PLoS ONE* **2014**, *9*, e96086. [CrossRef] [PubMed]
37. De Vries, F.T.; Griffiths, R.I.; Knight, C.; Nicolitch, O.; Williams, A.H. Harnessing rhizosphere microbiomes for drought-resilient crop production. *Science* **2020**, *368*, 270–274. [CrossRef] [PubMed]
38. Khan, W.; Zhu, Y.; Khan, A.; Zhao, L.; Yang, Y.; Wang, N.; Hao, M.; Ma, Y.; Nepal, J.; Ullah, F.; et al. Above-and below-ground feedback loop of maize is jointly enhanced by plant growth-promoting rhizobacteria and arbuscular mycorrhizal fungi in drier soil. *Sci. Total Environ.* **2024**, *917*, 170417. [CrossRef]
39. Cong, G.; Yin, C.; He, B.; Li, L.; Gao, K. Effect of the endophytic fungus *Chaetomium globosum* ND35 on the growth and resistance to drought of winter wheat at the seedling stage under water stress. *Acta Ecol. Sin.* **2015**, *35*, 6120–6128. [CrossRef]
40. Arzanesh, M.H.; Alikhani, H.A.; Khavazi, K.; Rahimian, H.; Miransari, M. Wheat (*Triticum aestivum* L.) growth enhancement by *Azospirillum* sp. under drought stress. *World J. Microbiol. Biotechnol.* **2011**, *27*, 197–205. [CrossRef]
41. Barnawal, D.; Bharti, N.; Pandey, S.S.; Pandey, A.; Chanotiya, C.S.; Kalra, A. Plant growth-promoting rhizobacteria enhance wheat salt and drought stress tolerance by altering endogenous phytohormone levels and *TaCTR1/TaDREB2* expression. *Physiol. Plant.* **2017**, *161*, 502–514. [CrossRef]
42. Zhang, B.; Chang, S.X.; Anyia, A.O. Mycorrhizal inoculation and nitrogen fertilization affect the physiology and growth of spring wheat under two contrasting water regimes. *Plant Soil* **2015**, *398*, 47–57. [CrossRef]
43. Getahun, A.; Muleta, D.; Assefa, F.; Kiros, S. Plant growth-promoting rhizobacteria isolated from degraded habitat enhance drought tolerance of acacia (*Acacia abyssinica* hochst. ex benth.) seedlings. *Int. J. Microbiol.* **2020**, *29*, 8897998. [CrossRef] [PubMed]
44. Tiepo, A.N.; Constantino, L.V.; Madeira, T.B.; Gonçalves, L.S.A.; Pimenta, J.A.; Bianchini, E.; De Oliveira, A.L.M.; Oliveira, H.C.; Stolf-Moreira, R. Plant growth-promoting bacteria improve leaf antioxidant metabolism of drought-stressed Neotropical trees. *Planta* **2020**, *251*, 83. [CrossRef]
45. Van Zelm, E.; Zhang, Y.; Testerink, C. Salt tolerance mechanisms of plants. *Annu. Rev. Plant Biol.* **2020**, *71*, 403–433. [CrossRef]
46. Navarro-Torre, S.; Barcia-Piedras, J.M.; Mateos-Naranjo, E.; Redondo-Gómez, S.; Camacho, M.; Caviedes, M.A.; Pajuelo, E.; Rodríguez-Llorente, I.D. Assessing the role of endophytic bacteria in the halophyte arthrocnemum macrostachyumsalt tolerance. *Plant Biol.* **2016**, *19*, 249–256. [CrossRef]
47. Singh, S.; Singh, U.P.; Trivedi, M.; Sahu, P.K.; Paul, S.; Sivanesan, I.; Saxena, A.K. Seed biopriming with salt-tolerant endophytic pseudomonas geniculata-modulated biochemical responses provide ecological fitness in maize (*Zea mays* L.) grown in saline sodic soil. *Int. J. Environ. Res. Public Health* **2019**, *17*, 253. [CrossRef]
48. Hameed, A.; Egamberdieva, D.; Abd-Allah, E.F.; Hashem, A.; Kumar, A.; Ahmad, P. Salinity stress and arbuscular mycorrhizal symbiosis in plants. In *Use of Microbes for the Alleviation of Soil Stresses*; Miransari, M., Ed.; Springer: New York, NY, USA, 2014; pp. 139–159. [CrossRef]
49. Afridi, M.S.; Van Hamme, J.D.; Bundschuh, J.; Khan, M.N.; Salam, A.; Waqar, M.; Chaudhary, H.J. Biotechnological approaches in agriculture and environmental management-bacterium *Kocuria rhizophila* 14ASP as heavy metal and salt tolerant plant growth-promoting strain. *Biologia* **2021**, *76*, 3091–3105. [CrossRef]
50. Moon, Y.; Ali, S. Possible mechanisms for the equilibrium of ACC and role of ACC deaminase-producing bacteria. *Appl. Microbiol. Biotechnol.* **2022**, *106*, 877–887. [CrossRef]

51. Kumari, S.; Vaishnav, A.; Jain, S.; Varma, A.; Choudhary, D.K. Bacterial-mediated induction of systemic tolerance to salinity with expression of stress alleviating enzymes in soybean (*Glycine max* L. Merrill). *J. Plant Growth Regul.* **2015**, *34*, 558–573. [CrossRef]
52. Mukherjee, P.; Mitra, A.; Roy, M. Halomonas rhizobacteria of avicennia marina of indian sundarbans promote rice growth under saline and heavy metal stresses through exopolysaccharide production. *Front. Microbiol.* **2019**, *10*, 1207. [CrossRef]
53. Ren, C.; Dai, C. Jasmonic acid is involved in the signaling pathway for fungal endophyte-induced volatile oil accumulation of *Atractylodes lancea* plantlets. *BMC Plant Biol.* **2012**, *12*, 128. [CrossRef] [PubMed]
54. Wang, X.M.; Yang, B.; Ren, C.G.; Wang, H.W.; Wang, J.Y.; Dai, C.C. Involvement of abscisic acid and salicylic acid in signal cascade regulating bacterial endophyte-induced volatile oil biosynthesis in plantlets of *Atractylodes lancea*. *Physiol. Plant.* **2015**, *153*, 30–42. [CrossRef]
55. Shao, M.; Zhao, W.; Su, Z.; Dong, L.; Guo, Q.; Ma, P. Effect of *Bacillus subtilis* NCD-2 on the growth of tomato and the microbial community structure of rhizosphere soil under salt stress. *Sci. Agric. Sin.* **2021**, *54*, 4573–4584.
56. Mahmud-Ur-Rahman, N.; Naser, I.B.; Abd-Elsalam, K.A.; Sarker, A.; Hoque, M.; Islam, T. A highly salt-tolerant bacterium brevibacterium sediminis promotes the growth of rice (*Oryza sativa* L.) seedlings. *Stresses* **2022**, *2*, 275–289. [CrossRef]
57. Akhtar, S.S.; Andersen, M.N.; Naveed, M.; Zahir, Z.A.; Liu, F. Interactive effect of biochar and plant growth-promoting bacterial endophytes on ameliorating salinity stress in maize. *Funct. Plant Biol.* **2015**, *42*, 770–781. [CrossRef] [PubMed]
58. Ali, S.; Charles, T.C.; Glick, B.R. Amelioration of high salinity stress damage by plant growth-promoting bacterial endophytes that contain ACC deaminase. *Plant Physiol. Biochem.* **2014**, *80*, 160–167. [CrossRef]
59. Sharma, S.; Kulkarni, J.; Jha, B. Halotolerant rhizobacteria promote growth and enhance salinity tolerance in peanut. *Front. Microbiol.* **2016**, *7*, 1600. [CrossRef] [PubMed]
60. Lastochkina, O.; Pusenkova, L.; Yuldashev, R.; Babaev, M.; Garipova, S.; Blagova, D.; Khairullin, R.; Aliniaeifard, S. Effects of Bacillus subtilis on some physiological and biochemical parameters of *Triticum aestivum* L. (wheat) under salinity. *Plant Physiol. Biochem.* **2017**, *121*, 80–88. [CrossRef]
61. Shahzad, R.; Khan, A.L.; Bilal, S.; Waqas, M.; Kang, S.M.; Lee, I.J. Inoculation of abscisic acid-producing endophytic bacteria enhances salinity stress tolerance in Oryza sativa. *Environ. Exp. Bot.* **2017**, *136*, 68–77. [CrossRef]
62. Islam, F.; Yasmeen, T.; Arif, M.; Ali, S.; Ali, B.; Hameed, S.; Zhou, W. Plant growth promoting bacteria confer salt tolerance in Vigna radiata by up-regulating antioxidant defense and biological soil fertility. *Plant Growth Regul.* **2016**, *80*, 23–36. [CrossRef]
63. Vega, C.; Rodríguez, M.; Llamas, I.; Béjar, V.; Sampedro, I. Silencing of phytopathogen communication by the halotolerant PGPR Staphylococcus equorum strain EN21. *Microorganisms* **2019**, *8*, 42. [CrossRef] [PubMed]
64. Baltruschat, H.; Fodor, J.; Harrach, B.D.; Niemczyk, E.; Barna, B.; Gullner, G.; Janeczko, A.; Kogel, K.; Schäfer, P.; Schwarczinger, I.; et al. Salt tolerance of barley induced by the root endophyte Piriformospora indica is associated with a strong increase in antioxidants. *New Phytol.* **2008**, *180*, 501–510. [CrossRef]
65. Sen, S.; Chandrasekhar, C.N. Effect of PGPR on growth promotion of rice (*Oryza sativa* L.) under salt stress. *Asian J. Plant Sci. Res.* **2014**, *4*, 62–67.
66. Conway, J.; Walton, W.G.; Salas-González, I.; Law, T.F.; Lindberg, C.A.; Crook, L.; Kosina, S.M.; Fitzpatrick, C.R.; Lietzan, A.; Northen, T.; et al. Diverse MarR bacterial regulators of auxin catabolism in the plant microbiome. *Nat. Microbiol.* **2022**, *7*, 1817–1833. [CrossRef]
67. Birkenbihl, R.P.; Kracher, B.; Ross, A.; Kramer, K.; Finkemeier, I.; Somssich, I.E. Principles and characteristics of the Arabidopsis WRKY regulatory network during early MAMP-triggered immunity. *Plant J.* **2018**, *96*, 487–502. [CrossRef]
68. Meena, K.K.; Sorty, A.M.; Bitla, U.M.; Choudhary, K.; Gupta, P.; Pareek, A.; Singh, D.P.; Prabha, R.; Sahu, P.K.; Gupta, V.K.; et al. Abiotic stress responses and microbe-mediated mitigation in plants: The omics strategies. *Front. Plant Sci.* **2017**, *8*, 172. [CrossRef]
69. Zhalnina, K.; Louie, K.; Hao, Z.; Mansoori, N.; Da Rocha, U.N.; Shi, S.; Cho, H.; Karaöz, U.; Loqué, D.; Bowen, B.P.; et al. Dynamic root exudate chemistry and microbial substrate preferences drive patterns in rhizosphere microbial community assembly. *Nat. Microbiol.* **2018**, *3*, 470–480. [CrossRef]
70. Preece, C.; Peñuelas, J. Rhizodeposition under drought and consequences for soil communities and ecosystem resilience. *Plant Soil* **2016**, *409*, 1–17. [CrossRef]
71. Williams, A.; De Vries, F.T. Plant root exudation under drought: Implications for ecosystem functioning. *New Phytol.* **2019**, *225*, 1899–1905. [CrossRef] [PubMed]
72. Zheng, Y.; Cao, X.; Zhou, Y.; Ma, S.; Wang, Y.; Li, Z.; Zhao, D.; Yang, Y.; Zhang, H.; Meng, C.; et al. Purines enrich root-associated Pseudomonas and improve wild soybean growth under salt stress. *Nat. Commun.* **2024**, *15*, 3520. [CrossRef] [PubMed]
73. Khokhar-Voytas, A.; Shahbaz, M.; Maqsood, M.F.; Zulfiqar, U.; Naz, N.; Iqbal, U.; Sarà, M.; Aqeel, M.; Khalid, N.; Noman, A.; et al. Genetic modification strategies for enhancing plant resilience to abiotic stresses in the context of climate change. *Funct. Integr. Genom.* **2023**, *23*, 283. [CrossRef]
74. Olejnik, P.; Madrzak, C.J.; Nuc, K. Cyclophilins and their functions in abiotic stress and Plant–Microbe interactions. *Biomolecules* **2021**, *11*, 1390. [CrossRef]

75. Nikolić, B.; Schwab, H.; Sessitsch, A. Metagenomic analysis of the 1-aminocyclopropane-1-carboxylate deaminase gene (acdS) operon of an uncultured bacterial endophyte colonizing *Solanum tuberosum* L. *Arch. Microbiol.* **2011**, *193*, 665–676. [CrossRef]
76. Ghorbanpour, A.; Salimi, A.; Ghanbary, M.A.T.; Pirdashti, H.; Dehestani, A. The effect of Trichoderma harzianum in mitigating low temperature stress in tomato (*Solanum lycopersicum* L.) plants. *Sci. Hortic.* **2018**, *230*, 134–141. [CrossRef]
77. Abuqamar, S.F.; Luo, H.; Laluk, K.; Mickelbart, M.V.; Mengiste, T. Crosstalk between biotic and abiotic stress responses in tomato is mediated by the AIM1 transcription factor. *Plant J.* **2009**, *58*, 347–360. [CrossRef]
78. Sziderics, A.H.; Rasche, F.; Trognitz, F.; Sessitsch, A.; Wilhelm, E. Bacterial endophytes contribute to abiotic stress adaptation in pepper plants (*Capsicum annuum* L.). *Can. J. Microbiol.* **2007**, *53*, 1195–1202. [CrossRef] [PubMed]
79. Rosenblueth, M.; Martínez-Romero, E. Rhizobium etli maize populations and their competitiveness for root colonization. *Arch. Microbiol.* **2004**, *181*, 337–344. [CrossRef] [PubMed]
80. Arora, N.K.; Fatima, T.; Mishra, I.; Verma, S. Microbe-based inoculants: Role in next Green Revolution. In *Environmental Concerns and Sustainable Development*; Shukla, V., Kumar, N., Eds.; Springer: Singapore, 2020; pp. 191–246. [CrossRef]
81. Waadt, R.; Seller, C.A.; Hsu, P.K.; Takahashi, Y.; Munemasa, S.; Schroeder, J.I. Plant hormone regulation of abiotic stress responses. *Nat. Rev. Mol. Cell Biol.* **2022**, *23*, 680–694. [CrossRef]
82. Bastías, D.A.; Balestrini, R.; Pollmann, S.; Gundel, P.E. Environmental interference of plant−microbe interactions. *Plant Cell Environ.* **2022**, *45*, 3387–3398. [CrossRef] [PubMed]
83. Hammerbacher, A.; Coutinho, T.A.; Gershenzon, J. Roles of plant volatiles in defence against microbial pathogens and microbial exploitation of volatiles. *Plant Cell Environ.* **2019**, *42*, 2827–2843. [CrossRef]
84. Sorty, A.M.; Meena, K.K.; Choudhary, K.; Bitla, U.M.; Minhas, P.S.; Krishnani, K.K. Effect of plant growth promoting bacteria associated with halophytic weed (*Psoralea corylifolia* L.) on germination and seedling growth of wheat under saline conditions. *Appl. Biochem. Biotechnol.* **2016**, *180*, 872–882. [CrossRef] [PubMed]
85. Vinale, F.; Sivasithamparam, K.; Ghisalberti, E.L.; Marra, R.; Barbetti, M.J.; Li, H.; Woo, S.L.; Lorito, M. A novel role for *Trichoderma* secondary metabolites in the interactions with plants. *Physiol. Mol. Plant Pathol.* **2008**, *72*, 80–86. [CrossRef]
86. Glick, B.R.; Penrose, D.M.; Li, J. A model for the lowering of plant ethylene concentrations by plant growth-promoting bacteria. *J. Theor. Biol.* **1998**, *190*, 63–68. [CrossRef] [PubMed]
87. Shinozaki, K.; Yamaguchi-Shinozaki, K. Gene networks involved in drought stress response and tolerance. *J. Exp. Bot.* **2006**, *58*, 221–227. [CrossRef]
88. Zhou, C.; Li, F.; Xie, Y.; Zhu, L.; Xiao, X.; Ma, Z.; Wang, J. Involvement of abscisic acid in microbe-induced saline-alkaline resistance in plants. *Plant Signal. Behav.* **2017**, *12*, e1367465. [CrossRef]
89. Shah, S.; Chen, C.; Sun, Y.; Wang, D.; Nawaz, T.; El-Kahtany, K.; Fahad, S. Mechanisms of nitric oxide involvement in plant-microbe interaction and its enhancement of stress resistance. *Plant Stress* **2023**, *10*, 100191. [CrossRef]
90. Ghosh, D.; Gupta, A.; Mohapatra, S. Dynamics of endogenous hormone regulation in plants by phytohormone secreting rhizobacteria under water-stress. *Symbiosis* **2019**, *77*, 265–278. [CrossRef]
91. Mendes, R.E.; Garbeva, P.; Raaijmakers, J.M. The rhizosphere microbiome: Significance of plant beneficial, plant pathogenic, and human pathogenic microorganisms. *FEMS Microbiol. Rev.* **2013**, *37*, 634–663. [CrossRef]
92. Contreras-Cornejo, H.A.; Macías-Rodríguez, L.; Cortés-Penagos, C.; López-Bucio, J. *Trichoderma virens*, a plant beneficial fungus, enhances biomass production and promotes lateral root growth through an auxin-dependent mechanism in arabidopsis. *Plant Physiol.* **2009**, *149*, 1579–1592. [CrossRef]
93. He, X.; Shao, C.; Wu, A.; Xia, L.; Li, T.; Pei, J.; Zhang, N.; Wang, Y. Arbuscular mycorrhizal fungi enhance nutrient acquisition and reduce aluminum toxicity in Lespedeza formosa under acid rain. *Environ. Sci. Pollut. Res.* **2022**, *29*, 29904–29916. [CrossRef]
94. Riquelme, M.; Aguirre, J.; Bartnicki-García, S.; Braus, G.H.; Feldbrügge, M.; Fleig, U.; Hansberg, W.; Herrera-Estrella, A.; Kämper, J.; Kück, U.; et al. Fungal morphogenesis, from the polarized growth of hyphae to complex reproduction and infection structures. *Microbiol. Mol. Biol. Rev.* **2018**, *82*, e00068-17. [CrossRef]
95. Zhu, X.; Song, F.; Liu, S.; Liu, T. Effects of arbuscular mycorrhizal fungus on photosynthesis and water status of maize under high temperature stress. *Plant Soil* **2011**, *346*, 189–199. [CrossRef]
96. Karthik, C.; Elangovan, N.; Kumar, T.S.; Govindharaju, S.; Barathi, S.; Oves, M.; Arulselvi, P.I. Characterization of multifarious plant growth promoting traits of rhizobacterial strain AR6 under Chromium (VI) stress. *Microbiol. Res.* **2017**, *204*, 65–71. [CrossRef]
97. Mathesius, U.; Mulders, S.; Gao, M.; Teplitski, M.; Caetano-Anollés, G.; Rolfe, B.G.; Bauer, W. Extensive and specific responses of a eukaryote to bacterial quorum-sensing signals. *Proc. Natl. Acad. Sci. USA* **2003**, *100*, 1444–1449. [CrossRef] [PubMed]
98. Rodríguez, H.; Fraga, R. Phosphate solubilizing bacteria and their role in plant growth promotion. *Biotechnol. Adv.* **1999**, *17*, 319–339. [CrossRef] [PubMed]
99. Collavino, M.M.; Sansberro, P.A.; Mroginski, L.A.; Aguilar, O.M. Comparison of in vitro solubilization activity of diverse phosphate-solubilizing bacteria native to acid soil and their ability to promote *Phaseolus vulgaris* growth. *Biol. Fertil. Soils* **2010**, *46*, 727–738. [CrossRef]

100. Liu, F.; Ma, H.; Peng, L.; Du, Z.; Ma, B.; Liu, X. Effect of the inoculation of plant growth-promoting rhizobacteria on the photosynthetic characteristics of *Sambucus williamsii* Hance container seedlings under drought stress. *AMB Express* **2019**, *9*, 169. [CrossRef]
101. Ratajczak, K.; Sulewska, H.; Błaszczyk, L.; Basińska-Barczak, A.; Mikołajczak, K.; Salamon, S.; Szymańska, G.; Dryjański, L. Growth and photosynthetic activity of selected spelt varieties (*Triticum aestivum* ssp. *spelta* L.) cultivated under drought conditions with different endophytic core microbiomes. *Int. J. Mol. Sci.* **2020**, *21*, 7987. [CrossRef]
102. Gou, W.; Li, T.; Ruan, Z.; Zheng, P.; Chen, F.; Zhang, L.; Zhiyan, C.; Zheng, P.; Zheng, L.; Gao, M.; et al. Accumulation of choline and glycinebetaine and drought stress tolerance induced in maize (*Zea mays*) by three plant growth promoting rhizobacteria (PGPR) strains. *Pak. J. Bot.* **2015**, *47*, 581–586.
103. Zareyan, M.; Mockevičiūtė, R.; Jurkonienė, S.; Gavelienė, V.; Paškevičius, A.; Šveikauskas, V. Physiological, biochemical, and genetic reactions of winter wheat to drought under the influence of plant growth promoting microorganisms and calcium. *Microorganisms* **2025**, *13*, 1042. [CrossRef]
104. Heidarianpour, M.B.; Aliasgharzad, N.; Olsson, P.A. Positive effects of co-inoculation with *Rhizophagus irregularis* and *Serendipita indica* on tomato growth under saline conditions, and their individual colonization estimated by signature lipids. *Mycorrhiza* **2020**, *30*, 455–466. [CrossRef] [PubMed]
105. Cao, L.; Li, Y.W.; Ling, K.J.; Wang, J.Q.; Yan, H.; Mou, L.J.; Jiang, J.L.; Fang, F.; Liu, R.Q. Effects of arbuscular mycorrhizal fungi inoculation on salt-tolerance of tomato plants. *Fujian J. Agric. Sci.* **2022**, *37*, 188–196.

plants

MDPI

Article

Investigating the Mechanism of Cadmium-Tolerant Bacterium *Cellulosimicrobium* and Ryegrass Combined Remediation of Cadmium-Contaminated Soil

Jiaqi Li [1], Xiaoyang Xu [1], Lanping Song [1], Meng Na [1], Shangqi Xu [1], Jie Zhang [1], Yongjie Huang [1], Xiaoping Li [2], Xianqing Zheng [3,*] and Jihai Zhou [1,*]

1 School of Ecology and Environment, Anhui Normal University, Wuhu 241002, China; krieben@163.com (J.L.); xxy18856023887@163.com (X.X.); songlanping98@163.com (L.S.); nameng1991@ahnu.edu.cn (M.N.); shangqixu@ahnu.edu.cn (S.X.); zhjie1211@sina.com (J.Z.); yongjiehuang0108@163.com (Y.H.)

2 Collaborative Innovation Center of Southern Modern Forestry, Nanjing Forestry University, Nanjing 210037, China; xpli@njfu.edu.cn

3 Institute of Eco-Environment and Plant Protection, Shanghai Academy of Agricultural Sciences, Shanghai 201403, China

* Correspondence: zxqfxf@163.com (X.Z.); zhoujihai2006@163.com (J.Z.)

Abstract: Cadmium (Cd) pollution has been rapidly increasing due to the global rise in industries. Cd not only harms the ecological environment but also endangers human health through the food chain and drinking water. Therefore, the remediation of Cd-polluted soil is an imminent issue. In this work, ryegrass and a strain of Cd-tolerant bacterium were used to investigate the impact of inoculated bacteria on the physiology and biochemistry of ryegrass and the Cd enrichment of ryegrass in soil contaminated with different concentrations of Cd (4 and 20 mg/kg). The results showed that chlorophyll content increased by 24.7% and 41.0%, while peroxidase activity decreased by 56.7% and 3.9%. In addition, ascorbic acid content increased by 16.7% and 6.3%, whereas glutathione content decreased by 54.2% and 6.9%. The total Cd concentration in ryegrass increased by 21.5% and 10.3%, and the soil's residual Cd decreased by 86.0% and 44.1%. Thus, the inoculation of Cd-tolerant bacteria can improve the antioxidant stress ability of ryegrass in Cd-contaminated soil and change the soil's Cd form. As a result, the Cd enrichment in under-ground and above-ground parts of ryegrass, as well as the biomass of ryegrass, is increased, and the ability of ryegrass to remediate Cd-contaminated soil is significantly improved.

Keywords: Cd pollution; biological co-remediation; antioxidant enzyme system; none-enzymatic antioxidant system; microorganisms

1. Introduction

Heavy metal pollution has been turning into a serious environmental concern across the world. About 19.4% of China's arable land (2.6×10^7 hm^2 area) is polluted by heavy metals [1]. According to the Report of the Ministry of Ecology and Environment on the Status of Ecological Environment in China in 2021, the main pollutants that affect the soil's environmental quality in agricultural land are heavy metals, among which cadmium is the primary pollutant [2]. Various heavy metals, such as cadmium, lead, chromium, and mercury, are toxic to most organisms at higher concentrations [3]. According to the literature, exposure to heavy metals has been linked to a variety of human diseases [4–6], such as emphysema, kidney failure, osteoporosis, and cancer [7–9]. Therefore, heavy metal remediation is emerging as a serious problem that needs to be effectively addressed.

Compared with physical remediation and chemical remediation, soil biological remediation has the comparative advantages of optimizing and improving the environment at a low cost. For this reason, it has become the main remediation method for cadmium-contaminated soil [10]. Phytoremediation refers to planting a plant on soil polluted by heavy metals to absorb heavy metals in the soil and achieve the purpose of soil remediation [11]. Ryegrass is often used in the remediation of heavy metal-contaminated soil, as it is an excellent tolerant herb with a fast growth rate, high yield, long growing season, and wide adaptability [12,13]. It has been demonstrated that ryegrass can effectively accumulate heavy metals, such as Cd [14,15]. However, the scope of simple phytoremediation is limited, and excessive absorption of heavy metals will have a large number of negative effects on plants and even lead to plant death [16,17]. On the contrary, the inoculation of microorganisms can promote plant growth and improve the efficiency of bioremediation. During the combined plant–microbial remediation, two aspects mainly govern the mechanism of microbial action. Microorganisms can regulate the absorption and accumulation of heavy metals by plants by changing the form of the heavy metals and the growth of plant roots [18]. Thus, the extraction capacity of heavy metals by plants can be improved and the removal of heavy metals from soils can be enhanced. On the other hand, the regionalization, chelation, and fixation of heavy metals in microbial cells can reduce the migration ability of the heavy metals [19], which is conducive to the fixation of heavy metals in plants. At the same time, the growth of plant roots could secrete protein, sugar, and other organic matter as a source of nutrients and energy for microorganisms and, at the same time, significantly improve the activity and adaptability of microorganisms [20,21]. Plants and microorganisms form symbionts for mutual benefit, which strengthens their respective roles in the remediation of heavy metal-contaminated soil and improves the removal efficiency. The presence of Cd in soil can cause damage to plants, such as damage to the antioxidant system, changes in membrane lipid catalase, enzyme activity, etc. [22]. Therefore, the physiological and biochemical characteristics of plants can be used as indicators to evaluate plant growth and repair efficiency. Soil microorganisms can regulate the decomposition of both soil organic matter and harmful compounds and participate in the biochemical cycle and the formation of the soil structure. In parallel, they are sensitive to external interference. Therefore, soil enzyme activity is considered a good index to detect soil quality [23].

To explore the impact of plant–microbial remediation, in this work, the ryegrass and Cd-tolerant bacteria were selected as remediation materials for cadmium-contaminated soil. Ryegrass and Cd-tolerant bacteria were used to study the changes in the physiological and biochemical characteristics of ryegrass, including the biomass, chlorophyll content, and antioxidant enzymes, as well as the changes in soil microbial activity and cadmium enrichment in ryegrass in an indoor pot experiment. The physiological response of ryegrass and the changes in the soil quality due to Cd pollution were thoroughly investigated to provide theoretical guidance and a practical basis for the remediation of Cd-contaminated soil.

2. Results

2.1. Changes in Growth, Physiology, and Biochemistry of Ryegrass

2.1.1. Changes in Biomass and Chlorophyll Content of Ryegrass

The above-ground and under-ground biomass of C1 and C2 were 55.4% and 46.8%, 91.3% and 88.3% lower than CK, respectively. Under the treatment of 4 mg/kg Cd, the biomass of above-ground and under-ground in C1B increased by 44.1% and 159.6%, respectively, compared to that of C1. For the concentration of 20 mg/kg Cd, the above-ground and under-ground biomass of C2B significantly increased by 111.1% and 765.2%, respectively, compared to C2 (Figure 1A). Compared to CK, the chlorophyll content of C1 and C2 decreased by 19.5% and 24.3%, respectively. In general, the inoculation of Cd-tolerant bacteria could increase the chlorophyll content of ryegrass. Under the treatment of 4 mg/kg Cd, the chlorophyll content of C1B significantly increased by 24.7% compared to that of C1. Under the treatment of 20 mg/kg Cd, the chlorophyll content of C2B greatly increased by 41.0% compared to that of C2 (Figure 1B).

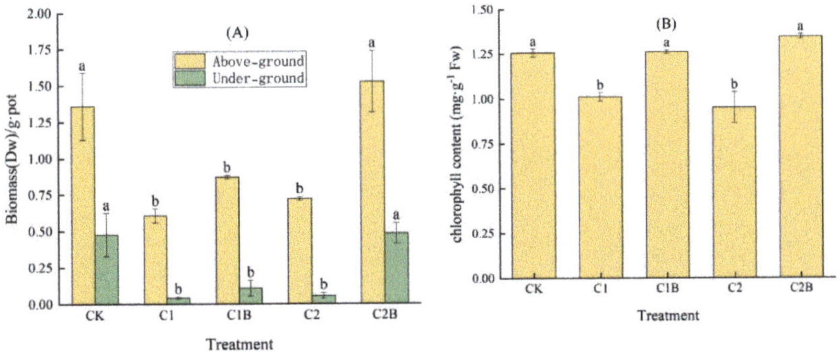

Figure 1. Changes in biomass (**A**) and chlorophyll content (**B**) of *Lolium perenne*.

CK is the control treatment. C1 is the treatment with 4 mg/kg Cd pollution. C2 is the treatment with 20 mg/kg Cd pollution. C1B is the treatment with 4 mg/kg Cd pollution and Cd-tolerant bacteria. C2B is the treatment with 20 mg/kg Cd pollution and Cd-tolerant bacteria. The different letter indicates significant differences between treatments ($p < 0.05$).

2.1.2. Changes in SOD and POD Activities of Ryegrass

The SOD activity of C1 ryegrass decreased by 7.9% compared to CK, while C2 increased by 2.1%. Under the application of a 4 mg/kg Cd treatment, C1B increased by 7.1% compared to C1. For the concentration of 20 mg/kg Cd, the SOD activity of C2B significantly declined (34.2%) compared to that of C2 (Figure 2A). The POD activity of C1 significantly increased by 81.1% compared to CK, while that of C2 increased by 16.2%. The POD activity of ryegrass declined due to the inoculation of Cd-tolerant bacteria. For a 4 mg/kg Cd concentration, C1B significantly decreased by 56.7% compared to C1, whereas for a 20 mg/kg Cd concentration, C2B reduced by about 3.9% compared to C2 (Figure 2B).

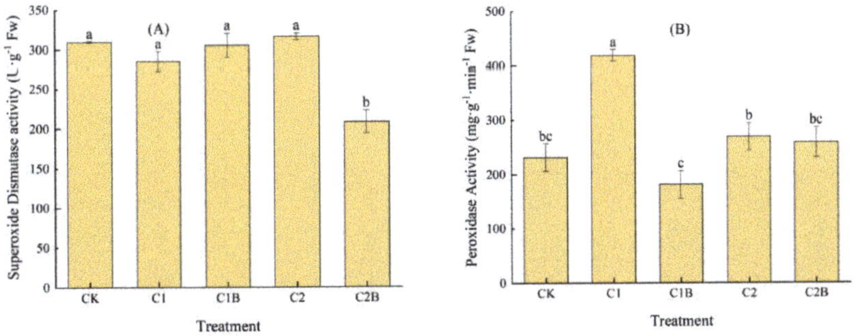

Figure 2. Changes in SOD (**A**) and POD (**B**) activity of ryegrass.

2.1.3. Changes in ASA and GSH Contents of Ryegrass

Under Cd stress, the ASA content significantly decreased by 45.2% in C1 and 24.2% in C2 compared to CK. The inoculation with Cd-tolerant bacteria could increase the ASA content in ryegrass. At a 4 mg/kg Cd concentration, the ASA content in C1B increased by about 16.7% compared to C1 with no significant difference. At a 20 mg/kg Cd concentration, C2B increased by 6.3% compared to C2 with no significant difference (Figure 3A). The Cd stress can also reduce the GSH content. More specifically, the GSH contents of C1 and C2 were significantly reduced by 13.5% and 52.6% compared to CK. The GSH content in ryegrass decreased by inoculation with the Cd-tolerant bacteria. Under a 4 mg/kg Cd treatment, the GSH content in C1B significantly decreased by 54.2% compared to C1. At 20 mg/kg Cd concentration, C2B decreased by 6.9% compared to C2 (Figure 3B).

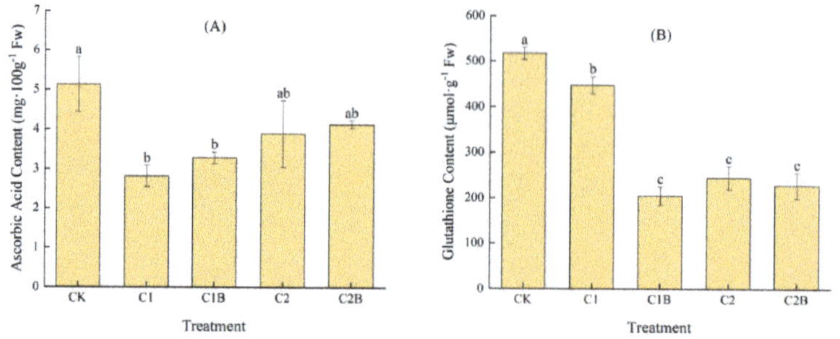

Figure 3. Changes in ASA (**A**) and GSH (**B**) content of ryegrass.

2.2. Changes in Soil Basal Respiration and Enzyme Activity

The Cd stress can reduce the intensity of soil basal respiration. Particularly, the soil basal respiration of C1 and C2 significantly decreased by 45.0% and 55.9% compared to CK. The inoculation of Cd-tolerant bacteria could reduce the soil basal respiration intensity. Under 4 mg/kg Cd treatment, the soil basal respiration intensity of C1B significantly reduced by about 36.9% compared to C1. For the 20 mg/kg Cd concentration, C2B significantly reduced by 65.6% compared to C2 (Figure 4A).

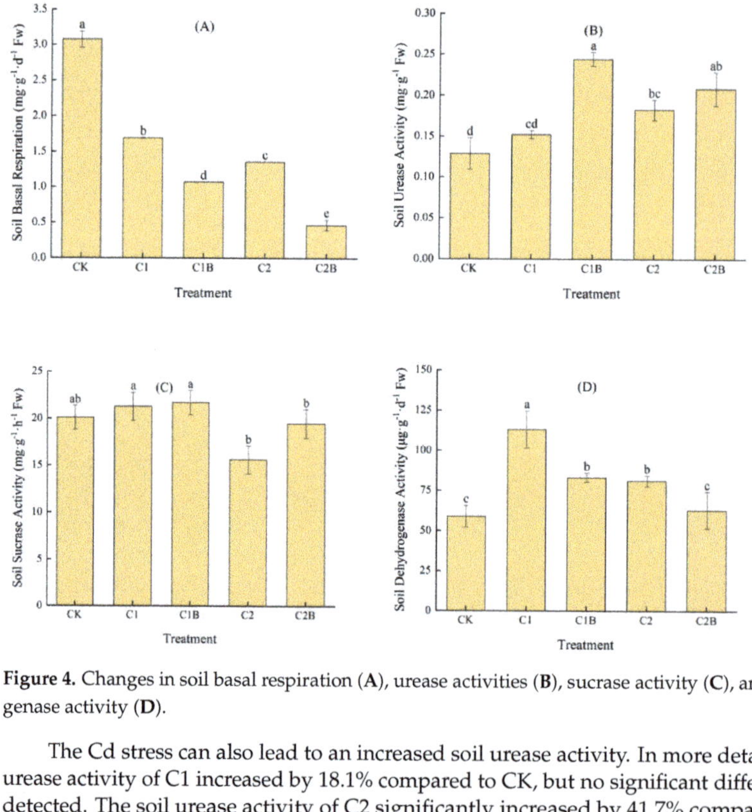

Figure 4. Changes in soil basal respiration (**A**), urease activities (**B**), sucrase activity (**C**), and dehydrogenase activity (**D**).

The Cd stress can also lead to an increased soil urease activity. In more detail, the soil urease activity of C1 increased by 18.1% compared to CK, but no significant difference was detected. The soil urease activity of C2 significantly increased by 41.7% compared to CK. The inoculation of Cd-tolerant bacteria can improve soil urease activity. At a 4 mg/kg Cd concentration, the soil urease activity of C1B considerably increased by 60.4% compared to C1. At a 20 mg/kg Cd concentration, the soil urease activity of C2B increased by 14.0% compared to C2 while having no significant difference (Figure 4B).

Soil sucrase activity of C1 increased by 5.9% compared to CK and C2 decreased by 22.3%. The inoculation of Cd-tolerant bacteria could improve soil sucrase activity. At a 4 mg/kg Cd concentration, the soil sucrase activity of C1B was 2.0% higher than C1 while having no significant difference. At a 20 mg/kg Cd treatment, the soil sucrase activity of C2B was 24.8% higher than C2 with no significant difference (Figure 4C).

The Cd stress can increase soil dehydrogenase activity. The soil dehydrogenase activity of C1 and C2 significantly increased by 92.2% and 37.6% compared to CK. The inoculation of Cd-tolerant bacteria decreased the soil dehydrogenase activity. At a 4 mg/kg Cd concentration, the soil dehydrogenase activity of C1B significantly decreased by 26.6% compared to C1. At a 20 mg/kg Cd concentration, the soil dehydrogenase activity of C2B significantly reduced by 22.3% compared to C2 (Figure 4D).

2.3. *Effects of Cd-Tolerant Bacteria on Soil Cd Forms and Cd Accumulation in Ryegrass*

2.3.1. The Content and Changes in Soil Cd Forms

The inoculation of Cd-tolerant bacteria could change the soil's Cd forms (Table 1). Under a 4 mg/kg Cd treatment, the proportion of exchangeable Cd in C1B increased by 4.19%, the proportion of potentially bioavailable Cd increased by 15.3%, and the proportion of residual Cd decreased by 86.0% compared to C1. The inoculation of Cd-tolerant bacteria could increase the contents of exchangeable and potentially bioavailable Cd in the soil to some extent and reduce the residual Cd content. Under the treatment of 20 mg/kg Cd, the proportion of exchangeable Cd in C2B decreased by about 4.56% and the proportion of residual Cd decreased by about 44.1% compared to C2. In striking contrast, the proportion of potentially bioavailable Cd increased by 26.5%.

Table 1. Contents of various forms of cadmium in soil (mg/kg).

Treatment	Exchangeable	Potential Bioavailability	Residual
C1	2.501 b	0.295 b	0.186 ab
C1B	2.637 b	0.340 b	0.026 b
C2	13.522 a	2.868 a	0.571 a
C2B	12.038 a	3.627 a	0.319 ab

Different letters in the same column indicated significant differences in different treatments ($p < 0.05$).

2.3.2. Changes in the Cd Content of Ryegrass

The inoculation of Cd-tolerant bacteria could increase the concentration of Cd accumulation in ryegrass, which leads to a reduced content of residual Cd in the soil. More specifically, under the application of a 4 mg/kg Cd concentration, the Cd content above-ground in C1B was 31.9% higher than that in C1 and 20.3% higher than that in C1 underground. Under the application of a 20 mg/kg Cd concentration, the Cd content of C2B in the above-ground significantly increased by 22.2% compared to C2 and the under-ground part increased by 8.0% (Figure 5).

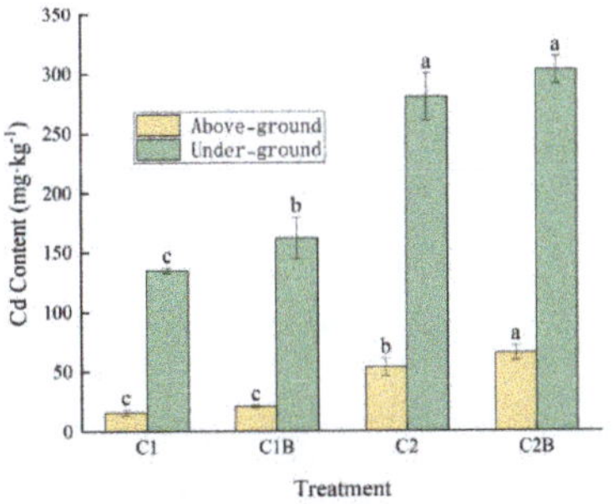

Figure 5. Cd Enrichment in Ryegrass.

2.4. Correlation of Cd Accumulation in Ryegrass with Microorganisms, Physiology of Ryegrass

The inoculation of Cd-tolerant bacteria remarkably increased the total amount of Cd enriched in the under-ground of ryegrass and improved the enrichment in the above-ground (Figure 6). The concentration of Cd was found to be positively correlated with the biomass of ryegrass. The content of chlorophyll was found to be positively correlated with above-ground and under-ground biomass. On the contrary, the SOD activity was negatively correlated with the concentration of cadmium. ASA was positively correlated with Cd enrichment, while GSH was inversely correlated. Moreover, the soil basal respiration and soil dehydrogenase activity were negatively correlated with Cd enrichment.

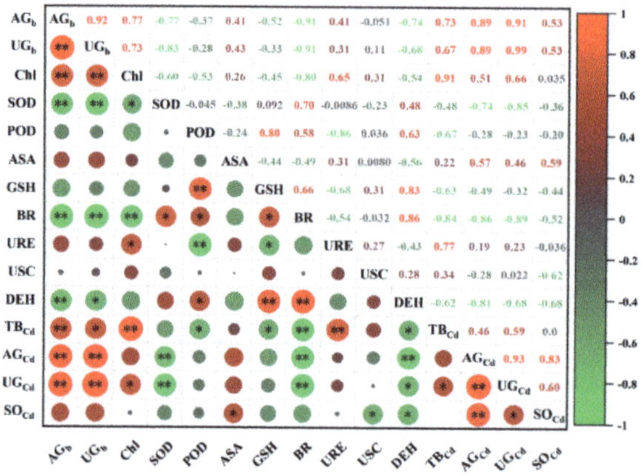

Figure 6. Correlations between the accumulation amount of Cd, physiological characteristics of ryegrass, and soil enzyme activities. AG_b—above-ground biomass, UG_b—under-ground biomass, Chl—chlorophyll content, SOD—superoxide dismutase, POD—peroxidase, ASA—ascorbic acid, GSH—glutathione, BR—soil basal respiration, URE—soil urease, USC—soil sucrase, DEH—soil dehydrogenase, TB_{Cd}—tolerant bacteria, AG_{Cd}—the accumulation amount of Cd in above-ground, UG_{Cd}—the accumulation amount of Cd in under-ground, SO_{Cd}—soil Cd content. * indicates $p \leq 0.05$, ** indicates $p \leq 0.01$.

3. Discussion

3.1. Effects of Cd-Tolerant Bacteria Inoculation on the Physiology and Biochemistry of Ryegrass

Biomass is a comprehensive index to measure the strength of plant tolerance. The increase in plant biomass in a polluted environment can be a direct indicator that plants have a stronger tolerance to pollution [24]. It has been reported in the literature that a certain degree of heavy metal pollution could directly lead to a decline in biomass. The existence of a higher concentration of heavy metals in the soil leads to a bigger toxicity to plants and greatly affects the plant biomass. The total biomass of organisms under the presence of the Cd stress could be significantly decreased [25]. Zhang et al. [26] found that inoculation with different PGPRs could promote the growth and photosynthesis of *P. praeruptorum* to different degrees, thereby improving its tolerance to the environment. Hu et al. [27] pointed out that arbustic mycorrhizal fungi could improve Alfred stonecrop (*Sedum alfredii* Hance) and fast-growing perennial ryegrass (*Lolium perenne* L.) under Cd-contaminated conditions biomass. In this work, the biomass of ryegrass was increased by the inoculation of Cd-tolerant bacteria. The origins of this effect could be ascribed to the fact that the inoculation of Cd-tolerant bacteria alleviated the toxic effect of Cd on ryegrass. As a result, the chlorophyll content increased, and the photosynthesis improved.

Chlorophyll is considered the basis of plant photosynthesis, and its content is closely related to plant growth and development. It has been demonstrated in the literature that the existence of a Cd stress has a negative impact on chlorophyll content, which is gradually decreased with the increase in the Cd concentration in the soil [28,29]. In this work, the chlorophyll content under Cd stress decreased compared to the case where there was no Cd stress. This effect indicated that Cd had an adverse impact on photosynthesis because the inflow of Cd^{2+} to plant cells could replace the cations related to chlorophyll synthesis in chloroplasts (ex: Fe^{2+}, Zn^{2+}, and Mg^{2+}). It could also bind to proteins containing SH in chloroplasts, inhibit the synthesis of chlorophyll precursors, destroy chloroplast structure, and ultimately reduce chlorophyll content [30,31]. El-Saadony et al. [32] found that inoculation with PGPR increased the chlorophyll content of pepper (*Capsicum annuum* L.) in Cd-contaminated soil. In this work, the inoculation of Cd-tolerant bacteria increased the chlorophyll content of ryegrass leaves, which could be attributed to the fact that Cd-tolerant bacteria enhanced the absorption of K and Fe in ryegrass and promoted the synthesis of chlorophyll. This assumption is also consistent with the results of previously reported works [33].

The antioxidant activity of various enzymes, such as SOD, plays a crucial role in protecting plants from oxidative stress-induced damage caused by abiotic factors, such as Cd stress [34]. It has been proven that the relative increase or decrease in antioxidant enzyme activity may vary based on species and the level of heavy metal pollution in the soil [35–37]. SOD is considered an important part of the plant's antioxidant system. To this end, it has been shown that the SOD activity of tobacco leaves decreases under the existence of a low Cd concentration [38], and a high Cd concentration can inhibit the enzyme activity or protein expression of SOD [39]. In this experiment, the SOD activity was not significantly increased under stress, and the activity was relatively stable. This effect could be attributed to the fact that the Cd concentration in the soil environment did not reach the threshold value of stress on ryegrass, and the SOD activity did not significantly change. This outcome is also in direct line with the literature [40]. The inoculation of Cd-tolerant bacteria under high Cd concentrations can reduce the SOD activity. The latter stems from the inoculation of Cd-tolerant bacteria under Cd stress, which can further alleviate the toxic effect of Cd on ryegrass and reduce the accumulation of reactive oxygen species in ryegrass, resulting in a decline in the SOD activity. POD is a highly active enzyme that can decompose H_2O_2 into O_2 and H_2O in plants, which is related to photosynthesis, respiration, and the auxin oxidation reaction. It is a common reference index in plant aging and stress. It has been reported that, under the existence of Cd stress, the POD content of Lemna minor could be increased and then decreased with the increase in the Cd concentration [41]. The increase in the POD content at low Cd concentrations further protected plants from peroxidation damage. At high Cd concentrations, the POD activity decreased, possibly because the high Cd content inhibited the enzyme synthesis or subunit assembly of enzymes [42]. The reduction in the POD activity after inoculation with Cd-tolerant bacteria could be ascribed to the decrease in free radical accumulation in plants after the inoculation with Cd-tolerant bacteria. As a result, the degree of oxidative damage to plants was reduced, and the POD activity was reduced to regulate the adaptive ability of plants.

The ascorbate–glutathione cycle is also considered an important component of the antioxidant system in plants. Ascorbic acid reduces H_2O_2 to H_2O via the catalysis of ascorbate peroxidase and removes excess H_2O_2 from plants. In this experiment, the ASA content decreased under the existence of Cd stress, which may be due to oxidative stress in plants under Cd stress [43]. The ASA consumption could lead to a decrease in the ASA content [44]. The increase in the ASA content by the inoculation of Cd-tolerant bacteria could be due to promoting the synthesis of ascorbic acid and the elimination of H_2O_2 in plants. Hence, the adverse impact of Cd stress on plants could be alleviated [45]. Glutathione can exert antioxidant effects through the ascorbate–glutathione cycle [46]. Zhang et al. [47] found that, with the increase in Cd concentration, the GSH content of ultra-black glutinous maize showed a decreasing trend, which was similar to our experiment. The decrease in glutathione content in this experiment could be attributed to the Cd-tolerant bacteria, which could not prevent the induced expression of Cd on the HO-1 gene, leading to the depletion of GSH [48]. It is also possible that Cd-tolerant bacteria alleviated the stress of Cd on ryegrass through other pathways. The response of glutathione to Cd weakened and the glutathione content decreased [49].

3.2. Effects of Cd-Tolerant Bacteria on Microbial Activity

Soil basal respiration is a major attribute related to soil fertility [50], which is commonly used to measure soil quality. It has been reported in the literature that soil basal respiration is decreased with the increase in Cd concentration [51]. This effect could originate from the fact that Cd has an adverse impact on soil microorganisms and reduces the metabolic activities of soil microorganisms. Soil respiration significantly decreased in the late stage of inoculation with *R. erythropolis* in Cd-contaminated soil, possibly because the *R. erythropolis* inoculants increased the bioavailability of toxic compounds. The latter was achieved by producing surface-active compounds that enhanced the toxicity of the contaminant environment and affected soil microorganisms [52]. In this experiment, the inoculated Cd-tolerant bacteria reduced the soil basal respiration under Cd stress. This effect stems from the competition of soil microorganisms induced by inoculation of Cd-tolerant bacteria and the decrease in the soil microbial abundance, as well as the weakening of the soil respiration intensity.

Soil enzyme activity can accurately reflect the response of soil microorganisms to heavy metal exposure during phytoremediation [53]. Soil urease could affect the content of soil organic matter and nitrogen conversion [54], which can catalyze the hydrolysis of urea into ammonia and provide nitrogen for soil microorganisms. According to the literature, urease activity in the soil increases for a certain range of Cd concentration compared to unpolluted conditions [55]. This effect is in direct line with our results, suggesting that plant-root exudates promote the improvement of soil enzyme activity, nutrient cycling, and the survival of microorganisms in soil [56]. He et al. [57] pointed out that the inoculation with EE2-degrading bacteria (*Hyphomicrobium* sp.) could increase the soil urease activity. The latter originates from the promotion of the growth of nitrogen-transforming bacteria in soil.

Sucrase is a hydrolase that catalyzes the hydrolysis of sucrose into glucose and fructose. It is widely distributed in plants, microorganisms, and soils. Its substrates and products are the most abundant soluble sugars in soil, which provide the necessary energy for microorganisms, animals, and plants. It also plays an important role in the conversion of soil carbon and nitrogen [58]. The activities of sucrase and dehydrogenase first increased and then decreased under the existence of the Cd stress. This may be due to the excitatory effect of toxicants, which takes place when the organism is stimulated by external pollutants. As a result, a compensation process in the organism could possibly exist. When the impact of pollutants on the organism is low, the compensation effect is higher than the impact of pollutants, which promotes the growth of the organism. With the increase in the pollutant concentration, the inhibitory effect of the pollutants exceeds the compensatory effect of organisms, and the growth of organisms gets inhibited [59]. Studies have shown that inoculated bacteria strains can enhance soil sucrase activity [60,61], which is consistent with the results of this experiment. The increased sucrase activity via inoculating Cd-tolerant bacteria could be due to the decomposition of certain substances in the soil by the tolerant bacteria, which provide nutrients for the microbial community [62].

The dehydrogenase activity is an important index to measure the toxicity of heavy metals in contaminated soil [63]. Dehydrogenase is an essential enzyme in living cells, involved in redox reactions that promote the dehydrogenation of soil organic matter and transfer energy to hydrogen receptors through the respiratory chain [64]. The addition of low Cd can stimulate soil microbial biomass and its metabolic activity [65]. In this experiment, the inoculation of Cd-tolerant bacteria reduced the dehydrogenase activity because the Cd-tolerant bacteria changed the soil environment, the pH, and the redox property of the soil, as well as inhibited the dehydrogenase activity. The inoculation of Cd-tolerant bacteria weakened the soil oxidative stress response and declined the enzyme activity accordingly.

3.3. Effects of Cd-Tolerant Bacteria on Soil Cd Forms and Cd Accumulation of Ryegrass

In this experiment, the inoculation of Cd-tolerant bacteria changed the Cd morphology in soil. The adverse effect of heavy metals on plants is not only related to the content of heavy metals but also to the form of heavy metals in soil. It is generally believed that the bioavailable Cd and the exchangeable Cd have high biological activity, which could be easily absorbed by animals and plants. Residual Cd has low bioavailability and weak migration ability and, hence, cannot be readily absorbed by animals and plants [66]. Liu et al. [67] proved that the combined treatment of arbuscular mycorrhiza and biochar amendments had a synergistic effect on inducing soil Cd immobilization, reducing plant availability of Cd, and imposing the risk of post-harvest transfer. Under the treatment of 4 mg/kg Cd, the contents of exchangeable and potential bioavailable forms of Cd in soil increased by inoculating Cd-tolerant bacteria. This effect was conducive to Cd enrichment in plant roots and the reduction of Cd in soil, which was identical to that of Sun et al. [68]. Under the application of 20 mg/kg Cd contamination, the proportions of exchangeable Cd and residual Cd in the soil inoculated with Cd-tolerant bacteria were lower than those in the soil without inoculating with Cd-tolerant bacteria. The potential bioavailability of Cd increased, which may be because the inoculation of Cd-tolerant bacteria promoted the transformation between different forms of Cd. It has been reported that the inoculation of metal-resistant PGRP on rice reduces the concentration of Cd in the soil [69]. This result indicates that the inoculation of Cd-tolerant bacteria is conducive to the absorption of Cd in soil by plants.

With the increase in the cadmium concentration in soil, the accumulation of cadmium in ryegrass increased, which reflected the high tolerance and enrichment ability of ryegrass to Cd [70]. Hu et al. [71] suggested that the Cd absorbed by ryegrass was mainly stored and distributed in the roots, which is in direct line with the results of this experiment. The inoculation of Cd-tolerant bacteria can generally improve the enrichment of Cd in ryegrass, which is consistent with the results obtained by Guo et al. [72], who had inoculated *Burkholderia* sp. D54 on ryegrass. Yung et al. [73] also revealed that the inoculation with Pr27 and Pr30 strains increased the amount of Cd absorbed by plants by 90%. The strain could increase the bioavailability of metal in the soil. This effect could also originate from the increased biomass of ryegrass and heavy metal accumulation, which is positively correlated with plant biomass [74]. Therefore, the Cd accumulation of ryegrass increased, which was also consistent with the decrease in the soil's residual Cd content in this experiment.

4. Materials and Methods

4.1. Materials

The soil was collected from a field having a clean soil layer of 0–20 cm. Stones, plant residues, and large and medium-sized soil animals were removed after collecting fresh soil and then air-dried and screened through a 2 mm sieve. The soil pH was 5.51. The total nitrogen content was 0.81 g/kg. The total phosphorus content was 0.48 g/kg. The total potassium content was 1.63 g/kg. Moreover, Cd was not detected. When determining soil pH, 1 g of soil was extracted with 2.5 mL of distilled water and measured with a pH meter. The total N of soil was determined by using an elemental analyzer. The total K was extracted by NaOH and determined by performing flame photometry measurements [75]. The total P was extracted with hydrofluoric acid and perchloric acid, and its content was determined by using the colorimetric method [76].

The plant was ryegrass (*Lolium perenne* L.), and the ryegrass seeds were purchased from Jiangsu Suqian City Horticulture Company Limited (Jiangsu Suqian, China).

The bacterial strain was screened from the contaminated soil of Fenghuang Mountain (30°53′32″ N, 118°1′21″ E), Tongling City, Anhui Province. It was identified as *Cellulosimicrobium* sp. Meanwhile, the strain was deposited in the China General Microbiological Culture Collection Centre as strain number CGMCC No. 25219, and the sequence of the partial 16S rRNA gene of that strain was deposited in the GenBank database under accession number PP204244 (please visit https://www.ncbi.nlm.nih.gov/nuccore/PP204244), accessed on 30 January 2024.

4.2. Experimental Design

A total of 5 treatments (Table 2) were set up in this experiment, with 3 replicates per treatment and a total of 15 pots. The test soil was contaminated by adding Cd in the form of a $CdCl_2{\cdot}2.5H_2O$ salt solution, and 0.5 g/L Cd^{2+} were accurately prepared as the mother liquid and evenly poured into the potting soil. According to the soil environmental quality standard of China (GB15618-2018 [77]), the Cd concentration (measured by Cd^{2+}) reached 0, 4 mg/kg (control value of farmland soil), and 20 mg/kg (screening value of Class I construction land) after aging for 60 days. Each pot (8 cm × a8 cm × 10 cm) was filled with 500 g (dry weight) of soil, and 50 ryegrass seeds were evenly planted in each pot. According to the conventional application amount of N (300 kg/hm^2) and P (150 kg/hm^2), the water-soluble fertilizer was diluted and applied as a base fertilizer. Each pot was inoculated with 9 mL of Cd-resistant bacteria (10^9 CFU/mL), illuminated for 14 h/d, and darkened for 10 h/d. The indoor culture experiment was carried out for 40 days. According to the soil-moisture condition of each pot, distilled water was weighed and added every day to maintain the potting soil's moisture content at 60% of the soil's saturated water content. Sampling was performed on day 40, and then, the plant physiology and biochemistry, soil microbial activity, and heavy metal content were analyzed.

Table 2. Experimental design.

Treatment	Cd Pollution Concentration (mg·kg^{-1})	Cd Tolerant Bacteria
CK	0	Without inoculation
C1	4	Without inoculation
C1B	4	Inoculation
C2	20	Without inoculation
C2B	20	Inoculation

4.3. Methods

4.3.1. Determination of Physiological and Biochemical Characteristics of Ryegrass

The biomass was measured by the drying method [78]. The samples were placed in an oven at 105 °C for 30 min, then baked at 80 °C to a constant weight and weighed to obtain the biomass.

The chlorophyll content was determined by using the ethanol-grinding method [79]. Initially, 0.2 g of the cut fresh sample were weighed, and a small amount of quartz sand and calcium carbonate powder and 2–3 mL 95% ethanol were added and ground until it homogenized. Then, it was filtered with 95% ethanol into a 25 mL brown volumetric bottle and, finally, filled with ethanol to 25 mL and shaken well. Next, the chlorophyll was extracted into a colorimetric cup with a light diameter of 1 cm. The absorbance was determined at 665 nm and 649 nm, with 95% ethanol as a blank.

Superoxide dismutase (SOD) was determined by nitrogen blue tetrazole colorimetry [80]. First, 0.5 g of plant leaves (remove coarse leaf veins) in a pre-cooled mortar were taken, and 1 mL of pre-cooled phosphoric acid buffer on an ice bath was added to grind into pulp. Subsequently, the mixture was centrifuged at 4000 r/min for 10 min, and the supernatant was the crude extract of SOD. In a 10 mL centrifuge tube, 1.5 mL 0.05 mol/L pH = 7.8 phosphate buffer, 0.3 mL 130 mmol/L methionine solution, and 0.3 mL 750 µmol/L NBT solution were injected into the sample group successively. The tube contained 0.3 mL 100 µmol/L EDTA-Na_2, 0.3 mL 20 µmol/L riboflavin solution, 0.05 mL enzyme solution, and 0.25 mL ultra-pure water. The enzyme solution was replaced by slow liquid in the control group. After mixing, half of the control group's measuring tubes were placed in the dark. The absorbance of the other tubes was measured at 560 nm after the reaction under 4000 lx sunlight for 20 min.

The peroxidase (POD) activity was determined by guaiacol colorimetry [81]. First, 0.2 g of leaves were taken, cut into pieces, and ground until homogenized with ultra-pure water, The volume was set to 25 mL and were shaken well. Next, they were centrifuged for 1000 r/min for 20 min, and the supernatant was the enzyme liquid to be measured. A clean glass tube was taken, and 1.0 mL 0.1% guaiacol, 1.0 mL 0.18% H_2O_2, 1.0 mL enzyme solution, and blank control plus 1.0 mL ultra-pure water were added. The mixture was shaken well to accurately react at 25 °C for 10 min. Finally, 0.2 mL of 5% metaphosphoric acid was added to terminate the reaction. Absorbance was measured at 470 nm.

Glutathione (GSH) was determined by the 2-nitrobenzoic acid method [82]. First, 1 g of leaves was taken and placed in a mortar. Next, 50 g/L TCA extract pre-cooled at 4 °C in batches were added and ground into homogenate in an ice bath. The mixture was centrifuged at 12,000 r/min for 20 min, and the supernatant was the liquid to be measured. In the test tube, 1 mL supernatant, 1 mL 0.1 mol/L pH = 7.7 phosphoric acid buffer, and 1 mL 4 mmol/L DTNB solution were successively added, and the absorbance was measured at 412 nm after the reaction at 25 °C for 10 min.

Ascorbic acid (ASA) was determined by the colorimetric method [83]. Initially, 1.0 g of leaves was placed in a mortar, and 50 g/L TCA extract pre-cooled at 4 °C in batches were added. The mixture was ground into a homogenate in an ice bath, and 50 g/L TCA at a constant volume of 10 mL were used. Next, it was well mixed and filtered, and the collected filtrate was used. In the test tube, 1 mL filtrate, 1 mL 50 g/L TCA, 1 mL anhydrous ethanol, 0.5 mL 0.4% phospho-ethanol solution, 1 mL 5 g/L Rhodophenanthroline–ethanol solution, and 0.5 mL 0.3 g/L $FeCl_3$-ethanol solution were successively added. When the reaction at 30 °C was completed after 60 min, the absorbance was measured at 534 nm.

4.3.2. Determination of Soil Microbial Activity

The soil basal respiration was measured by the alkali absorption method [84]. The ampoule containing a 10 mL 0.05 mol/L NaOH solution was placed in a glass jar, sealed tightly, cultured at 25 °C for 24 h, and then, removed. Then, the NaOH solution was washed into a 100 mL conical bottle with CO_2-free water, and 2 mL barium chloride was added. And, 2 drops of phenolphthalein indicator were titrated with 0.05 mol/L until the purplish red color disappeared. The soil basal respiration was expressed as CO_2 released per gram of soil after 24 h (mg/g/d).

The soil sucrase was determined by the 3,5-dinitrosalicylic acid color development method [85]. First, 2.50 g of fresh soil was accurately weighed into a triangular bottle. Next, 0.5 mL toluene, 15 mL 8% sucrose solution, and 5 mL phosphoric acid buffer with pH = 5.5 were added, mixed well, and cultured in a constant temperature incubator at 37 °C for 24 h. Then, the mixture was rapidly filtered and absorbed 1 mL of filtrate. Then, 3 mL DNS reagent were added. The water bath was boiled for 5 min and a water-cooling termination for 3 min with a distilled water fixed volume of 50 mL was performed. The absorbance was determined at 508 nm. No matrix control was set for each soil sample, and no soil control was performed for the whole experiment to eliminate the error caused by sucrose originally existing in the soil. The soil sucrase activity was expressed as the number of milligrams of glucose produced per gram of soil after 24 h (mg/g/d).

The soil dehydrogenase was determined by the 2,3,5-triphenyltetrazolium chloride reduction method [86]. First, 2.00 g of fresh soil were accurately weighed into the colorimetric tube. Next, 2 mL TTC solution, 2 mL glucose solution, and 2 mL Tris buffer were added, shaken well, and cultured in an incubator (37 °C) for 24 h. A small amount of methanol was extracted many times, filtered in the colorimetric tube, and methanol was used to reach 50 mL. The absorbance was measured at 485 nm.

The soil urease was determined by the phenol-blue colorimetric method [87]. Initially, 2.50 g of fresh soil were accurately weighed into a triangular bottle. Then, 0.5 mL toluene first were added and reacted for 10 min. Next, 10 mL 10% urea and 20 mL citrate buffer were added, mixed well, cultured in a 37 °C incubator for 24 h, and filtered, whereas 1 mL filtrate was added into a colorimetric tube. Four mL sodium phenol solution and 3 mL sodium hypochlorite solution were successively added, and after color development for 20 min, the absorbance was determined at 578 nm by shaking with distilled water to 50 mL. No matrix control was set for each soil sample, and the same volume of ultra-pure water was used to replace the matrix. The other operations were the same as the sample test. At the same time, no soil-sample control should be done for the whole test to eliminate the error caused by the original ammonia in the soil. Soil urease activity was expressed as mg of NH_3-N produced per gram of soil after 24 h (mg/kg/d).

4.3.3. Determination of Cd Content

The content of Cd in the soil samples was determined by the HCl-HNO_3-HF-$HClO_4$ system. First, the soil sample of 0.2500 g (over 100 mesh sieve) was weighed and dissolved in a Teflon crucible on a controlled electric heating plate in the fume hood (containing 0.5 mL water, 5 mL hydrochloric acid, and 5 mL nitric acid). After heating at 110 °C for 1 h, it was taken off and slightly cooled. Next, 4 mL hydrofluoric acid and 1.5 mL perchloric acid were added, and the mixture was heated at 200 °C for 1.5 h. Then, the cover of the crucible was removed, and the heating was continued for 1.5 h. After that, it was heated at the temperature value of 380 °C to drive the acid until the white smoke was gone. Fifty percent of the nitric acid solution was added to warm and dissolve the residue, and the constant volume was 50 mL. Double filter paper was used to filter, and the cadmium content was determined by atomic absorption spectrophotometer. The Cd from the soil was extracted by the Tesseier multi-stage continuous extraction method [88].

The Cd content of the plant was digested by HNO_3-$HClO_4$. The cadmium content was determined by the atomic absorption spectrophotometer [89]. Initially, 0.2 g of dried 60-mesh sieve plant samples were weighed and put into a 100 mL triangle bottle, and 15 mL of mixed acid (HNO_3:$HClO_4$ = 4:1) were added. The bottle was sealed with plastic wrap and allowed to rest overnight. The stepwise heating method was used to digest on an electric heating plate the next day. The plastic wrap was removed, and the funnel was covered and heated up to 120 °C. Then, it was heated up to 180 °C after 30 min and to 240 °C after 30 min. The solution was digested until it became clear and transparent and cooled to room temperature. Finally, a volume of distilled water of 50 mL was to be measured.

4.3.4. Identification of Tolerant Bacteria

The bacterial genome DNA of the strain was extracted by using the high salt method [90]. More specifically, 16S rDNA cloning, sequencing, and comparison processes concerning strains were carried out. Using the extracted total bacterial DNA as a template, the 16S rDNA universal primer was extracted. The forward primer was 5′-AGA GTT TGA TCC TGG CTC AG-3′ (27F). The reverse primer 5′-TAC CTT GTT ACG ACT T-3′ (1492R) was used to perform a 50 μL PCR reaction. Next, the PCR product was recovered, and the enzyme was connected to the pMD19-T vector (Takala, Japan) and, then, cloned for sequencing [91]. The sequencing was completed by Shenggong BioEngineering (Shanghai, China) Company Limited. Homology analysis was performed between the sequencing results and the corresponding 16SrDNA sequences on GenBank and its RDP. The 16SrDNA sequence of the type strain with close homology to the tested strain was compared from the NCBI database and identified as *Cellulosimicrobium* sp.

4.4. Instruments

A tissue grinder, microplate reader, graphite digester, TAS-990 AFG atomic absorption spectrophotometer (Beijing Puxi General Instrument Company Limited, Beijing, China), element analyzer, pH meter, centrifuge, thermostat, oscillating mixer, and MG48+ PCR instrument, BG-Power 600i Electrophoresis Instrument were used.

4.5. Data Analysis

SPSS27 was used for conducting a one-way ANOVA analysis of variance, and significant differences were determined by the Duncan test ($p < 0.05$). The Excel2016 and Origin2021 software packages were used for data statistics and plotting. The correlation between Cd accumulation and inoculated microorganisms, physiology, biochemistry of ryegrass, and microbial activity was evaluated by a Pearson correlation analysis.

5. Conclusions

By performing a set of experiments, it was found that the inoculation of Cd-tolerant bacteria successfully alleviated the stress of Cd pollution on ryegrass. Therefore, the peroxidase activity and glutathione content in ryegrass leaves increased. The Cd-tolerant bacteria enhanced the chlorophyll content in ryegrass leaves, improved the photosynthesis ability of plants, increased the ascorbic acid content, and removed excess H_2O_2 in ryegrass. The Cd-tolerant bacteria can change soil microbial activity, increase soil urease and sucrase activities, decrease soil basic respiration and dehydrogenase activities, and change soil Cd morphology. The enrichment of Cd in both above-ground and under-ground parts of ryegrass could be also increased. Therefore, the inoculation of Cd-tolerant bacteria can remarkably improve the ability of ryegrass to repair Cd-contaminated soil.

Author Contributions: J.L., J.Z. (Jihai Zhou) and L.S.: Data curation, Writing—Original draft preparation. J.L., M.N. and S.X.: Investigation. J.L., M.N., X.L. and S.X.: Visualization. J.L., X.X. and J.Z. (Jie Zhang): Software, Validation. X.Z., J.Z. (Jihai Zhou) and Y.H.: Conceptualization, Methodology, Supervision, Writing—Reviewing and Editing. All authors have read and agreed to the published version of the manuscript.

Funding: This work was funded by the National Natural Science Foundation of China (41660615), the National Agricultural Experimental Station for Agricultural Environment, Fengxian (NAES035AE03), the Natural Science Foundation of Anhui Province (2108085MD128), and the Key Research and Development Project of Wuhu City (2022yf56).

Data Availability Statement: All data generated or analyzed in this study are included in the published article.

Conflicts of Interest: The authors declare that they have no known competing financial interests or personal relationships that could have appeared to influence the work reported in this paper.

References

1. Zhao, F.-J.; Ma, Y.; Zhu, Y.-G.; Tang, Z.; McGrath, S.P. Soil contamination in China: Current status and mitigation strategies. *Environ. Sci. Technol.* **2014**, *49*, 750–759. [CrossRef] [PubMed]
2. Ministry of Ecology and Environment of the People's Pepublic of China. *Bulletin on Ecological Environment in China*; Ministry of Ecology and Environment of the People's Pepublic of China: Beijing, China, 2021. (In Chinese)
3. Zou, M.; Zhou, S.; Zhou, Y.; Jia, Z.; Guo, T.; Wang, J. Cadmium pollution of soil-rice ecosystems in rice cultivation dominated regions in China: A review. *Environ. Pollut.* **2021**, *280*, 116965. [CrossRef] [PubMed]
4. Genchi, G.; Sinicropi, M.S.; Lauria, G.; Carocci, A.; Catalano, A. The effects of cadmium toxicity. *Int. J. Environ. Res. Public Health* **2020**, *17*, 3782. [CrossRef] [PubMed]
5. Gao, J.; Peng, R.; Zhu, B.; Tian, Y.; Xu, J.; Wang, B.; Fu, X.; Han, H.; Wang, L.; Zhang, F.; et al. Enhanced phytoremediation of TNT and cobalt co-contaminated soil by AfSSB transformed plant. *Ecotoxicol. Environ. Saf.* **2021**, *220*, 112407. [CrossRef] [PubMed]
6. Oladoye, P.O.; Olowe, O.M.; Asemoloye, M.D. Phytoremediation technology and food security impacts of heavy metal contaminated soils: A review of literature. *Chemosphere* **2022**, *288*, 132555. [CrossRef] [PubMed]
7. Chen, H.; Zhang, W.; Yang, X.; Wang, P.; Mcgrath, S.P.; Zhao, F. Effective methods to reduce cadmium accumulation in rice grain. *Chemosphere* **2018**, *207*, 699–707. [CrossRef] [PubMed]
8. Huang, Y.; Chen, Q.; Deng, M.; Japenga, J.; Li, T.; Yang, X.; He, Z. Heavy metal pollution and health risk assessment of agricultural soils in a typical peri-urban area in southeast China. *J. Environ. Manag.* **2018**, *207*, 159–168. [CrossRef] [PubMed]
9. Hamid, Y.; Tang, L.; Sohail, M.I.; Cao, X.; Hussain, B.; Aziz, M.Z.; Usman, M.; He, Z.L.; Yang, X. An explanation of soil amendments to reduce cadmium phytoavailability and transfer to food chain. *Sci. Total Environ.* **2019**, *660*, 80–96. [CrossRef] [PubMed]
10. Liu, L.; Li, W.; Song, W.; Guo, M. Remediation techniques for heavy metal-contaminated soils: Principles and applicability. *Sci. Total Environ.* **2018**, *633*, 206–219. [CrossRef]

11. Lu, H.; Xu, D.; Kong, T.; Wang, D. Characteristics of enzyme activities during phytoremediation of Cd-contaminated soil. *Sustainability* **2022**, *14*, 9350. [CrossRef]
12. Lou, Y.; Lou, H.; Hu, T.; Li, H.; Fu, J. Toxic effects, uptake, and translocation of Cd and Pb in perennial ryegrass. *Ecotoxicology* **2013**, *22*, 207–214. [CrossRef] [PubMed]
13. Zhou, J.; Cheng, K.; Song, L.; Li, W.; Jiang, H.; Huang, G. Exogenous indoleacetic acid induces cadmium accumulation and growth in *Cinnamomum camphora. Sci. Hortic.* **2024**, *323*, 112518. [CrossRef]
14. He, S.; Wu, Q.; He, Z. Synergetic effects of DA-6/GA3 with EDTA on plant growth, extraction and detoxification of Cd by *Lolium perenne. Chemosphere* **2014**, *117*, 132–138. [CrossRef] [PubMed]
15. He, S.; Guo, H.; He, Z.; Wang, L. Effects of a new-type cleaning agent and a plant growth regulator on phytoextraction of cadmium from a contaminated soil. *Pedosphere* **2019**, *29*, 161–169. [CrossRef]
16. Fine, P.; Rathod, P.H.; Beriozkin, A.; Mingelgrin, U. Uptake of cadmium by hydroponically grown, mature Eucalyptus camaldulensis saplings and the effect of organic ligands. *Int. J. Phytoremed.* **2013**, *15*, 585–601. [CrossRef] [PubMed]
17. Vassilev, A.; Perez-Sanz, A.; Semane, B.; Carleer, R.; Vangronsveld, J. Cadmium accumulation and tolerance of two Salix genotypes hydroponically grown in presence of Cadmium. *J. Plant. Nutr.* **2005**, *28*, 2159–2177. [CrossRef]
18. Zhang, M.; Wang, X.; Yang, L.; Chu, Y. Research on progress in combined remediation technologies of heavy metal polluted sediment. *Int. J. Environ. Res. Public Health* **2019**, *16*, 5098. [CrossRef] [PubMed]
19. Jin, Y.; Luan, Y.; Ning, Y.; Wang, L. Effects and mechanisms of microbial remediation of heavy metals in soil: A critical review. *Appl. Sci.* **2018**, *8*, 1336. [CrossRef]
20. Hassan, M.K.; McInroy, J.A.; Kloepper, J.W. The interactions of rhizodeposits with plant growth-promoting rhizobacteria in the rhizosphere: A review. *Agriculture* **2019**, *9*, 142. [CrossRef]
21. Li, W.; Wang, D.; Hu, F.; Li, H.; Ma, L.; Xu, L. Exogenous IAA treatment enhances phytoremediation of soil contaminated with phenanthrene by promoting soil enzyme activity and increasing microbial biomass. *Environ. Sci. Pollut. Res.* **2016**, *23*, 10656–10664. [CrossRef]
22. Nagajyoti, P.C.; Lee, K.D.; Sreekanth, T.V.M. Heavy metals, occurrence and toxicity for plants: A review. Environ. *Chem. Lett.* **2010**, *8*, 199–216. [CrossRef]
23. Trasar-Cepada, C.; Leirós, M.C.; Seoane, S.; Gil-Sotres, F. Limitations of soil enzymes as indicators of soil pollution. *Soil Biol. Biochem.* **2000**, *32*, 867–875. [CrossRef]
24. Luo, J.; Qi, S.; Gu, X.W.S.; Hou, T.; Lin, L. Ecological risk assessment of EDTA-assisted phytoremediation of Cd under different cultivation systems. *Bull. Environ. Contam. Toxicol.* **2016**, *96*, 259–264. [CrossRef] [PubMed]
25. Zafar, Z.; Rasheed, F.; Khan, W.R.; Mohsin, M.; Rashid, M.Z.; Magiman, M.M.; Raza, Z.; Rosli, Z.; Afzal, S.; Bakar, F.A. The change in growth, osmolyte production and antioxidant ezymes activity explains the cadmium tolerance in four tree species at the saplings stage. *Forests* **2022**, *13*, 1343. [CrossRef]
26. Zhang, M.; Hua, M.; Guo, D.; Xue, Y.; Chen, X.; Rui, L.; Zhou, N. Effects of plant growth-promoting rhizobacteria on growth indicators and physiological characteristics of *Peucedanum praeruptorum* Dunn leaves. *Plant Signal. Behav.* **2023**, *18*, 2203571. [CrossRef] [PubMed]
27. Hu, J.; Wu, S.; Wu, F.; Leung, H.M.; Lin, X.; Wong, M.H. Arbuscular mycorrhizal fungi enhance both absorption and stabilization of Cd by Alfred stonecrop (*Sedum alfredii* Hance) and perennial ryegrass (*Lolium perenne* L.) in a Cd-contaminated acidic soil. *Chemosphere* **2013**, *93*, 1359–1365. [CrossRef] [PubMed]
28. Han, S.; Li, X.; Amombo, E.; Fu, J.; Xie, Y. Cadmium tolerance of perennial ryegrass induced by *Aspergillus aculeatus. Front Microbiol.* **2018**, *9*, 1579. [CrossRef] [PubMed]
29. Ali, B.; Qian, P.; Jin, R.; Ali, S.; Khan, M.; Aziz, R.; Tian, T.; Zhou, W. Physiological and ultra-structural changes in *Brassica napus* seedlings induced by cadmium stress. *Biol. Plant.* **2014**, *58*, 131–138. [CrossRef]
30. Andresen, E.; Küpper, H. Cadmium toxicity in plants. In *Cadmium: From Toxicity to Essentiality*; Springer: Dordrecht, The Netherlands, 2013; pp. 395–413.
31. Kalaji, H.M.; Jajoo, A.; Oukarroum, A.; Brestic, M.; Zivcak, M.; Samborska, I.A.; Cetner, M.D.; Łukasik, I.; Goltsev, V.; Ladle, R.J. Chlorophyll a fluorescence as a tool to monitor physiological status of plants under abiotic stress conditions. *Acta Physiol. Plant.* **2016**, *38*, 102. [CrossRef]
32. El-Saadony, M.T.; Desoky, E.S.M.; El-Tarabily, K.A.; AbuQamar, S.F.; Saad, A.M. Exploiting the role of plant growth promoting rhizobacteria in reducing heavy metal toxicity of pepper (*Capsicum annuum* L.). *Environ. Sci. Pollut. Res.* **2024**, *31*, 27465–27484. [CrossRef]
33. Zafar-ul-Hye, M.; Naeem, M.; Danish, S.; Khan, M.J.; Fahad, S.; Datta, R.; Brtnicky, M.; Kintl, A.; Hussain, G.S.; El-Esawi, M.A. Effect of cadmium-tolerant rhizobacteria on growth attributes and chlorophyll contents of bitter gourd under cadmium toxicity. *Plants* **2020**, *9*, 1386. [CrossRef] [PubMed]
34. Iori, V.; Pietrini, F.; Zacchini, M. Assessment of ibuprofen tolerance and removal capability in *Populus nigra* L. by in vitro culture. *J. Hazard. Mater.* **2012**, *229*, 217–223. [CrossRef] [PubMed]
35. Mohsin, M.; Kaipiainen, E.; Salam, M.M.A.; Evstishenkov, N.; Nawrot, N.; Villa, A.; Wojciechowska, E.; Kuittinen, S.; Pappinen, A. Biomass production and removal of nitrogen and phosphorus from processed municipal wastewater by *salix schwerinii*: A field trial. *Water* **2021**, *13*, 2298. [CrossRef]

36. Hussain, Z.; Rasheed, F.; Tanvir, M.A.; Zafar, Z.; Rafay, M.; Mohsin, M.; Pulkkinen, P.; Ruffner, C. Increased antioxidative enzyme activity mediates the phytoaccumulation potential of Pb in four agroforestry tree species: A case study under municipal and industrial wastewater irrigation. *Int. J. Phytorem.* **2021**, *23*, 704–714. [CrossRef] [PubMed]
37. Zhao, Z.; Xi, M.; Jiang, G.; Liu, X.; Bai, Z.; Huang, Y. Effects of IDSA, EDDS and EDTA on heavy metals accumulation in hydroponically grown maize (*Zea mays* L.). *J. Hazard. Mater.* **2010**, *181*, 455–459. [CrossRef] [PubMed]
38. Yang, F.; Zhang, H.; Wang, Y.; He, G.; Wang, J.; Guo, D.; Li, T.; Sun, G.; Zhang, H. The role of antioxidant mechanism in photosynthesis under heavy metals Cd or Zn exposure in tobacco leaves. *J. Plant Interact.* **2021**, *16*, 354–366. [CrossRef]
39. Wu, F.; Zhang, G.; Dominy, P. Four barley genotypes respond differently to cadmium: Lipid peroxidation and activities of antioxidant capacity. *Environ. Exp. Bot.* **2003**, *50*, 67–78. [CrossRef]
40. Wang, K.R.; Gong, H.; Wang, Y.; van der Zee, S.E.A.T.M. Toxic effects of cadmium on *Morus alba* L. and *Bombyx moril* L. *Plant Soil* **2004**, *261*, 171–180. [CrossRef]
41. Hou, W.; Chen, X.; Song, G.; Wang, Q.; Chang, C.C. Effects of copper and cadmium on heavy metal polluted waterbody restoration by duckweed (*Lemna minor*). *Plant Physiol. Biochem.* **2007**, *45*, 62–69. [CrossRef]
42. Wang, S.; Zhao, Y.; Guo, J.; Liu, Y. Antioxidative response in leaves and allelochemical changes in root exudates of *Ricinus communis* under Cu, Zn, and Cd stress. *Environ. Sci. Pollut. Res.* **2018**, *25*, 32747–32755. [CrossRef]
43. Zhu, T.; Li, L.; Duan, Q.; Liu, X.; Chen, M. Progress in our understanding of plant responses to the stress of heavy metal cadmium. *Plant Signal. Behav.* **2021**, *16*, 1836884. [CrossRef] [PubMed]
44. Maslennikova, D.; Nasyrova, K.; Chubukova, O.; Akimova, E.; Baymiev, A.; Blagova, D.; Ibragimov, A.; Lastochkina, O. Effects of *Rhizobium leguminosarum* Thy2 on the growth and tolerance to cadmium stress of wheat plants. *Life* **2022**, *12*, 1675. [CrossRef] [PubMed]
45. Song, L.; Xu, X.; Zheng, Y.; Hong, W.; Li, X.; Ai, Y.; Wang, Y.; Zhang, Z.; Chen, H.; Huang, Y.; et al. Dynamic mechanisms of cadmium accumulation and detoxification by *Lolium perenne* grown in soil inoculated with the cadmium-tolerant bacterium strain Cdq4-2. *Sci. Total Environ.* **2023**, *873*, 162314. [CrossRef] [PubMed]
46. Upadhyaya, C.P.; Venkatesh, J.; Gururani, M.A.; Asnin, L.; Sharma, K.; Ajappala, H.; Park, S.W. Transgenic potato overproducing Lascorbic acid resisted an increase in methylglyoxal under salinity stress via maintaining higher reduced glutathione level and glyoxalase enzyme activity. *Biotechnol. Lett.* **2011**, *33*, 2297–2307. [CrossRef] [PubMed]
47. Zhang, G.; Ba, Q.; Chen, S.; Liu, F.; Li, G. Exogenous application of glycine betaine alleviates cadmium toxicity in super black waxy maize by improving photosynthesis, the antioxidant system and glutathione-ascorbic acid cycle metabolites. *Cereal Res. Commun.* **2020**, *48*, 449–458. [CrossRef]
48. Horikawa, S.; Yoneya, R.; Nagashima, Y.; Hagiwara, K.; Ozasa, H. Prior induction of heme oxygenase 1 with glutathione depletor ameliorates the renal ischemia and reperfusion injury in the rat. *FEBS Lett.* **2002**, *510*, 221–224. [CrossRef] [PubMed]
49. Chen, L.; Beiyuan, J.; Hu, W.; Zhang, Z.; Duan, C.; Cui, Q.; Zhu, X.; He, H.; Huang, X.; Fang, L. Phytoremediation of potentially toxic elements (PTEs) contaminated soils using alfalfa (*Medicago sativa* L.): A comprehensive review. *Chemosphere* **2022**, *293*, 133577. [CrossRef] [PubMed]
50. Niemeyer, J.C.; Lolata, G.B.; de Carvalho, G.M.; Da Silva, E.M.; Sousa, J.P.; Nogueira, M.A. Microbial indicators of soil health as tools for ecological risk assessment of a metal contaminated site in Brazil. *Appl. Soil. Ecol.* **2012**, *59*, 96–105. [CrossRef]
51. Kızılkaya, R.; Aşkın, T.; Bayraklı, B.; Sağlam, M. Microbiological characteristics of soils contaminated with heavy metals. Eur. *J. Soil Biol.* **2004**, *40*, 95–102. [CrossRef]
52. Pacwa-Płociniczak, M.; Czapla, J.; Płociniczak, T.; Piotrowska-Seget, Z. The effect of bioaugmentation of petroleum-contaminated soil with *Rhodococcus erythropolis* strains on removal of petroleum from soil. *Ecotoxicol. Environ. Saf.* **2019**, *169*, 615–622. [CrossRef]
53. Cao, Y.; Ma, C.; Chen, H.; Chen, G.; White, J.C.; Xing, B. Copper stress in flooded soil: Impact on enzyme activities, microbial community composition and diversity in the rhizosphere of *Salix integra*. *Sci. Total Environ.* **2020**, *704*, 135350. [CrossRef] [PubMed]
54. Antonious, G.F.; Turley, E.T.; Dawood, M.H. Monitoring soil enzymes activity before and after animal manure application. *Agriculture* **2020**, *10*, 166. [CrossRef]
55. Xie, J.; Xu, X.; Zhang, S.; Yang, Z.; Wang, G.; Li, T.; Pu, Y.; Zhou, W.; Xu, C.; Lv, G.; et al. Activation and tolerance of *Siegesbeckia orientalis* L. rhizosphere to Cd stress. *Front. Plant Sci.* **2023**, *14*, 1145012. [CrossRef] [PubMed]
56. Li, Z.; Liang, Y.; Hu, H.; Shaheen, S.M.; Zhong, H.; Tack, F.M.G.; Wu, M.; Li, Y.-F.; Gao, Y.; Rinklebe, J.; et al. Speciation, transportation, and pathways of cadmium in soil-rice systems: A review on the environmental implications and remediation approaches for food safety. *Environ. Int.* **2021**, *156*, 106749. [CrossRef] [PubMed]
57. He, S.; Li, Y.; Guo, H.; Lu, L.; Yang, C. Combined effect of ryegrass and *Hyphomicrobium sp.* GHH on the remediation of EE2-Cd co-contaminated soil. *J. Soils Sediments* **2019**, *20*, 425–434. [CrossRef]
58. Wu, Y.; Zhang, H.; Li, J.; Wang, Z.; Jiang, Y. Adsorption of soil invertase to goethite, gibbsite and their organic complexes and the effects on enzyme catalytic performance. *Colloids Surf. B* **2023**, *222*, 113073. [CrossRef] [PubMed]
59. Calabrese, E.J.; Baldwin, L.A. Toxicology rethinks its central belief. *Nature* **2003**, *421*, 691–692. [CrossRef] [PubMed]
60. Jin, J.; Mi, R.; Li, Q.; Lang, J.; Lan, Y.; Huang, N.; Yang, G. Bacillus Thuringiensis Enhances the Ability of Ryegrass to Remediate Cadmium-Contaminated Soil. *Sustainability* **2023**, *15*, 5177. [CrossRef]

61. Zhang, S.; Li, Y.; Wang, P.; Zhang, H.; Ali, E.F.; Li, R.; Shaheen, S.M.; Zhang, Z. Lactic acid bacteria promoted soil quality and enhanced phytoextraction of Cd and Zn by mustard: A trial for bioengineering of toxic metal contaminated mining soils. *Environ. Res.* **2023**, *216*, 114646. [CrossRef]
62. Zhang, C.; Tao, Y.; Li, S.; Ke, T.; Wang, P.; Wei, S.; Chen, L. Bioremediation of cadmium-trichlorfon co-contaminated soil by Indian mustard (*Brassica juncea*) associated with the trichlorfon-degrading microbe *Aspergillus sydowii*: Related physiological responses and soil enzyme activities. *Ecotoxicol. Environ. Saf.* **2020**, *188*, 109756. [CrossRef]
63. Tan, X.; Liu, Y.; Yan, K.; Wang, Z.; Lu, G.; He, Y.; He, W. Differences in the response of soil dehydrogenase activity to Cd contamination are determined by the different substrates used for its determination. *Chemosphere* **2017**, *169*, 324–332. [CrossRef] [PubMed]
64. Visser, S.; Parkinson, D. Soil biological criteria as indicators of soil quality: Soil microorganisms. *Am. J. Altern. Agric.* **1992**, *7*, 33–37. [CrossRef]
65. Liao, M.; Luo, Y.; Zhao, X.; Huang, C. Toxicity of cadmium to soil microbial biomass and its activity: Effect of incubation time on Cd ecological dose in a paddy soil. *J. Zhejiang. Univ. Sci. B* **2005**, *6*, 324–330. [CrossRef] [PubMed]
66. Gan, C.; Chen, T.; Yang, J. Remediation of vanadium contaminated soil by alfalfa (*Medicago sativa* L.) combined with vanadium-resistant bacterial strain. *Environ. Technol. Innov.* **2020**, *20*, 101090. [CrossRef]
67. Liu, L.; Li, J.; Yue, F.; Yan, X.; Wang, F.; Bloszies, S.; Wang, Y. Effects of arbuscular mycorrhizal inoculation and biochar amendment on maize growth, cadmium uptake and soil cadmium speciation in Cd-contaminated soil. *Chemosphere* **2018**, *194*, 495–503. [CrossRef] [PubMed]
68. Sun, J.; Jia, Q.; Li, Y.; Dong, K.; Xu, S.; Ren, Y.; Zhang, T.; Chen, J.; Shi, N.; Fu, S. Effect of arbuscular mycorrhiza fungus *Diversispora eburnea* inoculation on *Lolium perenne* and *Amorpha fruticosa* growth, cadmium uptake, and soil cadmium speciation in cadmium-contaminated soil. *Int. J. Environ. Res. Public Health* **2023**, *20*, 795. [CrossRef] [PubMed]
69. Liu, A.; Wang, W.; Zheng, X.; Chen, X.; Fu, W.; Wang, G.; Ji, J.; Jin, C.; Guan, C. Improvement of the Cd and Zn phytoremediation efficiency of rice (*Oryza sativa*) through the inoculation of a metal-resistant PGPR strain. *Chemosphere* **2022**, *302*, 134900. [CrossRef] [PubMed]
70. Zhang, J.; Yang, N.; Geng, Y.; Zhou, J.; Lei, J. Effects of the combined pollution of cadmium, lead and zinc on the phytoextraction efficiency of ryegrass (*Lolium perenne* L.). *RSC Adv.* **2019**, *9*, 20603–20611. [CrossRef] [PubMed]
71. Hu, Z.; Wang, Y.; Fang, Z.; Shi, G.; Lou, L.; Ren, K.; Cai, Q. Italian ryegrass–rice rotation system for biomass production and cadmium removal from contaminated paddy fields. *J. Soils Sediments* **2020**, *20*, 874–882. [CrossRef]
72. Guo, J.K.; Zhao, J.; Ren, X.H.; Jia, H.L.; Muhammad, H.; Lv, X.; Wei, T.; Hua, L. Effects of *Burkholderia* sp. D54 on growth and cadmium uptake of tomato, ryegrass and soybean plants. *Int. J. Environ. Sci. Technol.* **2020**, *17*, 1149–1158. [CrossRef]
73. Yung, L.; Blaudez, D.; Maurice, N.; Azou-Barré, A.; Sirguey, C. Dark Septate Endophytes Isolated from Non-hyperaccumulator plants can increase phytoextraction of Cd and Zn by the hyperaccumulator *Noccaea caerulescens*. *Environ. Sci. Pollut. Res.* **2021**, *28*, 16544–16557. [CrossRef] [PubMed]
74. Lago-Vila, M.; Arenas-Lago, D.; Rodríguez-Seijo, A.; Andrade, M.L.; Vega, F.A. Ability of *Cytisus scoparius* for phytoremediation of soils from a Pb/Zn mine: Assessment of metal bioavailability and bioaccumulation. *J. Environ. Manag.* **2019**, *235*, 152–160. [CrossRef]
75. Alfaro, M.A.; Alfaro, M.A.; Jarvis, S.C.; Gregory, P.J. Factors affecting potassium leaching in different soils. *Soil. Use Manag.* **2004**, *20*, 182–189. [CrossRef]
76. Xie, X.; Pu, L.; Wang, Q.; Zhu, M.; Xu, Y.; Zhang, M. Response of soil physicochemical properties and enzyme activities to long-term reclamation of coastal saline soil, Eastern China. *Sci. Total Environ.* 2017; 607–608, 1419–1427.
77. Ministry of Ecology and Environment of the People's Pepublic of China. *Soil environmental quality Risk control standard for soil contamination of agricultural land*; Ministry of Ecology and Environment of the People's Pepublic of China: Beijing, China, 2018. (In Chinese)
78. Yang, Q.; Zhao, Z.; Bai, Z.; Hou, H.; Yuan, Y.; Guo, A.; Li, Y. Effects of mycorrhizae and water conditions on perennial ryegrass growth in rare earth tailings. *RSC Adv.* **2019**, *9*, 10881–10888. [CrossRef]
79. Cvitković, D.; Lisica, P.; Zorić, Z.; Repajić, M.; Pedisić, S.; Dragović-Uzelac, V.; Balbino, S. Composition and antioxidant properties of pigments of Mediterranean herbs and spices as affected by different extraction methods. *Foods* **2021**, *10*, 2477. [CrossRef]
80. Fukami, J.; de la Osa, C.; Ollero, F.J.; Megías, M.; Hungria, M. Co-inoculation of maize with *Azospirillum brasilense* and *Rhizobium tropici* as a strategy to mitigate salinity stress. *Funct. Plant Biol.* **2018**, *45*, 328–339. [CrossRef] [PubMed]
81. Khan, M.; Samrana, S.; Zhang, Y.; Malik, Z.; Khan, M.D.; Zhu, S. Reduced glutathione protects subcellular compartments from Pb-induced ROS injury in leaves and roots of upland cotton (*Gossypium hirsutum* L.). *Front. Plant Sci.* **2020**, *11*, 412. [CrossRef]
82. Rahman, I.; Kode, A.; Biswas, S.K. Assay for quantitative determination of glutathione and glutathione disulfide levels using enzymatic recycling method. *Nat. Protoc.* **2006**, *1*, 3159–3165. [CrossRef]
83. Zhang, J.; Li, H.; Xu, B.; Li, J.; Huang, B. Exogenous melatonin suppresses dark-induced leaf senescence by activating the superoxide dismutase-catalase antioxidant pathway and down-regulating chlorophyll degradation in excised leaves of perennial ryegrass (*Lolium perenne* L.). *Front. Plant Sci.* **2016**, *7*, 1500. [CrossRef]
84. Xu, Y.; Seshadri, B.; Bolan, N.; Sarkar, B.; Ok, Y.S.; Zhang, W.; Rumpel, C.; Sparks, D.; Farrell, M.; Hall, T.; et al. Microbial functional diversity and carbon use feedback in soils as affected by heavy metals. *Environ. Int.* **2019**, *125*, 478–488. [CrossRef]

85. Yang, D.; Tang, L.; Cui, Y.; Chen, J.; Liu, L.; Guo, C. Saline-alkali stress reduces soil bacterial community diversity and soil enzyme activities. *Ecotoxicology* **2022**, *31*, 1356–1368. [CrossRef] [PubMed]
86. Filipović, L.; Romić, M.; Sikora, S.; Huić Babić, K.; Filipović, V.; Gerke, H.H.; Romić, D. Response of soil dehydrogenase activity to salinity and cadmium species. *J. Soil Sci. Plant Nutr.* **2020**, *20*, 530–536. [CrossRef]
87. Liu, G.; Zhang, X.; Wang, X.; Shao, H.; Yang, J.; Wang, X. Soil enzymes as indicators of saline soil fertility under various soil amendments. *Agric. Ecosyst. Environ.* **2017**, *237*, 274–279.
88. Tessier, A.; Campbell, P.G.C.; Bisson, M. Sequential extraction procedure for the speciation of particulate trace metals. *Anal. Chem.* **1979**, *51*, 844–851. [CrossRef]
89. Shi, G.; Yan, Y.; Yu, Z.; Zhang, L.; Cheng, Y.; Shi, W. Modification-bioremediation of copper, lead, and cadmium-contaminated soil by combined ryegrass (*Lolium multiflorum* Lam.) and Pseudomonas aeruginosa treatment. *Environ. Sci. Pollut. Res.* **2020**, *27*, 37668–37676. [CrossRef] [PubMed]
90. Guillemaut, P.; Maréchal-Drouard, L. Isolation of plant DNA: Afast, inexpensive, and reliable method. *Plant Mol. Biol. Rep.* **1992**, *10*, 60–65. [CrossRef]
91. Sambrook, J.; Fritch, E.F.; Maniatis, T. *Molecular Cloning: A Laboratory Manual*, 2nd ed.; Cold Spring Harbor Laboratory Press: New York, NY, USA, 1989.

International Journal of
Plant Biology

MDPI

Review

Synergism or Antagonism: Do Arbuscular Mycorrhizal Fungi and Plant Growth-Promoting Rhizobacteria Work Together to Benefit Plants?

Noah Savastano * and Harsh Bais *

Department of Plant and Soil Sciences, University of Delaware, 357 AP Biopharma, 590 Avenue 1743, Newark, DE 19713, USA
* Correspondence: nsav@udel.edu (N.S.); hbais@udel.edu (H.B.)

Abstract: In agriculture, abiotic and biotic stress reduce yield by 51–82% and 10–16%, respectively. Applications of biological agents such as plant growth-promoting rhizobacteria (PGPR) and arbuscular mycorrhizal fungi (AMF) can improve plant growth. Applications of lone PGPR and AMF also help plants resist abiotic and biotic stressors. The reports for dual inoculation of AMF and PGPR to benefit plants and tackle stressors are largely unknown. It is speculated that PGPR colonization in plants enhances AMF infection during dual AMF and PGPR application, although increased AMF colonization does not always correlate with the increased benefits for the plant hosts. Further research is needed regarding molecular mechanisms of communication during dual inoculations, and dual-inoculation enhancement of induced systemic resistance under pathogen stress, to understand how dual inoculations can result in enhanced plant benefits. The influence of application timing of AMF and PGPR dual inoculations on mitigating abiotic and biotic stress is also not well understood. This review documents the factors that govern and modulate the dual application of AMF and PGPR for plant benefits against stress responses, specifically abiotic (drought) stress and stress from pathogen infection.

Keywords: microorganisms; plant beneficial microorganisms; abiotic stress; biotic stress; dual inoculation

1. Introduction

Each year, plant pathogens reduce global agricultural yields by 10–16% [1]. Additionally, abiotic stress decreases yield by 51–82% [2]. Agricultural yield loss from biotic and abiotic stress contributes to global food insecurity. Therefore, agricultural yield improvements are needed [1].

One approach to mitigating abiotic and biotic stress to improve yield is the application of biological agents. Biological agents have several advantages in comparison to more commonly used chemical agents, for plant pathogen control. Biological agent applications do not contaminate the environment or pose a risk to human health, unlike chemical pesticides. Also, biological agents increase plant growth under both ideal and stress conditions, improving crop yields [3].

Biological agents such as plant growth-promoting rhizobacteria (PGPR) and arbuscular mycorrhizal fungi (AMF) can improve plant pathogen resistance by acting antagonistically towards pathogens, mainly via competition and antimicrobial compound production [4]. Biological agents such as PGPR and AMF can also mitigate abiotic stress, specifically drought stress, by reducing cell damage, producing antioxidants, and improving water and nutrient uptake [5]. Additionally, dual inoculations of AMF and PGPR have enhanced plant benefits, compared to singular or no inoculation. PGPRs stimulate AMF colonization of plant roots, enhancing plant biomass and nutrient uptake [6–8]. Dual inoculations can also further enhance abiotic and biotic stress resistance, specifically by mitigating drought stress and improving plant pathogen defense [9–13].

However, there are several unknowns regarding the application of AMF and PGPR, both singularly and dually. Specifically, interactions between AMF, PGPR, and plants are not yet fully understood [14,15]. Also, it is known that the high diversity of the microbial community in the soil surrounding plant roots, the rhizosphere, is necessary for maintaining plant and soil health, both in general and under stress conditions [4,5]. However, a greater understanding is needed of how biological-control agent applications influence native soil microbes, and how the native microbial community affects the efficacy of biological agents [4].

This review will document several mechanisms of PGPR and AMF drought- and pathogen-stress mitigation, and the potential benefit of dual PGPR and AMF applications in mitigating plant stress.

2. Plant Growth-Promoting Rhizobacteria (PGPR)

Rhizobacteria colonize the area of soil surrounding plant roots known as the rhizosphere [16]. "Plant growth-promoting rhizobacteria" (PGPR) refers to 2–5% of rhizobacteria colonizing the rhizosphere that promote plant growth [16,17]. Plants and PGPR exchange chemical metabolites and create a symbiotic relationship with one another [18]. Within this relationship, PGPR promote plant growth by producing plant growth hormones including gibberellin and indole-3-acetic acid, to improve plant nutrient uptake. PGPR also improve nutrient availability and uptake by solubilizing insoluble phosphate, solubilizing iron via siderophore production and fixing atmospheric nitrogen. PGPR also mitigate plant pathogen stress, to promote plant growth in the presence of pathogens [16]. Additionally, PGPR aid plants in abiotic stress tolerance, such as tolerance to nutrient deficiency, heavy metals, salt, and drought [18]. While many studies have reported abiotic- and biotic-stress reduction via PGPR application, due to the vast number of plant and PGPR species, many species-specific PGPR–plant interactions under abiotic- and biotic-stress conditions are yet to be investigated. Some PGPR species' growth promotion is specific to plant species, cultivars, or genotypes. However, the specifications of PGPR–plant compatibility has not been well studied [16]. In addition, while previous studies have identified PGPR root colonization, they do not indicate which regions within the roots are colonized by PGPR. The specific site for PGPR root localization is key to deducing the interplay a PGPR colonization may have with AMF infection in plants.

2.1. Mechanisms of Mitigating Plant Drought Stress

2.1.1. Increased Proline Production

Reactive oxygen species (ROS) cause oxidative damage in plant cells, specifically affecting the DNA, lipids, and proteins within cells. Cellular damage from ROS occurs when plants experience abiotic stress, such as drought stress. Osmolytes such as proline act as ROS scavengers that neutralize ROS in plant cells. PGPR synthesize osmolytes, including proline, to mitigate oxidative damage from ROS during drought stress conditions [19]. Increased proline concentration correlates with the mitigation of drought stress in plants [20]. For cucumber (*Cucumis sativus* L.) under drought stress, inoculation with PGPR consortia including *Bacillus cereus* AR156, *Bacillus subtilis* SM21, and *Serratia* sp. XY21 increased proline content in leaves, as well as improving leaf chlorophyll content, and reduced wilt symptoms, compared to uninoculated plants [21]. Additionally, when maize (*Zea mays* L.) was exposed to drought stress, plants inoculated with PGPR *Bacillus* spp. produced more proline and had an increase in biomass compared to the control [22]. For *Thymus vulgaris*, *Santolina chamaecyparissus*, *Lavandula dentata*, and *Salvia officinalis* shrub species under drought conditions, PGPR inoculations improved nutrition and morphological traits in comparison to uninoculated plants. PGPR, especially *Bacillus* sp. and *Enterobacter* sp., also produced more proline to help remove free radicals from the plant when under drought stress. Bacterial inoculation increased proline content in *L. dentata* leaves [23]. Also, for wheat (*Triticum aestivum* L.) exposed to drought stress, *Bacillus megaterium* inoculation increased proline content, as well as improving relative water content, protein content, and chlorophyll a, b, and carotenoids [24].

2.1.2. Antioxidant Enzyme Production

PGPR have been found to regulate antioxidant enzyme activity in plants under drought stress. Antioxidant enzymes act as ROS scavengers, mitigating oxidative damage that normally occurs during drought stress. PGPR inoculants *Bacillus altitudinis* FD48 and *Bacillus methylotrophicus* RABA6 increased antioxidant enzyme activity, specifically superoxide dismutase (SOD), peroxidase (POD), catalase (CAT), and ascorbate reductase (APX), and mitigated drought stress for rice (*Oryza sativa*) exposed to drought conditions [25]. PGPR *B. megaterium* (MU2) increased SOD, POD, APX, CAT, and glutathione reductase (GR) activity and mitigated drought stress for wheat under drought conditions, in comparison to the control [24].

2.1.3. ABA Production, Regulating Stomatal Closure

PGPR inoculation increases abscisic acid (ABA) production in plants under drought conditions, resulting in a decrease in water loss and amelioration of drought stress. Increased ABA production from PGPR inoculation signals the plant to increase calcium levels in the cytosol of guard cells on plant leaves. Increased calcium levels enhance the rate of potassium and anion efflux out of the cell, causing water to exit the guard cell and turgor pressure to be lost. Reduction in turgor pressure triggers guard cells to close and decreases water loss via stomata [26]. PGPR inoculant *Azospirillum brasilense* Sp 245 enhanced ABA levels in *Arabidopsis thaliana* exposed to drought conditions compared to uninoculated plants, triggering the plants to close their stomata and conserve water. The PGPR inoculation and enhanced ABA production increased plant biomass and seed yield compared to the control [27]. Inoculants *Bacillus licheniformis* Rt4M10 and *Pseudomonas fluorescens* Rt6M10 also increased ABA production and minimized water loss for grapevine (*Vitis vinifera*) under water stress, leading to improved water content and turgidity [28]. Though previous studies found PGPR increase in ABA plant production to influence stomatal activity, they did not investigate other mechanisms of PGPR regulating pathways that control stomata opening and closing during drought stress [20].

2.1.4. Reduced Ethylene Overproduction

Plants respond differently to both biotic and abiotic stress, and ethylene as a signaling molecule plays a potential role in modulating root and shoot growth [29]. It is known that various PGPR mitigate ethylene production by synthesizing ACC deaminase (1-aminocyclopropane-1-carboxylate deaminase) to maintain plant growth [20,30]. During ethylene synthesis, ACC produced is a precursor of ethylene. ACC deaminase enzyme, produced by PGPR, deteriorates ACC, and therefore decreases ethylene production [29]. ACC deaminase production by *Enterobacter* sp., *Bacillus thuringiensis*, *Bacillus* sp., and *B. megaterium* inoculants mitigated drought stress for *T. vulgaris*, *S. chamaecyparissus*, *L. dentata*, and *S. officinalis* shrub species [23]. PGPR *B. licheniformis* inoculation produced ACC deaminase, and increased plant growth of pepper (*Capsicum annuum* L.) exposed to drought conditions in comparison to uninoculated plants [31]. Inoculants *Pseudomonas aeruginosa*, *Enterobacter cloacae*, *Achromobacter xylosoxidans* and *Leclercia adecarboxylata* increased shoot and root length for maize under drought stress, in comparison to uninoculated plants. It was suggested that ACC deaminase production decreased ethylene levels and led to increased plant growth [32].

2.1.5. Volatile Organic Compound (VOC) Production

PGPR produce volatile organic compounds (VOC) that induce stomatal closure, conserving water, and mitigating drought stress [33–35]. PGPR production of VOCs has been found to promote plant growth. Two PGPR strains produced VOC 2,3-butanediol and acetoin and increased growth in *Arabidopsis* in comparison to PGPR mutants that did not produce 2,3-butanediol and acetoin [35]. In drought conditions, PGPR *Pseudomonas chlororaphis* produced VOC 2,3-butanediol, which regulated stomatal closure and improved drought tolerance in *Arabidopsis* [33]. The VOC 2,3-butanediol, produced by PGPR, also induces nitric oxide and hydrogen peroxide production in plants. Nitric oxide and hydrogen peroxide both influence ABA signaling to close stomata and improve drought tolerance when plants are exposed to drought stress [34].

2.1.6. Extracellular Polymeric Substance (EPS) Production

Extracellular polymeric substances (EPS) are biopolymers that form the biofilm matrix [36]. EPS increase soil water retention and decrease soil drying rate [37]. For plants exposed to drought stress, EPS produced by PGPR help to create hydrophilic biofilms at the rhizosphere, improving drought tolerance [38]. When PGPR-inoculated maize was exposed to water stress, *Bacillus* spp. produced more EPS under drought conditions, improving water and nutrient uptake and increasing biomass in comparison to uninoculated plants [22]. PGPR production of EPS also mitigated drought stress for grapevine [38]. However, previous studies did not investigate EPS effect on plant physiology [39].

2.1.7. Summary

PGPR mitigate drought stress via several mechanisms. PGPR produce the osmolyte proline and regulate antioxidant enzyme activity, both of which act as ROS scavengers and neutralize ROS [19,24,25]. PGPR increase ABA production and production of VOCs to decrease water loss via stomata [26,34]. ACC deaminase production by PGPR reduces ethylene production to maintain plant growth under drought conditions [20,30]. Moreover, EPS production by PGPR produces hydrophilic biofilms surrounding plant roots, which improves drought tolerance [38]. Although several mechanisms of drought stress mitigation by PGPR have been identified, more efforts are needed to understand other mechanisms of improved drought tolerance by PGPR application. Examples of PGPR plant-growth promotion under drought stress are summarized in Table 1.

Table 1. Role of PGPR in plant growth enhancement and drought stress mitigation under drought stress.

PGPR	Plant	Response	Reference
Bacillus cereus AR156, *Bacillus subtilis* SM21, and *Serratia* sp. XY21	Cucumber	Increased leaf proline and chlorophyll content, decreased wilt symptoms	[21]
Bacillus spp.	Maize	Increased proline content, increased water and nutrient uptake, increased biomass	[22]
Enterobacter sp., *Bacillus thuringiensis*, *Bacillus* sp., and *Bacillus megaterium*	Shrub species	Improved nutrition, improved morphological traits, increased proline production, increased ACC deaminase production	[23]
Bacillus megaterium	Wheat	Increased proline content, relative water content, protein content, and chlorophyll a, b, carotenoids, and antioxidant enzyme activity	[24]
Bacillus altitudinis FD48 and *Bacillus methylotrophicus* RABA6	Rice	Increased antioxidant enzyme activity, mitigated drought stress	[25]
Azospirillum brasilense Sp 245	*Arabidopsis*	Increased ABA production, plant biomass, and seed yield	[27]
Bacillus licheniformis Rt4M10 and *Pseudomonas fluorescens* Rt6M10	Grapevine	Increased ABA production, water content, and turgidity	[28]
Bacillus licheniformis	Pepper	Increased ACC deaminase production and plant growth	[31]

Table 1. *Cont.*

PGPR	Plant	Response	Reference
Pseudomonas aeruginosa, Enterobacter cloacae, Achromobacter xylosoxidans, Leclercia adecarboxylata	Maize	Increased ACC deaminase production, shoot and root length	[32]
Bacillus subtilis GB03, *Bacillus amyloliquefaciens* IN937a	*Arabidopsis*	VOC 2,3-butanediol and acetoin production, increased plant growth	[35]
Pseudomonas chlororaphis	*Arabidopsis*	VOC 2,3-butanediol production, improved drought tolerance	[33]
Pseudomonas sp. S1, *Acinetobacter* sp. S2, *Pseudomonas* sp. S3, *Bacillus* sp. S4, *Delftia* sp. S5 and *Sphingobacterium* sp. S6	Grapevine	Increased plant growth, mitigated drought stress	[38]

2.2. *PGPR and Plant Defense Response*

In addition to improving plant abiotic stress resistance, PGPR inoculants enhance the ability of plants to resist stress from pathogens. Plants release root exudates that are metabolized by PGPR, while PGPR produce compounds to trigger ISR, and produce antifungal and antibacterial secondary metabolites that enhance pathogen resistance. When PGPR induce systemic resistance, parts of the plant not previously exposed to a given pathogen are protected from future infection via plant signaling pathways that mediate pathogen resistance [18]. When avocado (*Persea americana*) was inoculated with *B. subtilis*, the PGPR produced antifungal secondary metabolites that improved resistance to fungal pathogens. More *B. subtilis* was present on the rhizosphere of healthy avocado compared to avocado showing signs of fungal pathogen infection, implying that the PGPR presence improved fungal pathogen resistance [40]. PGPR can slow disease progression and prevent disease symptoms. Inoculant *B. subtilis* slowed powdery mildew disease progression in the beginning stages of development for strawberry (*Fragaria vesca* 'Elvira') [41]. Additionally, *B. subtilis* inoculation suppressed the *Rhizoctonia solani* pathogen and prevented wilting, a disease symptom, of cotton (*Gossypium hirsutum*) [42]. While previous studies have shown PGPR ability to slow and prevent disease, they did not fully investigate the mechanisms of pathogen resistance by PGPR [40].

2.3. *PGPR Mitigating Plant Drought and Pathogen Stress*

Since previous studies have identified PGPR mechanisms of pathogen and drought stress mitigation, it is thought that PGPR inoculation can improve abiotic- and biotic-stress resistance when plants are simultaneously exposed to drought stress and pathogen stress [43–45]. Additionally, PGPR application can indirectly improve abiotic- and biotic-stress resistance by increasing plant nutrient uptake and can further mitigate stress from simultaneous drought and pathogen stress [44,45]. However, few studies have been performed regarding PGPR mitigation of simultaneous drought and pathogen stress [44,45]. For bean (*Phaseolus vulgaris*), tomato (*Solanum lycopersicum*), and zucchini (*Cucurbita pepo* cv. Xara) inoculated with *Bacillus amyloliquefaciens* and under drought and pathogen stress, the PGPR induced an ISR response. Drought stress did not affect the ability of the PGPR to improve pathogen resistance [44]. Also, for rice under drought stress, PGPR *P. fluorescens* inoculation increased plant defense against six different pathogens by increasing abiotic- and biotic-defense enzyme production [45]. Although a few studies have previously demonstrated simultaneous pathogen and drought stress mitigation by PGPR, there is a lack of studies investigating species-specific interactions and mechanisms responsible for stress mitigation under dual abiotic and biotic stress conditions.

3. Arbuscular Mycorrhizal Fungi

Arbuscular mycorrhizal fungi are a type of endomycorrhizal fungi that colonize plant roots by entering the plant cell wall and moving into the plant cell membrane [46]. AMF form a mutually beneficial association with plants at the root level. AMF colonization enhances plant nutrient and water uptake, growth, and disease resistance. Meanwhile, AMF attain organic carbon from the plant, for their growth [47]. Mechanisms of plant growth promotion by AMF include hyphae improvement of nutrient and water uptake by plants, and production of glomalin to improve soil aggregation and nutrient availability. AMF also improve the plant defense response to pathogens. While some mechanisms of AMF plant growth promotion have been identified, more efforts are needed to understand enhancement of plant growth by AMF colonization [46]. AMF also improve plant resistance to abiotic stresses, including drought stress [47]. Previous studies found that the species AMF, associated with a certain plant species, influences the potential benefit of AMF colonization. However, studies identifying the ideal combination of AMF and plant species for enhanced plant benefit are yet to be performed [48].

3.1. AMF Mechanisms of Mitigating Plant Drought Stress

3.1.1. Hyphae Improve Plant Water Uptake

AMF hyphal networks increase surface area for water absorption to improve plant water uptake, which aids plants in drought resistance [49]. For trifoliate orange (*Poncirus trifoliata*) inoculated with *Funneliformis mosseae* and *Paraglomus occultum*, the water absorption rate of hyphae increased under drought conditions, indicating the significance of hyphal water uptake for plants under drought stress [50]. Moreover, *F. mosseae* and *Glomus constrictum* inoculation onto *Sophora davidii* under drought stress improved plant biomass, root length, and water use efficiency. The improved water use efficiency was attributed to AMF hyphae increasing plant water uptake [51].

3.1.2. Increased Reactive Oxygen Species (Hydrogen Peroxide) Efflux

AMF inoculation mitigates drought stress by increasing the efflux of hydrogen peroxide out of plant roots. Hydrogen peroxide presence, a reactive oxygen species, otherwise damages plants, via oxidative burst, when there is low hydrogen peroxide efflux. Two different studies of trifoliate orange under drought stress found that inoculation with *F. mosseae* enhanced drought stress tolerance by increasing hydrogen peroxide root efflux, resulting in increased biomass and plant growth promotion [52,53].

3.1.3. Increased Osmolyte Production

Similar to PGPR, AMF increase osmolyte levels that reduce oxidative stress from ROS during drought conditions. For shrubby horsetail (*Ephedra foliate*) under drought conditions, AMF *Glomus etunicatum*, *Rhizophagus irregularis*, and *F. mosseae* increased osmolyte production, mitigating oxidative damage from drought stress. The inoculated plants, as a result, had increased plant growth [54]. Also, when pistachio (*Pistacia vera*) was exposed to drought conditions, AMF *G. etunicatum* increased plant proline content, reducing oxidative damage and improving plant growth [55].

3.1.4. Increased Antioxidant Enzyme Production

AMF also increase antioxidant enzyme activity, similar to PGPR, to neutralize ROS and improve drought tolerance. *Rhizophagus intraradices* and *F. mosseae* AMF inoculants increased antioxidant enzyme activity for caucasian hackberry (*Celtis caucasica*) under drought conditions. The increased enzyme activity decreased hydrogen peroxide presence and correlated with an increase in plant growth [56]. *G. etunicatum* inoculant enhanced antioxidant enzymes of pistachio under drought conditions, increasing plant growth [55]. Additionally, AMF *G. etunicatum*, *R. irregularis*, and *F. mosseae* increased the activity of SOD, CAT, APX, and GR antioxidant enzymes for shrubby horsetail under drought stress, increasing plant growth [54].

3.1.5. Regulation of Aquaporins to Control Water Movement

Aquaporins are proteins that control water movement in cells. AMF regulate aquaporin activity in plants and in the fungus itself, to control water movement and improve drought tolerance [57,58]. The exposure to drought increased the expression of GintAQPF1 and GintAQPF2 aquaporin genes in *R. irregularis*, improving water transport to maize plants and increasing drought tolerance [58]. AMF *R. intraradices* inoculation also regulated aquaporin genes for maize under drought conditions, increasing plant growth [57]).

3.1.6. Summary

AMF reduce drought stress via multiple mechanisms. AMF production of hyphae increases the surface area for water and nutrient uptake [49]. AMF reduce oxidative stress by increasing hydrogen peroxide root efflux [52,53]. AMF also mitigate oxidative stress by producing osmolytes and increasing antioxidant enzyme activity to neutralize ROS during drought conditions [54,55]. Additionally, AMF regulate aquaporins to control water movement and reduce drought stress [57,58]. Studies identifying additional mechanisms of AMF drought stress mitigation are lacking. Examples of AMF plant growth promotion under drought conditions are summarized in Table 2.

Table 2. Role of AMF in plant growth enhancement and drought stress mitigation under drought stress.

AMF	Plant	Response	Reference
Funneliformis mosseae, Paraglomus occultum	Trifoliate orange	Increased water absorption rate via AMF hyphae	[50]
Funneliformis mosseae, Glomus constrictum	*Sophora davidii*	Increased plan biomass, root length, and water use efficiency via AMF hyphae	[51]
Funneliformis mosseae	Trifoliate orange	Increased hydrogen peroxide root efflux, biomass, and plant growth promotion	[52,53]
Glomus etunicatum, Rhizophagus irregularis, Funneliformis mosseae	Shrubby horsetail	Increased osmolyte production, antioxidant enzyme activity, and plant growth	[54]
Glomus etunicatum	Pistachio	Increased proline content, antioxidant enzyme production, and plant growth	[55]
Rhizophagus intraradices, Funneliformis mosseae	Caucasian hackberry	Increased antioxidant enzyme activity and plant growth	[56]
Rhizophagus irregularis	Maize	Increased expression of aquaporin genes, improved water transport	[58]
Rhizophagus intraradices	Maize	Improved aquaporin regulation, increased plant growth	[57]

3.2. *AMF and Plant Defense Response*

In addition to mitigating drought stress, AMF increase the plant defense response to pathogens. Improved plant nutrient uptake by AMF symbiosis enhances the ability of plants to resist disease. *Glomus* sp. colonization of tomato increased nitrogen, phosphorus, and potassium content and improved disease resistance to *Fusarium oxysporum* f. sp. *lycopersici* [59]. AMF *R. irregularis* provided nitrogen and carbon to inoculated soybean (*Glycine max*) infected with *Phytophthora sojae*, influencing the production of nitric oxide (NO), which signals pathogen resistance. The increased nitrogen and carbon uptake also enhanced the plant defense response [60]. AMF colonization on diseased plants also influences defense-related gene expression, mitigating disease. *Rhizoglomus irregulare* colonization upregulated defense-related gene expression in tomato and decreased tomato bushy stunt virus and tomato mosaic pathogen infection in young leaves [61]. *R. irregularis* increased jasmonic acid levels in soybean, which is involved in plant defense gene expression, mitigating *P. sojae* infection [60]. AMF colonization additionally enhances plant pathogen resistance by decreasing oxidative stress that occurs when ROS are produced during pathogen infection. *R. irregularis* decreased hydrogen peroxide buildup during *P. sojae* infection by increasing antioxidant enzyme activity, to improve pathogen resistance for soybean [60]. AMF *Glomus fasciculatum* colonization also decreased ROS presence in tomato, to enhance resistance to *F. oxysporum* f. sp. *lycopersici* [62]. While previous studies determined AMF plant pathogen mitigation, they did not deeply investigate AMF interactions with the host plant that resulted in improved pathogen resistance [60].

3.3. AMFs Improve Drought and Pathogen Resistance

AMF mitigation of abiotic and biotic stress similarly benefits plants affected by both drought stress and pathogen stress, simultaneously. Cowpea (*Vigna unguiculata*) infected with *Macrophomina phaseolina* (charcoal rot disease) and inoculated with *Glomus deserticola* and *Gigaspora gigantea* achieved normal growth under drought conditions [63]. AMF *Glomus monosporus* and *Glomus clarum* also reduced plant death and enhanced plant growth for date palm (*Phoenix dactylifera*) infected with *F. oxysporum* and exposed to drought stress [64]. Moreover, *Verticillium dahliae* pathogen had an especially negative effect on pepper under drought conditions, when not inoculated with *G. deserticola*, indicating the significance of AMF inoculation in mitigating abiotic and biotic stress [65]. However, few studies have previously investigated AMF reduction of simultaneous drought and pathogen stress and the mechanisms involved in abiotic and biotic stress reduction by AMF [63–65].

4. AMF and PGPR Dual Inoculation

4.1. Interactions of AMF and PGPR during Dual Inoculation

Although negative interactions are possible, dual inoculations with both PGPR and AMF most often have increased benefits for plants in comparison to singular or no inoculation. PGPR enhance plant growth when co-inoculated with AMF by stimulating AMF colonization, further benefiting the plant [6–8]. The inoculation of PGPR *B. subtilis* increased AMF *F. mosseae* colonization for rose-scented geranium (*Pelargonium graveolens*), increasing shoot biomass and nutrient uptake in comparison to singular or no inoculation [6]. *B. subtilis* also enhanced AMF *R. irregularis* colonization for onion (*Allium cepa*), increasing plant biomass and nutrient acquisition [7]. Additionally, *Bacillus amyloliquefaciens* increased the number of arbuscules and the glomalin production of AMF *Rhizophagus intraradices* when dually inoculated on winter wheat, indicating an increase in AMF biomass [8]. Notably, a meta-analysis found that although PGPR and AMF dual inoculation did have enhanced benefits, PGPR inoculation did not always increase AMF colonization, suggesting that other mechanisms may be responsible for the enhanced benefits of dual inoculation [15]. Few studies have investigated the interactions of specific AMF and PGPR species and the role of nutrients in plant–AMF–PGPR relationships [14,15]. Additionally, the influence of application timing on enhanced benefits of PGPR and AMF dual inoculations has yet to be studied.

4.2. Dual AMF and PGPR Inoculation Mitigates Plant Drought Stress

Dual inoculation of AMF and PGPR has been found to enhance drought stress tolerance for plants, compared to singular or no inoculation [9–11,13]. PGPR inoculation improves AMF growth and enhances the benefits of AMF inoculation for plants exposed to drought conditions. *Bacillus* sp. inoculation enhanced AMF *F. mosseae* and *R. irregularis* growth, mitigating drought stress for lettuce (*Lactuca sativa*) [13]. Dual inoculation also improves nutrient uptake, water uptake and photosynthetic activity, and reduces oxidative stress [9–11]. Co-inoculation of AMF, *Glomus versiforme*, and PGPR, *Bacillus methylotrophicus*, increased nitrogen, phosphorus, and potassium uptake and decreased oxidative stress by improving antioxidant enzyme activity for tobacco (*Nicotiana tabacum*). Dual inoculation also improved plant biomass and height [10]. Dual AMF and PGPR inoculation applied to date palm also improved nitrogen and phosphorus nutrient uptake and reduced oxidative stress. The dual inoculation additionally improved stomatal conductance, plant growth, and morphological traits, in comparison to singular or no inoculation. It was suggested that increased water and nutrient absorption by AMF hyphae improved plant nutrient uptake [9]. Dual inoculation of *R. irregularis* and *Bacillus thuringiensis* also increased water uptake for *Retama sphaerocarpa*, due to AMF hyphal water absorption [11]. Although previous studies found that dual AMF and PGPR mitigated drought stress, they did not investigate communication between plants, AMF, and PGPR at the molecular level, during drought conditions [12]. A summary of the mechanisms PGPR and AMF use independently to mitigate plant drought stress, and the benefits of dual PGPR and AMF applications for drought stress mitigation are shown in Figure 1.

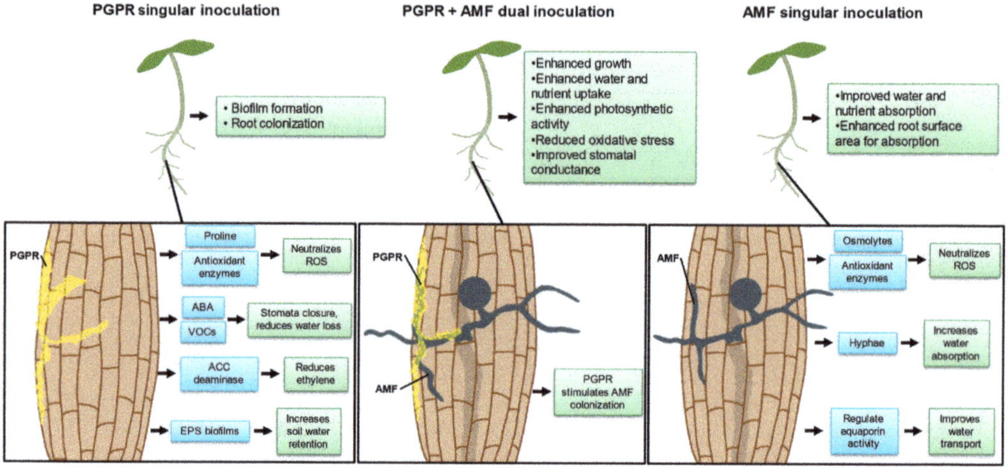

Figure 1. PGPR–AMF–plant interactions under drought stress. Schematic showing physiological and beneficial traits in plants inoculated with PGPR, AMF and a dual inoculation of PGPR and AMF.

4.3. AMF and PGPR Dual Inoculation Improves Pathogen Resistance

AMF and PGPR can provide enhanced benefits to plants under biotic stress when dually inoculated [12]. AMF and PGPR can both antagonistically toward plant pathogens and induce systemic resistance to improve plant defense [41]. AMF can compete with pathogens for resources to mitigate the negative impacts of the pathogen. AMF *Glomus* spp. increased competition with the white rot fungi pathogen (*Sclerotium rolfsii*), which improved pathogen defense for the common bean [66]. PGPR can also produce antipathogenic compounds to deter pathogens. PGPR *B. subtilis* produced antipathogenic compounds that improved white rot fungi disease resistance for the common bean [66]. The co-inoculation of AMF and PGPR can also enhance the activity of enzymes involved in pathogen defense more than using singular or no inoculation. Dual inoculation of *Glomus* spp. and *B. subtilis* enhanced peroxidase and polyphenyl oxidase (PPO) enzyme activity, which are both involved in plant pathogen defense [66]. Dual inoculation can also increase phosphorus uptake to improve plant growth during pathogen infection [66]. PGPR inoculation can increase the ability of AMF to improve phosphorus uptake for plants, by providing phosphorus in plant-available forms [67]. Although PGPR have been found to increase AMF colonization during dual inoculation to improve colonization benefits [6–8], PGPR inoculation does not always stimulate AMF colonization [66]. Also, when *R. irregularis* and *B. subtilis* were dually inoculated to enhance *Aspergillus niger* pathogen defense for lettuce, an increase in AMF colonization did not correlate with increased pathogen defense [67]. A lack of correlation indicates that other mechanisms enhance pathogen resistance during dual inoculation, although previous studies have not investigated how microbe–microbe interactions improve pathogen resistance and the benefit of dual AMF-PGPR application in agriculture [68]. Mechanisms of pathogen stress mitigation via individual PGPR and AMF applications, and the benefit of dual PGPR and AMF application in improving the plant pathogen defense response are summarized in Figure 2. Representative images of singular and dual PGPR and AMF root colonization are included in Figure 3.

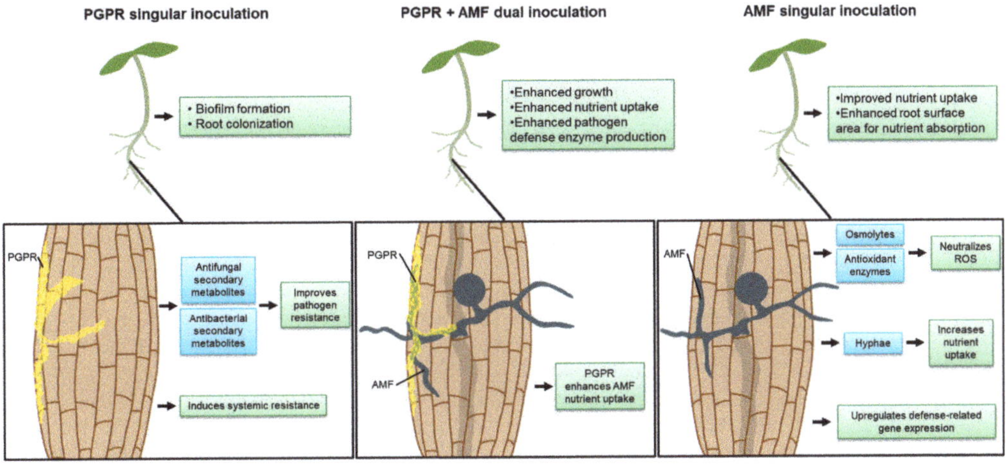

Figure 2. PGPR–AMF–plant interactions under biotic stress. Schematic showing physiological and beneficial traits in plants inoculated with PGPR and AMF and a dual inoculation of PGPR and AMF.

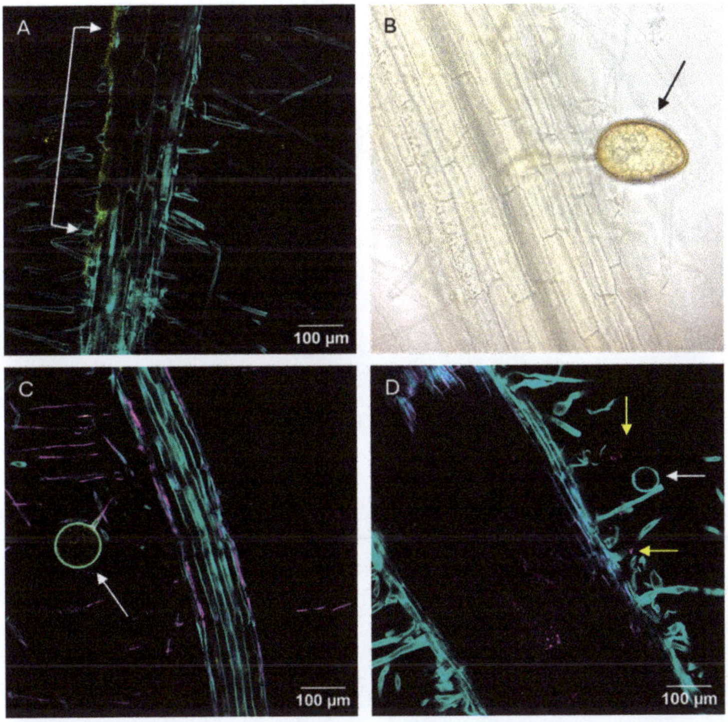

Figure 3. Representative micrographs depicting the single and dual colonization of PGPR (*B. subtilis*) and AMF (*R. irregularis*) on tomato roots. The arrow in (**A**) shows a fluorescent microscopy image of

B. subtilis root plane colonization. PGPR was stained with SYTO13 (yellow) fluorescent dye. (**B**) shows a phase image of an AMF spore on tomato root plane. (**C**) shows a fluorescent image of the single AMF colonization on tomato roots. AMF were stained with wheat germ agglutinin 594 (magenta) and calcofluor white (cyan) fluorescent dyes. (**D**) shows the dual inoculation of AMF and PGPR on tomato roots. The white arrow indicates an AMF spore, and yellow arrows show PGPR colonization on tomato roots.

5. Commercial AMF and PGPR Inoculums in the Agricultural Sector

Microorganisms were first applied agriculturally for plant benefit at the start of the 20th century, when rhizobial bacteria were applied to legume crops [69]. Availability of commercial PGPR and AMF inoculums arose more recently, in the 21st century [69]. Commercial PGPR products mainly include *Bacillus* and *Pseudomonas* bacterial strains [69]. Commercial AMF inoculums currently only include a select number of AMF strains in the *Glomeraceae* family [70]. Some of these products include AMF consortia of more than one AMF species [70]. However, products including multiple AMF strains may only enhance plant benefits if the strains included perform different beneficial effects [70]. While dual application can enhance plant benefits, few commercial products include both AMF and PGPR species [70]. Limitations to the commercial success of biological agents include short shelf life, variable benefits to plant growth, and a lack of understanding of how microbial inoculums may alter the native soil microbiota [46].

6. Conclusions

AMF and PGPR have several mechanisms for mitigating drought stress in plants. PGPR increase drought tolerance by increasing ABA production and producing VOCs to control stomata closure and decrease water loss. PGPR production of ACC deaminase also decreases ethylene production, to promote plant growth during drought conditions. PGPR also produce EPS to form hydrophilic biofilms and reduce drought stress. AMF colonization regulates aquaporin activity, to control water movement during drought conditions. AMF hyphae also increase plant water uptake. Both AMF and PGPR also reduce oxidative damage during drought by increasing osmolyte levels and increasing antioxidant enzyme activity. AMF also mitigate ROS damage by increasing the rate of hydrogen peroxide root efflux.

AMF and PGPR also promote pathogen defense. PGPR produce antipathogenic compounds and secondary metabolites, to reduce pathogen infection. PGPR additionally increase pathogen defense-related enzyme activity. PGPR also induce systemic resistance in plants. AMF promote plant health during pathogen infection by improving nutrient uptake. AMF also regulate defense-related gene expression. Additionally, AMF reduce oxidative stress during pathogen infection by increasing antioxidant enzyme activity. However, there has been little investigation into understanding the mechanisms of both PGPR and AMF in pathogen resistance.

PGPR and AMF dual inoculations promote abiotic- and biotic-stress tolerance in agriculture. Although previous studies found that dual inoculations increase water and nutrient uptake, enhance photosynthetic activity, and decrease oxidative stress in drought conditions, they did not fully investigate molecular-level communications of PGPR, AMF, and plants under drought stress. Dual inoculations also provide enhanced benefits under biotic stress. AMF compete with pathogens for resources to mitigate pathogen stress, while PGPR produce antipathogenic compounds to deter pathogens. Dual inoculation also increases phosphorus uptake, to improve plant health during pathogen infection. Additionally, co-inoculations enhance defense-related enzyme activity. D inoculation enhancement of ISR has been understudied.

PGPR can stimulate AMF colonization, to provide enhanced benefits during dual inoculation, although PGPR inoculation does not always result in increased AMF colonization, including during pathogen infection. Additionally, an increase in AMF colonization does not consistently correlate with improved pathogen defense, indicating that other mechanisms may enhance the benefits of dual AMF and PGPR associations. Though the enhanced plant benefits from AMF and PGPR dual inoculation have been studied previously, few studies identify mechanisms resulting in enhanced benefits.

There are several unknowns regarding interactions between PGPR, AMF, and plants. Mechanisms of pathogen defense, plant growth promotion, and the influence of species specificity are not well understood for singular PGPR or AMF inoculations. Additionally, the influence of species specificity, application timing, and nutrient availability on the enhanced benefits of dual AMF and PGPR inoculations is not well understood. More efforts are needed to understand the influence of dual inoculations on plants under abiotic and biotic stress, specifically molecular interactions under drought conditions and interactions that occur between microbes and plants, to improve pathogen resistance.

Author Contributions: Conceptualization, N.S. and H.B.; methodology, N.S. and H.B.; investigation, N.S. and H.B.; writing—original draft preparation, N.S. and H.B.; writing—review and editing, N.S. and H.B.; supervision, H.B. All authors have read and agreed to the published version of the manuscript.

Funding: The graduate assistantship to NS was funded through the UD-CANR support. The research was supported by the USDA-HATCH grant to HB. Microscopy was funded by grants from the NIH-NIGMS (P20 GM103446), the NIGMS (P20 GM139760) and the state of Delaware. Microscopy equipment was acquired with NIST (70NANB21H085).

Data Availability Statement: No new data were created or analyzed in this study. Data sharing is not applicable to this article. (Where no new data were created, or where data is unavailable due to privacy or ethical restrictions, a statement is still required. Please check if this statement is OK. Suggested Data Availability Statements are available in section "MDPI Research Data Policies" at https://www.mdpi.com/ethics.)

Acknowledgments: This manuscript is part of the thesis "The influence of plant growth-promoting rhizobacteria and arbuscular mycorrhizal fungi dual inoculation on tomato plants against abiotic stress regime" submitted to the University of Delaware by Noah Savastano for the fulfillment of a master's degree requirement. Microscopy access in the Bioimaging Facility at the Delaware Biotechnology Institute was supported by grants from the NIH-NIGMS (P20 GM103446), the NIGMS (P20 GM139760) and the state of Delaware. We thank Jeffrey Caplan and Tim Chaya in the Bio-Imaging Center for assistance with fluorescent microscopy imaging.

Conflicts of Interest: The authors declare no conflicts of interest.

References

1. Chakraborty, S.; Newton, A.C. Climate Change, Plant Diseases and Food Security: An Overview. *Plant Pathol.* **2011**, *60*, 2–14. [CrossRef]
2. Oshunsanya, S.O.; Nwosu, N.J.; Li, Y. Abiotic Stress in Agricultural Crops Under Climatic Conditions. In *Sustainable Agriculture, Forest and Environmental Management*; Jhariya, M.K., Banerjee, A., Meena, R.S., Yadav, D.K., Eds.; Springer: Singapore, 2019; pp. 71–100. ISBN 978-981-13-6830-1.
3. Usta, C. Microorganisms in Biological Pest Control—A Review (Bacterial Toxin Application and Effect of Environmental Factors). In *Current Progress in Biological Research*; Silva-Opps, M., Ed.; InTech: London, UK, 2013; ISBN 978-953-51-1097-2.
4. de Faria, M.R.; Costa, L.S.A.S.; Chiaramonte, J.B.; Bettiol, W.; Mendes, R. The Rhizosphere Microbiome: Functions, Dynamics, and Role in Plant Protection. *Trop. Plant Pathol.* **2021**, *46*, 13–25. [CrossRef]
5. Vidal, C.; González, F.; Santander, C.; Pérez, R.; Gallardo, V.; Santos, C.; Aponte, H.; Ruiz, A.; Cornejo, P. Management of Rhizosphere Microbiota and Plant Production under Drought Stress: A Comprehensive Review. *Plants* **2022**, *11*, 2437. [CrossRef] [PubMed]
6. Alam, M.; Khaliq, A.; Sattar, A.; Shukla, R.S.; Anwar, M.; Dharni, S. Synergistic Effect of Arbuscular Mycorrhizal Fungi and Bacillus Subtilis on the Biomass and Essential Oil Yield of Rose-Scented Geranium (*Pelargonium Graveolens*). *Arch. Agron. Soil Sci.* **2011**, *57*, 889–898. [CrossRef]
7. Toro, M.; Azcon, R.; Barea, J. Improvement of Arbuscular Mycorrhiza Development by Inoculation of Soil with Phosphate-Solubilizing Rhizobacteria To Improve Rock Phosphate Bioavailability ((Sup32)P) and Nutrient Cycling. *Appl. Environ. Microbiol.* **1997**, *63*, 4408–4412. [CrossRef]
8. Wilkes, T.I.; Warner, D.J.; Edmonds-Brown, V.; Davies, K.G. Species-Specific Interactions of Bacillus Innocula and Arbuscular Mycorrhizal Fungi Symbiosis with Winter Wheat. *Microorganisms* **2020**, *8*, 1795. [CrossRef]

9. Anli, M.; Baslam, M.; Tahiri, A.; Raklami, A.; Symanczik, S.; Boutasknit, A.; Ait-El-Mokhtar, M.; Ben-Laouane, R.; Toubali, S.; Ait Rahou, Y.; et al. Biofertilizers as Strategies to Improve Photosynthetic Apparatus, Growth, and Drought Stress Tolerance in the Date Palm. *Front. Plant Sci.* **2020**, *11*, 516818. [CrossRef]
10. Begum, N.; Wang, L.; Ahmad, H.; Akhtar, K.; Roy, R.; Khan, M.I.; Zhao, T. Co-Inoculation of Arbuscular Mycorrhizal Fungi and the Plant Growth-Promoting Rhizobacteria Improve Growth and Photosynthesis in Tobacco Under Drought Stress by Up-Regulating Antioxidant and Mineral Nutrition Metabolism. *Microb. Ecol.* **2022**, *83*, 971–988. [CrossRef]
11. Marulanda, A.; Barea, J.M.; Azcón, R. An Indigenous Drought-Tolerant Strain of Glomus Intraradices Associated with a Native Bacterium Improves Water Transport and Root Development in Retama Sphaerocarpa. *Microb. Ecol.* **2006**, *52*, 670–678. [CrossRef]
12. Nanjundappa, A.; Bagyaraj, D.J.; Saxena, A.K.; Kumar, M.; Chakdar, H. Interaction between Arbuscular Mycorrhizal Fungi and *Bacillus* spp. in Soil Enhancing Growth of Crop Plants. *Fungal Biol. Biotechnol.* **2019**, *6*, 23. [CrossRef]
13. Vivas, A.; Marulanda, A.; Ruiz-Lozano, J.M.; Barea, J.M.; Azcón, R. Influence of a *Bacillus* sp. on Physiological Activities of Two Arbuscular Mycorrhizal Fungi and on Plant Responses to PEG-Induced Drought Stress. *Mycorrhiza* **2003**, *13*, 249–256. [CrossRef] [PubMed]
14. Gopal, S.; Chandrasekaran, M.; Shagol, C.; Kim, K.-Y.; Sa, T.-M. Spore Associated Bacteria (SAB) of Arbuscular Mycorrhizal Fungi (AMF) and Plant Growth Promoting Rhizobacteria (PGPR) Increase Nutrient Uptake and Plant Growth Under Stress Conditions. *Korean J. Soil Sci. Fertil.* **2012**, *45*, 582–592. [CrossRef]
15. Primieri, S.; Magnoli, S.M.; Koffel, T.; Stürmer, S.L.; Bever, J.D. Perennial, but Not Annual Legumes Synergistically Benefit from Infection with Arbuscular Mycorrhizal Fungi and Rhizobia: A Meta-analysis. *New Phytol.* **2022**, *233*, 505–514. [CrossRef] [PubMed]
16. Katiyar, D. Plant Growth Promoting Rhizobacteria-an Efficient Tool for Agriculture Promotion. *Adv. Plants Agric. Res.* **2016**, *4*, 426–434. [CrossRef]
17. Kloepper, J.W.; Schroth, M.N. Plant growth-promoting rhizobacteria on radishes. In Proceedings of the Station de Pathologie, 4th International Conference on Plant Pathogenic Bacteria, Tours, France, 27 August–2 September 1978; Végétale et Phyto-Bactériologie, Ed.; pp. 879–882.
18. Borriss, R. Use of Plant-Associated Bacillus Strains as Biofertilizers and Biocontrol Agents in Agriculture. In *Bacteria in Agrobiology: Plant Growth Responses*; Maheshwari, D.K., Ed.; Springer: Berlin/Heidelberg, Germany, 2011; pp. 41–76. ISBN 978-3-642-20331-2.
19. Das, K.; Roychoudhury, A. Reactive Oxygen Species (ROS) and Response of Antioxidants as ROS-Scavengers during Environmental Stress in Plants. *Front. Environ. Sci.* **2014**, *2*, 53. [CrossRef]
20. Sati, D.; Pande, V.; Pandey, S.C.; Samant, M. Recent Advances in PGPR and Molecular Mechanisms Involved in Drought Stress Resistance. *J. Soil Sci. Plant Nutr.* **2023**, *23*, 106–124. [CrossRef]
21. Wang, C.-J.; Yang, W.; Wang, C.; Gu, C.; Niu, D.-D.; Liu, H.-X.; Wang, Y.-P.; Guo, J.-H. Induction of Drought Tolerance in Cucumber Plants by a Consortium of Three Plant Growth-Promoting Rhizobacterium Strains. *PLoS ONE* **2012**, *7*, e52565. [CrossRef]
22. Vardharajula, S.; Zulfikar Ali, S.; Grover, M.; Reddy, G.; Bandi, V. Drought-Tolerant Plant Growth Promoting *Bacillus* spp.: Effect on Growth, Osmolytes, and Antioxidant Status of Maize under Drought Stress. *J. Plant Interact.* **2011**, *6*, 1–14. [CrossRef]
23. Armada, E.; Barea, J.-M.; Castillo, P.; Roldán, A.; Azcón, R. Characterization and Management of Autochthonous Bacterial Strains from Semiarid Soils of Spain and Their Interactions with Fermented Agrowastes to Improve Drought Tolerance in Native Shrub Species. *Appl. Soil Ecol.* **2015**, *96*, 306–318. [CrossRef]
24. Rashid, U.; Yasmin, H.; Hassan, M.N.; Naz, R.; Nosheen, A.; Sajjad, M.; Ilyas, N.; Keyani, R.; Jabeen, Z.; Mumtaz, S.; et al. Drought-Tolerant Bacillus Megaterium Isolated from Semi-Arid Conditions Induces Systemic Tolerance of Wheat under Drought Conditions. *Plant Cell Rep.* **2022**, *41*, 549–569. [CrossRef]
25. Narayanasamy, S.; Thangappan, S.; Uthandi, S. Plant Growth-Promoting *Bacillus* sp. Cahoots Moisture Stress Alleviation in Rice Genotypes by Triggering Antioxidant Defense System. *Microbiol. Res.* **2020**, *239*, 126518. [CrossRef] [PubMed]
26. Schroeder, J.I.; Kwak, J.M.; Allen, G.J. Guard Cell Abscisic Acid Signalling and Engineering Drought Hardiness in Plants. *Nature* **2001**, *410*, 327–330. [CrossRef] [PubMed]
27. Cohen, A.C.; Bottini, R.; Pontin, M.; Berli, F.J.; Moreno, D.; Boccanlandro, H.; Travaglia, C.N.; Piccoli, P.N. *Azospirillum Brasilense* Ameliorates the Response of *Arabidopsis Thaliana* to Drought Mainly via Enhancement of ABA Levels. *Physiol. Plant.* **2015**, *153*, 79–90. [CrossRef] [PubMed]
28. Salomon, M.V.; Bottini, R.; De Souza Filho, G.A.; Cohen, A.C.; Moreno, D.; Gil, M.; Piccoli, P. Bacteria Isolated from Roots and Rhizosphere of *Vitis Vinifera* Retard Water Losses, Induce Abscisic Acid Accumulation and Synthesis of Defense-related Terpenes in in Vitro Cultured Grapevine. *Physiol. Plant.* **2014**, *151*, 359–374. [CrossRef]
29. Vandana, U.K.; Singha, B.; Gulzar, A.B.M.; Mazumder, P.B. Molecular Mechanisms in Plant Growth Promoting Bacteria (PGPR) to Resist Environmental Stress in Plants. In *Molecular Aspects of Plant Beneficial Microbes in Agriculture*; Elsevier: Amsterdam, The Netherlands, 2020; pp. 221–233. ISBN 978-0-12-818469-1.
30. Ravanbakhsh, M.; Sasidharan, R.; Voesenek, L.A.C.J.; Kowalchuk, G.A.; Jousset, A. Microbial Modulation of Plant Ethylene Signaling: Ecological and Evolutionary Consequences. *Microbiome* **2018**, *6*, 52. [CrossRef]
31. Lim, J.-H.; Kim, S.-D. Induction of Drought Stress Resistance by Multi-Functional PGPR Bacillus Licheniformis K11 in Pepper. *Plant Pathol. J.* **2013**, *29*, 201–208. [CrossRef]
32. Danish, S.; Zafar-Ul-Hye, M.; Hussain, S.; Riaz, M.; Qayyum, M.F. Mitigation of Drought Stress in Maize through Inoculation with Drought Tolerant ACC Deaminase Containing PGPR under Axenic Conditions. *Pak. J. Bot.* **2020**, *52*, 49–60. [CrossRef]

33. Cho, S.M.; Kang, B.R.; Han, S.H.; Anderson, A.J.; Park, J.-Y.; Lee, Y.-H.; Cho, B.H.; Yang, K.-Y.; Ryu, C.-M.; Kim, Y.C. 2R,3R-Butanediol, a Bacterial Volatile Produced by *Pseudomonas Chlororaphis* O6, Is Involved in Induction of Systemic Tolerance to Drought in *Arabidopsis Thaliana*. *Mol. Plant-Microbe Interact.* **2008**, *21*, 1067–1075. [CrossRef]
34. Cho, S.-M.; Kim, Y.H.; Anderson, A.J.; Kim, Y.C. Nitric Oxide and Hydrogen Peroxide Production Are Involved in Systemic Drought Tolerance Induced by 2R,3R-Butanediol in Arabidopsis Thaliana. *Plant Pathol. J.* **2013**, *29*, 427–434. [CrossRef]
35. Ryu, C.-M.; Farag, M.A.; Hu, C.-H.; Reddy, M.S.; Wei, H.-X.; Paré, P.W.; Kloepper, J.W. Bacterial Volatiles Promote Growth in *Arabidopsis*. *Proc. Natl. Acad. Sci. USA* **2003**, *100*, 4927–4932. [CrossRef]
36. Flemming, H.-C.; Wingender, J. The Biofilm Matrix. *Nat. Rev. Microbiol.* **2010**, *8*, 623–633. [CrossRef] [PubMed]
37. Roberson, E.B.; Firestone, M.K. Relationship between Desiccation and Exopolysaccharide Production in a Soil *Pseudomonas* sp. *Appl. Environ. Microbiol.* **1992**, *58*, 1284–1291. [CrossRef] [PubMed]
38. Rolli, E.; Marasco, R.; Vigani, G.; Ettoumi, B.; Mapelli, F.; Deangelis, M.L.; Gandolfi, C.; Casati, E.; Previtali, F.; Gerbino, R.; et al. Improved Plant Resistance to Drought Is Promoted by the Root-associated Microbiome as a Water Stress-dependent Trait. *Environ. Microbiol.* **2015**, *17*, 316–331. [CrossRef] [PubMed]
39. Morcillo, R.; Manzanera, M. The Effects of Plant-Associated Bacterial Exopolysaccharides on Plant Abiotic Stress Tolerance. *Metabolites* **2021**, *11*, 337. [CrossRef] [PubMed]
40. Cazorla, F.M.; Romero, D.; Pérez-García, A.; Lugtenberg, B.J.J.; Vicente, A.D.; Bloemberg, G. Isolation and Characterization of Antagonistic Bacillus Subtilis Strains from the Avocado Rhizoplane Displaying Biocontrol Activity: Characterization of Antagonistic Bacillus. *J. Appl. Microbiol.* **2007**, *103*, 1950–1959. [CrossRef]
41. Lowe, A.; Rafferty-McArdle, S.M.; Cassells, A.C. Effects of AMF- and PGPR-Root Inoculation and a Foliar Chitosan Spray in Single and Combined Treatments on Powdery Mildew Disease in Strawberry. *Agric. Food Sci.* **2012**, *21*, 28–38. [CrossRef]
42. Medeiros, F.H.V.; Souza, R.M.; Medeiros, F.C.L.; Zhang, H.; Wheeler, T.; Payton, P.; Ferro, H.M.; Paré, P.W. Transcriptional Profiling in Cotton Associated with Bacillus Subtilis (UFLA285) Induced Biotic-Stress Tolerance. *Plant Soil* **2011**, *347*, 327–337. [CrossRef]
43. Nadeem, S.M.; Ahmad, M.; Zahir, Z.A.; Javaid, A.; Ashraf, M. The Role of Mycorrhizae and Plant Growth Promoting Rhizobacteria (PGPR) in Improving Crop Productivity under Stressful Environments. *Biotechnol. Adv.* **2014**, *32*, 429–448. [CrossRef]
44. Pertot, I.; Puopolo, G.; Hosni, T.; Pedrotti, L.; Jourdan, E.; Ongena, M. Limited Impact of Abiotic Stress on Surfactin Production in Planta and on Disease Resistance Induced by *Bacillus Amyloliquefaciens* S499 in Tomato and Bean. *FEMS Microbiol. Ecol.* **2013**, *86*, 505–519. [CrossRef]
45. Prathuangwong, S.; Chuaboon, W.; Chatnaparat, T.; Kladsuwan, L.; Shoorin, M.; Kasem, S. Induction of Disease and Drought Resistance in Rice by Pseudomonas Fluorescens SP007s. *CMU J. Nat. Sci. Spec. Issue Agric. Nat. Res.* **2012**, *11*, 45–55.
46. Sagar, A.; Rathore, P.; Ramteke, P.W.; Ramakrishna, W.; Reddy, M.S.; Pecoraro, L. Plant Growth Promoting Rhizobacteria, Arbuscular Mycorrhizal Fungi and Their Synergistic Interactions to Counteract the Negative Effects of Saline Soil on Agriculture: Key Macromolecules and Mechanisms. *Microorganisms* **2021**, *9*, 1491. [CrossRef] [PubMed]
47. Kumar, A.; Dames, J.F.; Gupta, A.; Sharma, S.; Gilbert, J.A.; Ahmad, P. Current Developments in Arbuscular Mycorrhizal Fungi Research and Its Role in Salinity Stress Alleviation: A Biotechnological Perspective. *Crit. Rev. Biotechnol.* **2015**, *35*, 461–474. [CrossRef] [PubMed]
48. Hage-Ahmed, K.; Rosner, K.; Steinkellner, S. Arbuscular Mycorrhizal Fungi and Their Response to Pesticides. *Pest Manag. Sci.* **2019**, *75*, 583–590. [CrossRef] [PubMed]
49. Pagano, M.C. Drought Stress and Mycorrhizal Plant. In *Use of Microbes for the Alleviation of Soil Stresses, Volume 1*; Miransari, M., Ed.; Springer: New York, NY, USA, 2014; pp. 97–110. ISBN 978-1-4614-9465-2.
50. Zhang, F.; Zou, Y.-N.; Wu, Q.-S. Quantitative Estimation of Water Uptake by Mycorrhizal Extraradical Hyphae in Citrus under Drought Stress. *Sci. Hortic.* **2018**, *229*, 132–136. [CrossRef]
51. Gong, M.; Tang, M.; Chen, H.; Zhang, Q.; Feng, X. Effects of Two Glomus Species on the Growth and Physiological Performance of Sophora Davidii Seedlings under Water Stress. *New For.* **2013**, *44*, 399–408. [CrossRef]
52. Huang, Y.-M.; Zou, Y.-N.; Wu, Q.-S. Alleviation of Drought Stress by Mycorrhizas Is Related to Increased Root H_2O_2 Efflux in Trifoliate Orange. *Sci. Rep.* **2017**, *7*, 42335. [CrossRef]
53. Zou, Y.-N.; Huang, Y.-M.; Wu, Q.-S.; He, X.-H. Mycorrhiza-Induced Lower Oxidative Burst Is Related with Higher Antioxidant Enzyme Activities, Net H_2O_2 Effluxes, and Ca^{2+} Influxes in Trifoliate Orange Roots under Drought Stress. *Mycorrhiza* **2015**, *25*, 143–152. [CrossRef]
54. Al-Arjani, A.-B.F.; Hashem, A.; Abd_Allah, E.F. Arbuscular Mycorrhizal Fungi Modulates Dynamics Tolerance Expression to Mitigate Drought Stress in Ephedra Foliata Boiss. *Saudi J. Biol. Sci.* **2020**, *27*, 380–394. [CrossRef]
55. Abbaspour, H.; Saeidi-Sar, S.; Afshari, H.; Abdel-Wahhab, M.A. Tolerance of Mycorrhiza Infected Pistachio (*Pistacia vera* L.) Seedling to Drought Stress under Glasshouse Conditions. *J. Plant Physiol.* **2012**, *169*, 704–709. [CrossRef]
56. Sepahvand, T.; Etemad, V.; Matinizade, M.; Shirvany, A. Symbiosis of AMF with Growth Modulation and Antioxidant Capacity of Caucasian Hackberry (*Celtis caucasica* L.) Seedlings under Drought Stress. *Cent. Asian J. Environ. Sci. Technol. Innov.* **2021**, *2*, 20–35. [CrossRef]
57. Bárzana, G.; Aroca, R.; Bienert, G.P.; Chaumont, F.; Ruiz-Lozano, J.M. New Insights into the Regulation of Aquaporins by the Arbuscular Mycorrhizal Symbiosis in Maize Plants Under Drought Stress and Possible Implications for Plant Performance. *Mol. Plant Microbe Interact.* **2014**, *27*, 349–363. [CrossRef] [PubMed]

58. Li, T.; Hu, Y.; Hao, Z.; Li, H.; Wang, Y.; Chen, B. First Cloning and Characterization of Two Functional Aquaporin Genes from an Arbuscular Mycorrhizal Fungus *Glomus Intraradices*. *New Phytol.* **2013**, *197*, 617–630. [CrossRef] [PubMed]
59. Kumari, S.M.P.; Prabina, B.J. Protection of Tomato, Lycopersicon Esculentum from Wilt Pathogen, *Fusarium Oxysporum* f.sp. *Lycopersici* by Arbuscular Mycorrhizal Fungi, *Glomus* sp. *Int. J. Curr. Microbiol. App. Sci.* **2019**, *8*, 1368–1378. [CrossRef]
60. Li, Y.; Liu, Z.; Hou, H.; Lei, H.; Zhu, X.; Li, X.; He, X.; Tian, C. Arbuscular Mycorrhizal Fungi-Enhanced Resistance against Phytophthora Sojae Infection on Soybean Leaves Is Mediated by a Network Involving Hydrogen Peroxide, Jasmonic Acid, and the Metabolism of Carbon and Nitrogen. *Acta Physiol. Plant* **2013**, *35*, 3465–3475. [CrossRef]
61. Khoshkhatti, N.; Eini, O.; Koolivand, D.; Pogiatzis, A.; Klironomos, J.N.; Pakpour, S. Differential Response of Mycorrhizal Plants to Tomato Bushy Stunt Virus and Tomato Mosaic Virus Infection. *Microorganisms* **2020**, *8*, 2038. [CrossRef] [PubMed]
62. Math, S.; Arya, S.; Sonawane, H.; Patil, V.; Chaskar, M. Arbuscular Mycorrhizal (*Glomus fasciculatum*) Fungi as a Plant Immunity Booster against Fungal Pathogen. *Curr. Agri. Res. J.* **2019**, *7*, 99–107. [CrossRef]
63. Oyewole, B.O.; Olawuyi, O.J.; Odebode, A.C.; Abiala, M.A. Influence of Arbuscular Mycorrhiza Fungi (AMF) on Drought Tolerance and Charcoal Rot Disease of Cowpea. *Biotechnol. Rep.* **2017**, *14*, 8–15. [CrossRef]
64. Meddich, A.; Ait El Mokhtar, M.; Bourzik, W.; Mitsui, T.; Baslam, M.; Hafidi, M. Optimizing Growth and Tolerance of Date Palm (*Phoenix dactylifera* L.) to Drought, Salinity, and Vascular Fusarium-Induced Wilt (*Fusarium oxysporum*) by Application of Arbuscular Mycorrhizal Fungi (AMF). In *Root Biology*; Giri, B., Prasad, R., Varma, A., Eds.; Soil Biology; Springer International Publishing: Cham, Switzerland, 2018; Volume 52, pp. 239–258. ISBN 978-3-319-75909-8.
65. Garmendia, I.; Goicoechea, N.; Aguirreolea, J. Moderate Drought Influences the Effect of Arbuscular Mycorrhizal Fungi as Biocontrol Agents against Verticillium-Induced Wilt in Pepper. *Mycorrhiza* **2005**, *15*, 345–356. [CrossRef]
66. Mohamed, I.; Eid, K.E.; Abbas, M.H.H.; Salem, A.A.; Ahmed, N.; Ali, M.; Shah, G.M.; Fang, C. Use of Plant Growth Promoting Rhizobacteria (PGPR) and Mycorrhizae to Improve the Growth and Nutrient Utilization of Common Bean in a Soil Infected with White Rot Fungi. *Ecotoxicol. Environ. Saf.* **2019**, *171*, 539–548. [CrossRef]
67. Kohler, J.; Caravaca, F.; Carrasco, L.; Roldán, A. Interactions between a Plant Growth-Promoting Rhizobacterium, an AM Fungus and a Phosphate-Solubilising Fungus in the Rhizosphere of *Lactuca sativa*. *Appl. Soil. Ecol.* **2007**, *35*, 480–487. [CrossRef]
68. Niu, B.; Wang, W.; Yuan, Z.; Sederoff, R.R.; Sederoff, H.; Chiang, V.L.; Borriss, R. Microbial Interactions Within Multiple-Strain Biological Control Agents Impact Soil-Borne Plant Disease. *Front. Microbiol.* **2020**, *11*, 585404. [CrossRef] [PubMed]
69. Backer, R.; Rokem, J.S.; Ilangumaran, G.; Lamont, J.; Praslickova, D.; Ricci, E.; Subramanian, S.; Smith, D.L. Plant Growth-Promoting Rhizobacteria: Context, Mechanisms of Action, and Roadmap to Commercialization of Biostimulants for Sustainable Agriculture. *Front. Plant Sci.* **2018**, *9*, 1473. [CrossRef] [PubMed]
70. Basiru, S.; Hijri, M. The Potential Applications of Commercial Arbuscular Mycorrhizal Fungal Inoculants and Their Ecological Consequences. *Microorganisms* **2022**, *10*, 1897. [CrossRef] [PubMed]

International Journal of
Plant Biology

MDPI

Article

Antagonism and Synergism Characterize the Interactions between Four North American Potato Virus Y Strains

Prakash M. Niraula [1], Patricia Baldrich [2], Junaid A. Cheema [1], Hashir A. Cheema [1], Dejah S. Gaiter [1], Blake C. Meyers [2] and Vincent N. Fondong [1,*]

1 Department of Biological Sciences, Delaware State University, Dover, DE 19901, USA; pniraula@desu.edu (P.M.N.); j18@students.desu.edu (J.A.C.); hacheema20@students.desu.edu (H.A.C.); dsgaiter18@students.desu.edu (D.S.G.)
2 Donald Danforth Plant Science Center, St. Louis, MO 63132, USA; pbaldrich@danforthcenter.org (P.B.); bmeyers@danforthcenter.org (B.C.M.)
* Correspondence: vfondong@desu.edu

Abstract: Potato virus Y (PVY) is one of the most important constraints to potato production worldwide. There is an increasing occurrence of recombinant PVY strains PVY^{NTN} and $PVY^{N\text{-}Wi}$ and a decline in the incidence of the nonrecombinant PVY^{O}. We hypothesized that this may be due to the ability of these recombinant strains to antagonize and/or outcompete PVY^{O} in mixed infections. To determine this, we investigated interactions between PVY^{O} and three recombinant PVY strains common in North America: PVY^{NTN}, $PVY^{N\text{-}Wi}$, and $PVY^{N:O}$. Overall, our study showed that these interactions are tissue-dependent. Specifically, PVY^{NTN}, the main causal agent of potato tuber necrotic ringspot disease (PTNRD), was found to be more adaptable than PVY^{O}, especially in potato leaves due, at least in part, to the *Ny* gene that confers hypersensitive resistance (HR) to PVY^{O}. Furthermore, $PVY^{N\text{-}Wi}$ was found to repress PVY^{O} in potato tubers but act synergistically in potato leaves. The PVY^{O}-induced foliage necrosis in cultivar 'Ranger Russet' was observed to be more severe in plants co-infected by $PVY^{N\text{-}Wi}$ and $PVY^{N:O}$, respectively, resulting in plant death. Strikingly, this PVY^{O} -induced necrosis was suppressed by PVY^{NTN} in doubly infected plants. These interactions may, at least partially, explain the decreasing incidence of PVY^{O} in United States potato production regions, especially given that many cultivars contain the *Ny* gene, which likely limits PVY^{O} enabling PVY^{NTN} and $PVY^{N\text{-}Wi}$ to outcompete. We also found that replication and cell-to-cell movement of these PVY strains in tubers at 4 °C was similar to levels at ambient temperature.

Keywords: antagonism; infectivity; pathogenicity; Potato virus Y; RT-qPCR; superinfection exclusion; synergism

1. Introduction

Potato virus Y (PVY) is one of the most important pathogens of solanaceous plants, such as potato, tobacco, pepper, and ornamental plants, leading to significant losses of the crop [1]. In potato, diseases caused by PVY pose critical challenges to the production of the crop worldwide. PVY is disseminated in potato by aphids and through seed tubers, which are the main mode of potato propagation. The virus exists as a complex of strains that induce a wide variety of foliar and tuber symptoms in potato, leading to yield reductions and loss of tuber quality. PVY evolves through accumulation of mutations and rapidly through recombination between different strains, adapting to new potato cultivars across different environments [2]. Although many PVY recombinants with important genome differences have been identified, almost all these recombinants have PVY^{O} and PVY^{N} as parents [3,4]. In recent decades, new recombinant strains have emerged that have rapidly adapted to the potato ecosystem; these strains continue to dominate virus populations over vast geographical areas in North America [3,5,6]. These recombinants tend to cause milder foliar symptoms; thus, infected plants are less apparent than parent strains, but they cause more severe agricultural losses due to tuber necrosis, which reduces tuber quality. Usual foliar symptoms caused by PVY include leaf mosaic, crinkling, localized necrotic lesions, and leaf drop. Some PVY strains induce distinct ring patterns on the surface of tubers, causing the so-called potato tuber necrotic ringspot disease (PTNRD), which is one of the most damaging viral diseases in potatoes and poses a serious threat to seed and commercial potato production industries [2].

To date, at least nine recombination patterns of PVY^{O} and PVY^{N} sequences have been identified in potato-infecting PVY isolates; the three most common recombinant patterns characteristic of PVY strains are PVY^{NTN}, $PVY^{N:O}$, and $PVY^{N\text{-}Wi}$ [7,8]. $PVY^{N:O}$ and $PVY^{N\text{-}Wi}$ are believed to have acquired their recombinant PVY^{O} segments from two separate PVY^{O} lineages, while PVY^{NTN}, which has risen globally and overpopulated other strains worldwide, is thought to have acquired its recombinant PVY^{O} segment from the same lineage as $PVY^{N:O}$ [4].

Recently, the incidence and occurrence of the strain PVY^{NTN} have been increasing in the United States seed potato crops, while PVY^{O}, the ordinary, non-recombinant strain, has been decreasing [9]. PVY^{NTN} is the main cause of PTNRD in susceptible potato cultivars, leading to reductions in tuber quality and quantity [10]. $PVY^{N\text{-}Wi}$ has also been reported to cause PTNRD in the United States and Canada [3]. Interestingly, PVY^{NTN} and $PVY^{N\text{-}Wi}$ infections are usually either asymptomatic to the foliage or result in transient mild mosaic patterning [2]. On the other hand, PVY^{O} typically causes a strong hypersensitive resistance (HR), including localized necrosis and leaf drop and/or mild mosaic symptoms on the foliage of potato cultivars having the *Ny* resistance gene. PVY^{O}-induced HR is mediated by the *Ny* gene, which confers partial resistance in many North American potato varieties [8]. $PVY^{N:O}$ is less damaging than other strains, but causes mosaic symptoms in susceptible potato varieties [11].

Interactions between PVY strains and potato are defined in large part by resistance genes. Two types of PVY single dominant resistance genes have been identified in potato, namely, *R* genes that confer extreme resistance (ER) and *N* genes that confer HR [12]. Only a limited number of cultivated potatoes contain *R* genes, including *Rysto* from *S. stoloniferum* [13]. Unlike *R* genes, *N* genes occur widely in cultivated potato, and the ability of potatoes containing *N* genes to exhibit the HR upon PVY infection is strain-specific [14,15]. Thus, Nc_{tbr} and Nc_{spl} from *S. tuberosum* and *S. sparsipilum*, respectively, confer HR to PVY^{C} only, and the Ny_{tbr} from *S. tuberosum* confers HR resistance to PVY^{O} isolates only. Strain groups PVY^{C}, PVY^{O}, and PVY^{Z} elicit HR phenotypes *Nc*, *Ny*, and *Nz*, respectively, while PVY^{D} elicits HR phenotypes Ny an Nc [16]. In addition to PVY^{O}, the strains PVY^{C}, PVY^{D}, PVY^{N}, and PVY^{Z} are non-recombinant and serve as parents for many recombinant strains. PVY^{Z} was first described based on its ability to elicit HR in a potato cultivar carrying the *Nz* gene and the inability to induce necrosis in tobacco [17].

There are growing reports that PVY^{NTN} has been spreading rapidly in North America, and that the occurrence of PVY^{O}, the ordinary strain, has been decreasing [2,3,6,9,15]. We therefore hypothesized that antagonistic interactions between PVY^{O} and other PVY strains have led to a decline in importance of the former. In this study, we assessed the infectivity of PVY^{O}, PVY^{NTN}, $PVY^{N:O}$, and $PVY^{N\text{-}Wi}$ in *N. tabacum* and in tubers of potato cultivars 'Desiree', 'Ranger Russet', 'Russet Norkotah' and 'Eva', which have different levels of resistance to PVY. Our results indicate a high adaptability *of* PVY^{NTN} and that it outcompetes PVY^{O}, especially in leaves. Furthermore, $PVY^{N\text{-}Wi}$ was shown to antagonize PVY^{O} but act synergistically with PVY^{NTN} and $PVY^{N:O}$. These findings may, at least partially, explain the current PVY infection dynamics in North America.

2. Materials and Methods

2.1. Virus Inoculations

Virus inoculum was prepared by homogenizing systemically infected leaves of *N. tabacum* plants in a 0.01 M phosphate buffer, pH 7.0, at a dilution of 1/20 (*w*/*v*). Control plants were inoculated with sap from PVY-free *N. tabacum* cv. Samsun plants. Plants were placed in a growth room at 16 h light and 22 ± 3 °C temperature. At five days post-infection (dpi), whole inoculated leaves were collected for each of three biological replications, ground to powder in liquid nitrogen, and then stored at −80 °C until RNA extraction. Symptoms produced by each strain were recorded for up to 30 dpi.

The four PVY strains from the United States (kindly provided by Stewart Gray of Cornell University) investigated in this study were collected between 2004 and 2006. Three of the strains, PVY^{O} (isolate NY090031; GenBank # KY848009), PVY^{NTN} (isolate NY090029; GenBank # KY848008), and $PVY^{N:O}$ (isolate NY090004; GenBank # KY848007), were isolates from New York State (PVY^{O}, PVY^{NTN}, and $PVY^{N:O}$), while the fourth strain, $PVY^{N\text{-}Wi}$ (isolate MN15_G_52; GenBank # KY847981), was from the state of Minnesota [5]. These isolates were confirmed by RT-PCR of total RNA from tuber samples using strain-specific primers listed in Table 1. Virus inoculum was then prepared by homogenizing systemically infected leaves (at 14 to 21 dpi) of *N. tabacum* plants in a 0.01 M phosphate buffer, pH 7.0, at a dilution of 1/20 (w/v). The control was sap from PVY-free *N. tabacum* plants. PVY strain infectivity was investigated in potato foliage using the cultivar 'Ranger Russet'. The inoculum was prepared as described above. Plants were inoculated at a 4 to 5 leaf stage. Inoculated plants were placed in a room at 16 h light and 22 ± 3 °C temperature.

Table 1. List of primers and targets used in this study.

A/RT-PCR			
Target	**Designation**	**Sequence (5′-3′)**	**Length (bp)**
* PVY^{NTN}	NTN7350-F NTN8266-R	ACATCACCGATGAGCAGG GTACATACCCTCGATTAGCA	918
* PVY^{O}	O1962-F O2296-R	TCAACATTCTATCCACCAAC ACGTTTGAGTGTCATGGT	335
$PVY^{N:O}$	N:O1008-F N:O1703-R	GCACGTTCCAAGGTTACC TCGCTTAGCATGATATTCCCT	695
** $PVY^{N\text{-}Wi}$	N-Wi_YN5-1780-F N-Wi_YO3-2558-R	TCCGAATGGGACAAGAAAACTTG AGGCTCATCTAACAGCAACTGTC	778
B/RT-qPCR * **			
Target	**Designation**	**Sequence (5′-3′)**	**Length (bp)**
PVY^{NTN}	NTNq-6515-F NTNq-6631-R	TCCGAGCTCCAGTGCAGAAT AAGTGCTGCCCGGTACATTG	116
PVY^{O}	Oq-4-F Oq-138-R	CGCAAAAACACTCATAAAAGCTCA TGGTTGGAAGTGATGAAATTGCT	134
$PVY^{N:O}$	NOq-6444-F NOq-6574-R	GGATATCATCCTCATCAAATGCCG TCGACGATGCATACTTCTCCTG	130
$PVY^{N\text{-}Wi}$	N-Wiq-35-F N-Wiq-156-R	TTCCTTGCAATTCTCTTAAACGGT ACGAACCGAAACAGATTGTTGAC	121
All strains	Uni-q-9426-F Uni-q-9549-R	GTGGCAGGGTGATTTCGTCA AGAATCGCAACATCACCTAATCG	123
1-alpha EF1α	EF1-F EF1-R	TGGAGGCACTCCCTGGTGACA TGTGGCAGTCGAGCACTGGT	194

* [18]; ** [19]. *** Primer pairs with efficiencies ranged from 95 to 109%.

To characterize virus replication and cell-to-cell movement in potato tubers, we investigated three potato cultivars with different levels of resistance to PVY, namely, 'Desiree' (susceptible), 'Ranger Russet' (moderately susceptible) [3], and 'Eva' (extreme resistance). All tubers used in these experiments were at the sprouting stage and were tested with reverse transcription polymerase chain reaction (RT-PCR) and confirmed to be PVY-free. To inoculate the tubers, a 1 cm corkborer was used to produce a well in the tuber pith (inner medulla) at the center of approximately 2 cm thick sections of potato tubers. Corkborer wells were sealed at one end of the tuber section with an agarose gel in a Petri dish. Virus inoculum was prepared by homogenizing systemically infected leaves of *N. tabacum* plants in a tuber inoculation buffer (20 mM PBS, 2% PVFP, 20 mM Na_2SO_3, 2 mM EDTA) at a dilution of 1/20 (*w*/*v*), followed by centrifugation at 10,000 revolutions per minute (rpm) for 5 min. Approximately 150 µL of supernatant was placed in the well, and the inoculated tubers were placed in a closed box and incubated until sample collection, which was conducted using a 0.5 cm corkborer. Samples were taken in triplicate, frozen in liquid nitrogen, and stored at −80 °C until RNA extraction.

2.2. Isolation of RNA and Reverse Transcription Quantitative PCR (RT-qPCR)

Total RNA was isolated from tuber and leaf samples using the RNeasy Mini kit (Qiagen, Germantown, MD, USA) with modifications. Here, approximately 200 mg of ground leaf and tuber tissues, respectively, were aliquoted for total RNA extraction using a leaf extraction buffer (5 M guanidine thiocyanate, 50 mM Tris-HCl pH 8, 10 mM EDTA, 2% N-lauroylsarcosine) and a tuber RNA extraction buffer (6 M guanidine hydrochloride, 20 mM MES hydrate, 20 mM EDTA, pH 8, 10% β-mercaptoethanol). The quality and quantity of RNA were determined using a NanoDrop 1000 Spectrophotometer (Thermo Fisher Scientific, Waltham, MA, USA) and agarose gel electrophoresis. The Invitrogen SuperScript III Platinum™ SYBR™ green one-step RT-qPCR Kit (Thermo Fisher Scientific, USA) was used to quantify viral RNA in approximately 50 ng of total RNA on a Light Cycler 480 system (Roche Diagnostics, Indianapolis, IN, USA). The 10 µL reaction mix contained 5 µL SYBR Green supermix, 0.25 µL SuperScript III reverse transcriptase (RT)/Platinum® Taq DNA polymerase enzyme mix, 0.4 µL 100 µM forward and reverse primers, and 2 µL of total RNA. The RT-qPCR conditions were as follows: cDNA synthesis at 50 °C for 15 min; qPCR cycling at 95 °C for 5 min and 45 three-step cycles, each at 95 °C for 15 s, 60 °C for 30 s, and 72 °C for 10 s, followed by a final melting curve of 65–95 °C for 5 s. The specificities of primers used in RT-qPCR analyses were determined by running amplification products in an agarose gel and confirming amplification of a single band corresponding to the expected size. Furthermore, amplification efficiencies were determined in a RT-qPCR assay using duplicates of a 10-fold dilution series (four dilutions) as the template.

Mean threshold cycle (Ct) values of the dilution series were plotted against log (1/dilution factor), and the resulting slope value was used to calculate the amplification efficiency (E) using the equation: $E = 10^{-1/\text{slope}}$ (Rasmussen 2001). Primer pairs had efficiencies ranging from 95 to 109% (Table 1; Supplementary Figure S1), which were, therefore, within the recommended range [20].

2.3. Determination of Relative Viral RNA Titer

Here, we used elongation factor 1 (EF1) as the housekeeping gene. To compensate for potential variation between runs and plates, EF1 Ct values were normalized by subtracting the median Ct value from individual EF1 Ct values in each replicate. Normalized EF1 Ct values were then subtracted from individual target Ct values to obtain normalized target Ct values, which were used to calculate target ΔCt and ΔΔCt values. In the calculation of the relative viral RNA titer, we removed variation from background signals unrelated to viral RNA target by subjecting healthy tissues to the same amplification conditions as for each treatment [21]; this was followed by calculation of the mean ΔCt and ΔΔCt of the background noise. These background ΔCt and ΔΔCt values were then subtracted from experimental ΔCt and ΔΔCt values, respectively, the results of which were used to calculate individual $2^{-\Delta\Delta Ct}$ values [22] to obtain relative viral RNA levels. Error bars were computed to indicate the standard deviation of three different technical repeats from three biological replicates, and significant differences were determined using a *t*-test.

3. Results

3.1. Antagonistic and Synergistic Interactions between PVY Strains in Potato

There are several reports that PVY^{NTN} has been spreading rapidly in North America, and that the occurrence of PVY^{O}, the ordinary strain, has been decreasing [3,6,15,18]. We investigated whether this may be due to a PVY^{NTN} competitive advantage over PVY^{O}, during interactions between these viral strains in co-infected plants. We therefore examined coinfections in the potato cultivar 'Ranger Russet', which displays an HR upon PVY^{O} infection, due to the presence of the *Ny* gene [23]. The results showed that leaves of 'Ranger Russet' inoculated with PVY^{O} displayed severe local veinal necrosis within 5 dpi in inoculated leaves, and by 10 dpi, these symptoms had spread to systemically infected leaves, which displayed yellowing (Figure 1A). By 21 dpi, severe veinal and systemic necrosis had spread to all leaves of plants inoculated with PVY^{O} (Supplementary Figure S2A). This PVY^{O}-induced systemic necrosis, which is triggered by the corresponding *N* gene, resulted in extensive leaf-drop within 30 dpi (Figure 1B). We note that although the Hc-Pro avirulence factor of PVY^{O} restricts virus spread from inoculated leaves of potato cultivars with the *Ny* gene [23], it has been observed that this resistance is usually overcome, due to other host factors and/or environmental conditions [24–28], resulting in severe systemic necrosis and leaf-drop [15]. As discussed later in this text, these differential symptoms between PVY^{O}, PVY^{NTN}, and PVY$^{N\text{-}Wi}$ have been shown to be due to Hc-Pro, which has been identified as the PVY avirulence factor that corresponds to *Ny* tbr and *Nc* spl [29].

We then investigated whether phenotypes displayed by singly infected plants could be influenced by other PVY strains by doubly inoculating 5 to 6 plants and comparing them to symptoms displayed by single infections. Interestingly, in all inoculated plants, the systemic necrosis caused by PVY^{O} was more severe when the latter was co-inoculated with PVY$^{N\text{-}Wi}$ and PVY$^{N:O}$, and leaves of these plants displayed severe necrosis and leaf yellowing by 10 dpi (Figure 1A). These symptoms became severe and resulted in plant death within 30 dpi (Figure 1B; Supplementary Figure S2B). Plant death started from the lower stem and progressed to the apical meristem, which, although wilted, had not died at 30 dpi, when observations were discontinued. Also, plants inoculated with PVY$^{N\text{-}Wi}$ and PVY$^{N:O}$ displayed systemically infected leaves with a deep orange phenotype (Supplementary Figure S2A). Strikingly, the systemic necrosis and leaf drop caused by PVY^{O} was observed to be absent in plants doubly infected with PVY^{NTN} and PVY^{O}. These plants displayed a phenotype similar to the phenotype displayed by plants singly infected with PVY^{NTN}. Thus, the ability of PVY^{O} to cause systemic necrosis and leaf drop in 'Ranger Russet' is repressed by PVY^{NTN} (Figure 1B). Together, these symptoms show that PVY^{O} interacts synergistically with PVY$^{N\text{-}Wi}$ and PVY$^{N:O}$ but is antagonized by PVY^{NTN} in 'Ranger Russet'.

3.2. RT-qPCR Quantification of Viral RNA

To determine the effect of mixed infections on viral RNA levels, viral RNA was quantified in singly and doubly infected plants at 7 dpi from inoculated leaves and at 30 dpi from newly emerged leaves. RT-qPCR quantification of viral RNA was carried out using strain-specific primers. The specificity of the primers used in this study was confirmed by running RT-PCR amplification products on an agarose gel (Supplementary Figure S3). All experiments had three biological replicates and three technical replicates. The results showed that at 7 dpi (Figure 2A(a–d)), there were almost undetectable levels of PVY^{O}, likely due to the *Ny* gene in 'Ranger Russet' that confers HR, characterized by necrosis to PVY^{O}. Interestingly, in the presence of PVY$^{N\text{-}Wi}$, there was a significant increase in the level PVY^{O} RNA. Interactions between PVY$^{N:O}$ and PVY$^{N\text{-}Wi}$, as well as between PVY$^{N:O}$ and PVY^{O}, which resulted in significantly increased levels of PVY$^{N:O}$. These increases in viral RNA levels indicate synergism between PVY$^{N:O}$ and PVY$^{N\text{-}Wi}$, as well as between PVY$^{N:O}$ and PVY^{O}. The very high RNA levels in plants inoculated with PVY$^{N\text{-}Wi}$ and PVY^{O}, and with PVY$^{N:O}$ and PVY$^{N\text{-}Wi}$ may at least partly explain plant death in these plants. There were almost undetectable levels of viral RNA in samples collected from the apical meristem of plants inoculated with PVY^{NTN} at 30 dpi (Figure 2A(e–h)), consistent with the apparent absence of symptoms in these plants. This is likely due to a deficiency in long-distance movement of this strain. We note that although, by 30 dpi, most of the stem of all plants inoculated with PVY$^{N:O}$ and PVY^{O}, as well as PVY$^{N\text{-}Wi}$ and PVY^{O}, had died, the apical part of each plant was still in a condition (Figure 1B) for samples to be collected for RNA analysis.

Figure 1. Interactions between PVY strains PVY^{O}, PVY^{NTN}, PVY$^{N:O}$, and PVY$^{N\text{-}Wi}$ in potato cultivar 'Ranger Russet'. Systemically infected leaves of plants inoculated with PVY^{O} displayed severe local veinal necrosis at 10 dpi, while leaves of plants inoculated with PVY^{NTN}, PVY$^{N\text{-}Wi}$, and PVY$^{N:O}$ displayed necrotic spots (**A**). Leaves doubly infected with PVY$^{N\text{-}Wi}$ and PVY^{O}, as well as with PVY$^{N:O}$ and PVY^{O}, displayed severe necrosis leading leaf death (**B**).

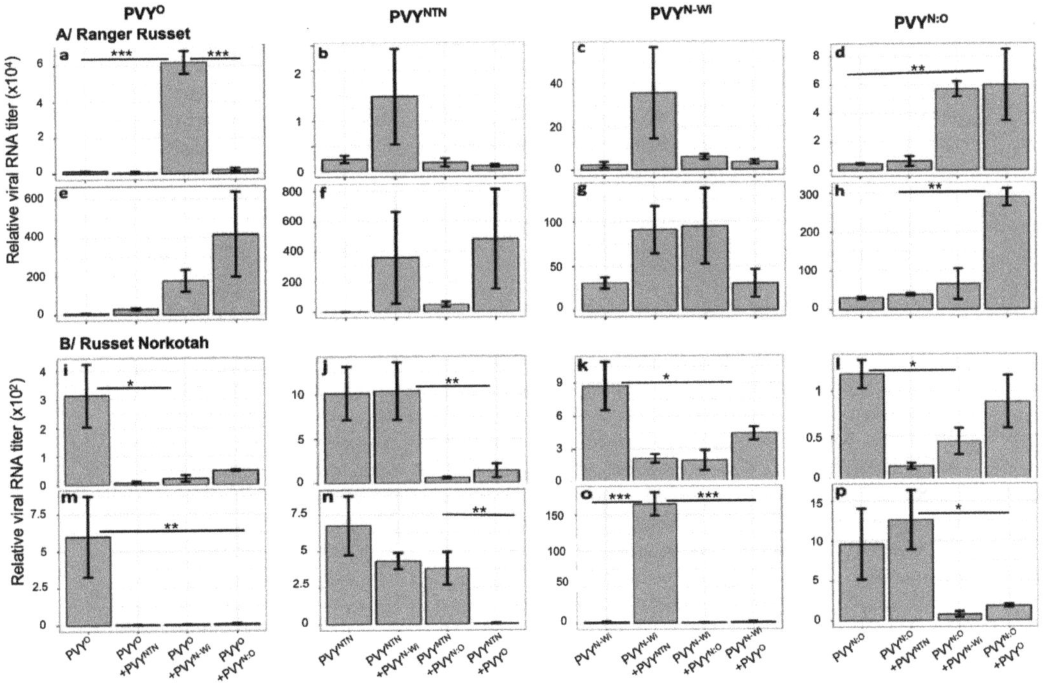

Figure 2. RT-qPCR quantification of viral RNA in leaves of potato cultivar 'Ranger Russet' (**A**) and 'Russet Norkotah' (**B**) using strain-specific primers. Viral titers were determined at 7 dpi (**a–d**,**i–l**) and 30 dpi (**e–h**,**m–p**). Each experiment had three biological replicates and three technical replicates. Details of the approach used to determine relative levels of viral RNA are provided in Materials and Methods. Mean differences were determined using a *t*-test (* $p < 0.05$; ** $p < 0.01$; *** $p < 0.001$).

Because 'Ranger Russet' has the *Ny* gene, which confers HR to PVY^{O}, we reasoned that the presence of this gene may have influenced interactions between PVY^{O} and other PVY strains. Thus, we further investigated these interactions in 'Russet Norkotah', which does not contain the *Ny* gene [8]. Samples were collected for viral RNA quantification at 7 and 30 dpi from inoculated and systemic leaves, respectively. Analysis of viral RNA from three biological replicates and three technical replicates at 7 dpi showed that, as is consistent with results recorded in 'Ranger Russet', PVY^{O} levels were significantly lower in 'Russet Norkotah' leaves when co-inoculated with PVY^{NTN}, PVY^{N-Wi}, and PVY$^{N:O}$, respectively (Figure 2B(i)). Similar results were recorded at 30 dpi (Figure 2B(m)). Strikingly, unlike the result obtained for 'Ranger Russet', PVY^{NTN} levels were significantly lower in mixed infections with PVY^{O} (Figure 2B(j,n)), indicating that the presence of the *Ny* gene likely inhibited the ability of PVY^{O} to compete with PVY^{NTN} in 'Ranger Russet'. Curiously, PVY^{N-Wi} showed very low levels in systemically infected leaves of 'Russet Norkotah', except in the presence of PVY^{NTN} at 30 dpi (Figure 2B(o)). Together, these results show differential influences on virus replication and movement in the presence of the *Ny* gene.

3.3. Replication and Cell-To-Cell Movement of PVY Strains

Potato-infecting viruses, as well as viruses of other tuberous root crops, tend to spend more time in tubers than elsewhere in the plant. Thus, here, we investigated the ability of all four PVY strains to replicate and move in tubers using tubers of three potato cultivars with different levels of resistance to PVY. Two of the cultivars, 'Desiree' and 'Ranger Russet', contained *N* genes [23], while the third cultivar, 'Eva', contained the Ry_{adg} extreme the inoculum well using a 0.5 cm corkborer and samples collected from positions a, b, and c (**A**) at 6, 24, 48, and 96 h post-inoculation. Viral RNA levels were analyzed using RT-qPCR as described in Materials and Methods. Tubers infected by PVY^{NTN} and PVY^{O} displayed lower levels of viral RNA than tubers singly infected by either of the strains, as indicated by RT-qPCR analysis using strain-specific primers (**B**,**C**). Mean differences were determined using a *t*-test (* $p < 0.05$; ** $p < 0.01$).

We further investigated the interaction of other PVY strains in tubers of 'Ranger Russet'. Thus, tubers were doubly inoculated with pairs of all four strains and viral RNA levels compared between single and co-infections using strain-specific as well as universal primers. In 'Ranger Russet' tubers, the results indicated that in tubers, mixed infection between $PVY^{N:O}$ and PVY^{O} appeared not to significantly affect levels of either strain (Figure 5A). Correspondingly, $PVY^{N\text{-}Wi}$ levels appeared not to be affected by PVY^{O}, which itself showed a significant decrease when co-infected with $PVY^{N\text{-}Wi}$ (Figure 5B). Also, mixed infection between $PVY^{N:O}$ and PVY^{NTN} caused no significant effect on PVY^{NTN}, but resulted in an increase in the level of $PVY^{N:O}$ (Figure 5C). Additionally, $PVY^{N\text{-}Wi}$ and PVY^{NTN} antagonized each other (Figure 5D). Moreover, except for tubers doubly infected with $PVY^{N\text{-}Wi}$ and PVY^{NTN}, all doubly infected tubers recorded higher viral RNA loads than singly infected tubers, as determined with universal primers (Supplementary Figure S5).

To determine whether interactions between PVY strains were influenced by the *Ny* gene, we investigated these interactions in tubers of 'Russet Norkotah', which does not contain the gene. The results showed that $PVY^{N\text{-}Wi}$ and PVY^{NTN} each acted synergistically with PVY^{O} [35,36]. Strikingly, unlike observations in 'Ranger Russet', in 'Russet Norkotah', there were significantly lower levels of PVY^{NTN} in leaves co-inoculated with PVY^{O}. (Supplementary Figure S6). Thus, PVY^{O} likely represses PVY^{NTN} in the absence of the Ny gene, to which the host exhibits a HR in the presence of PVY^{O}. Levels of $PVY^{N:O}$ were also repressed by PVY^{NTN} and $PVY^{N\text{-}Wi}$, and not by PVY^{O} in 'Russet Norkotah' leaves, but showed a significant increase in tubers co-infected with PVY^{NTN}.

3.5. Effect of Temperature on PVY Cell-to-Cell Movement in Tubers

To determine whether low seed storage temperature influences PVY replication and cell-to-cell movement in potato tubers, we quantified the levels of PVY^{O} and PVY^{NTN} RNA over time in tubers at 4 °C and at 22 °C, respectively. Samples were collected at 6, 24, and 48 hpi using a 0.5 cm corkborer at an approximate distance of 5 mm from the inoculation well, at position "a", as well as at positions "b" and "c", for a total distance of approximately 1.5 cm from the center of the well, as illustrated in Figure 4A. The results of RT-qPCR quantification indicated that under both temperature conditions, viral RNA was detected at position "a" at 6 hpi, thus indicating that both PVY strains were replicating and carrying out cell-to-cell movement. Interestingly, viral RNA levels declined in samples collected from the same position over time, which may be due to the tuber's antiviral response.

There were differences between the levels of viral RNA in the two strains. For instance, PVY^{NTN} RNA was significantly higher in 'Desiree' tubers at 4 °C compared with 22 °C from positions "a" through "c". No decline was observed in 'Ranger Russet' at position "c" (Figure 6A). In 'Eva', there was almost no detectable viral RNA beyond position "a", consistent with the fact that 'Eva' has PVY extreme resistance, which likely involves virus movement. This result contrasts with the high levels of viral RNA observed along the infection gradient of tubers of the more susceptible 'Desiree' and 'Ranger Russet'. Once again, the almost indetectable levels of viral RNA in 'Eva', compared with the less resistant cultivars is further evidence of viral replication and validates use of this approach to quantify virus replication in tubers. Consistent with results obtained in infectivity assessment above, we observed lower levels of PVY^{O} in tubers of all three cultivars at both incubation temperatures compared with levels of PVY^{NTN}, especially at 22 °C, where 'Eva' and 'Ranger Russet' showed very low levels of PVY^{O} RNA (Figure 6B). PVY^{O} RNA levels were more uniform in 'Desiree' tubers at 22 °C.

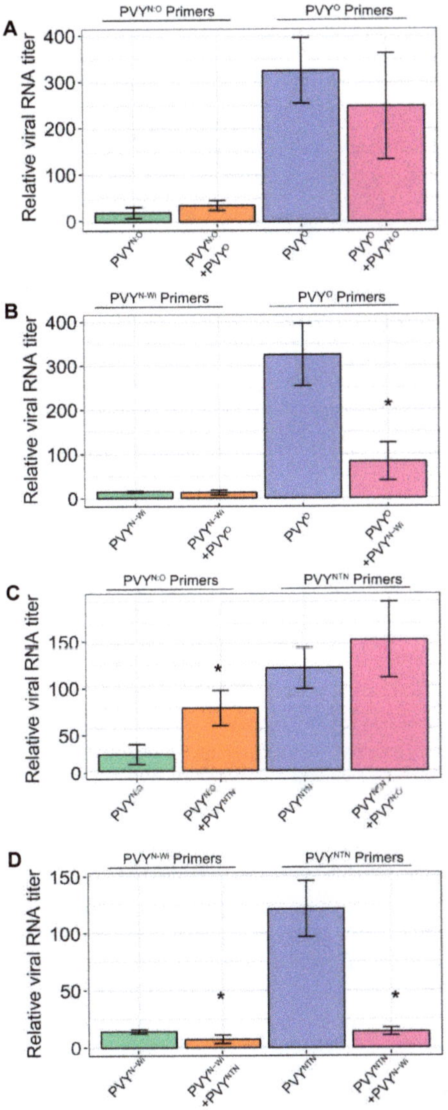

Figure 5. Antagonism and synergistic interactions between PVY strains PVY^{O}, PVY^{NTN}, PVY$^{N:O}$, and PVY$^{N\text{-}Wi}$ in potato tubers. PVY strains were singly and co-inoculated with (**A**) PVY$^{N:O}$ and PVY^{O}; (**B**) PVY$^{N\text{-}Wi}$ and PVY^{O}; (**C**) PVY$^{N:O}$ and PVY^{NTN}; and (**D**) PVY$^{N\text{-}Wi}$ and PVY^{NTN}, to determine the effect of mixed infection on replication, which was determined using RT-qPCR. Relative viral RNA levels were determined as described in Materials and Methods. PVY$^{N\text{-}Wi}$ was observed to repress PVY^{O} and PVY^{NTN}, but not PVY$^{N:O}$, and the latter in turn represses PVY$^{N\text{-}Wi}$ as indicated by strain-specific primers. Mean differences were determined using a *t*-test (* $p < 0.05$).

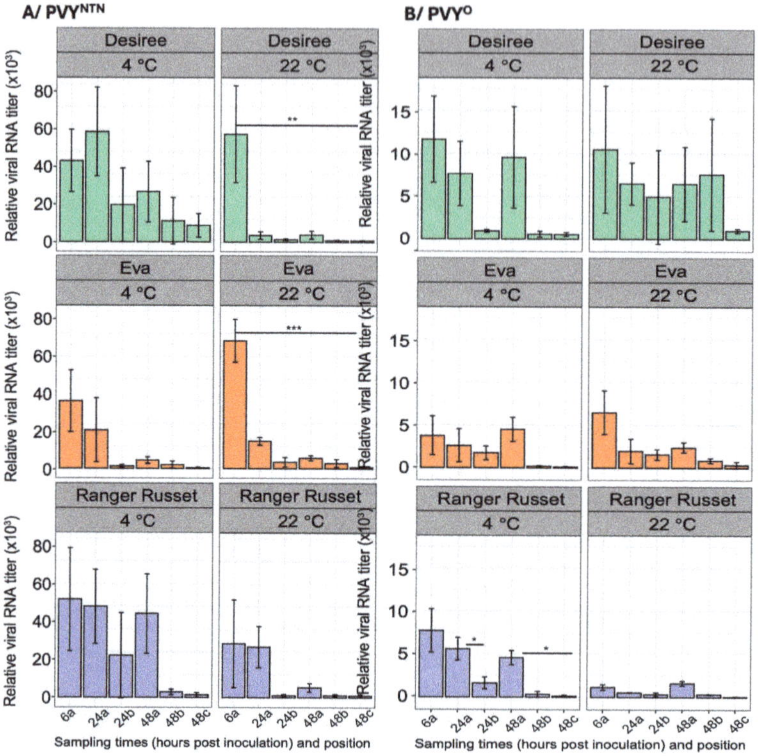

Figure 6. Effect of temperature on tuber infectivity of PVY^{NTN} (**A**) and PVY^{O} (**B**). Tubers of 'Desiree', Eva, and 'Ranger Russet' were inoculated and incubated at 4 °C and 22 °C, respectively. Samples were collected 6 hpi, 24 hpi, and 48 hpi along the infection gradient from the periphery at ~22 mm from the inoculation well to ~1.5 cm. Viral RNA levels were quantified using RT-qPCR as described in Materials and Methods. Mean differences were determined using a *t*-test test (* $p < 0.05$; ** $p < 0.01$; *** $p < 0.001$).

3.6. Evidence of Virus Replication in Potato Tubers

To confirm that the viral RNA detected in this study was from a replicating virus, we carried out an RT-PCR targeting each of PVY^{NTN} and PVY^{O} Hc-Pro cistron negative sense strand RNA [(−)ssRNA], which, unlike the (+)ssRNA, is not packaged in the virion. Here, samples were collected at 6, 24, 48, and 96 hpi, and total tuber RNA was extracted as described in Materials and Methods for reverse transcription of the (−)ssRNA of the Hc-Pro cistron. Reverse transcription was carried out using the Verso 1-step RT- PCR Reddymix kit (Thermo Fisher Scientific, Waltham, MA USA) at 50 °C for 20 min with a primer targeting the (−)ssRNA of the Hc-Pro cistron. The reverse transcriptase was then inactivated for 10 min at 95 °C prior to PCR amplification of the cDNA. Two microliters of the reverse transcription reaction were utilized for PCR amplification using the Dreamtaq PCR mastermix kit (Thermo Fisher Scientific, Waltham, MA USA) and both Hc-Pro forward and reverse primers (Table 1). Agarose gel analysis of the PCR products showed the presence of an amplicon corresponding to Hc-Pro cistron (Supplementary Figure S7). Given that the inoculum used in this study was crude sap, where viral RNA is naturally degraded by RNAases, the (−)ssRNA obtained in this analysis could only have come from a replicating virus. This result, together with low levels of viral RNA recorded in cultivar 'Eva' as compared to RNA levels in the less resistant 'Ranger Russet" and 'Desiree' observed above, indicates that the viral RNA detected in tubers in this study was from a replicating virus. Thus, the corkborer approach can be used to quantify replicating viral RNA in tubers.

4. Discussion

This study showed that the levels of PVY^{NTN} RNA were consistently higher in potato tubers than those of PVY^{O}. The results also showed that during co-infection, PVY^{NTN} outcompeted PVY^{O} in tubers of 'Ranger Russet' (Figure 6B) but not in those of 'Russet Norkotah'. This was likely due to the presence in 'Ranger Russet' of *Ny* gene, which confers resistance to PVY^{O}, but not to PVY^{NTN}. Given the increasing importance of potato cultivars with the *Ny* gene in commercial potatoes [8,23,37], these interactions may, at least partly, explain the increasing importance in North America of PVY^{NTN}, and to a lesser extent, $PVY^{N\text{-}Wi}$ [3,14,15]. Further, although synergistic interactions are generally between unrelated viruses [38–40] and relationships between related viruses are mostly antagonistic (competitive) [38,39], these data show that synergistic interactions do occur in very closely related PVY strains and that these interactions are genotype- and tissue-dependent. It is likely that there are host factors present in tubers with which proteins of either of the virus strains are interacting. Future studies will need to determine the mechanisms of action, including by analyzing differentially expressed genes. Such a study will determine the factors involved in PVY^{NTN} induction of potato tuber necrotic ringspot disease (PTNRD).

It has been shown that Hc-Pro is the PVY avirulence factor corresponding to Ny_{tbr} and Nc_{spl} [29]. The specific determinant was subsequently mapped to the Aspartic acid residue at position 419 (Asp419) in PVY^{O} Hc-Pro [41]. In PVY^{N} strains, there is Asp419 Glu mutation. However, breakdown of Ny_{tbr} resistance in PVY^{NTN} and $PVY^{N\text{-}Wi}$ was attributed to recombination rather than accumulation of point mutations [29]. The ability of PVY^{NTN} and $PVY^{N\text{-}Wi}$ to overcome resistance conferred by the *Ny* gene may, at least partially, explain the antagonism between these strains and PVY^{O}. Thus, PVY^{NTN} and $PVY^{N\text{-}Wi}$ overcome the resistance conferred by Nc_{spl} and outcompete PVY^{O}. It must be stressed that necrosis alone may not be sufficient or essential for the restriction of viral movement. Indeed, other host factors have been reported to be involved in the recognition of the avirulence protein or signaling for HR, as well as light intensity and/or quality, influencing the outcome of HR [24–28]. Hence, depending on the potato cultivar and environmental conditions, necrotic responses may vary phenotypically and, in addition to local necrosis, PVY may spread, causing systemic necrosis and leaf-drop in the plant, as observed on the 'Ranger Russet' cultivar in this study.

Epidemiologically, these data support the hypothesis that the global increase in occurrence of PVY^{NTN}, and to a limited extent, of $PVY^{N\text{-}Wi}$, is at least partially due to the ability of these strains to outcompete PVY^{O}, the hitherto predominant strain. Thus, in mixed infection situations, the fitness cost is more severe on PVY^{O} than on PVY^{NTN} or $PVY^{N\text{-}Wi}$. Furthermore, it has been indicated that, in potato cultivars with the *Ny* gene, such as 'Ranger Russet', the severe systemic necrosis observed in this study is beneficial in PVY^{O} management because it leads to developmentally impaired plants, which are unlikely to be used for seed tubers [2]. Given that a large proportion of cultivated potatoes have the *Ny* gene [8,23,37], this results in an increasing importance of PVY^{NTN} and $PVY^{N\text{-}Wi}$ at the expense of PVY^{O}. Moreover, the fact that there were no apparent viral symptoms in the foliage of potato plants infected by PVY^{NTN} supports the view that much of the spread is likely through tubers. Equally epidemiologically important is the fact that potato plants doubly infected by $PVY^{N\text{-}Wi}$ and PVY^{O}, and by $PVY^{N:O}$ and PVY^{O}, exhibited severe veinal and systemic necrosis, leading to plant death under our conditions. Thus, these plants are less likely to be used as seed tubers, thereby limiting these viral strains in mixed-infected seed tubers.

Differences in fitness have been suggested to be the driving force in mixed infections between pathogens, and underpin the natural selection process [42,43]. For viruses, fitness is the extent to which the virus adapts to the host and can produce infectious progeny (i.e., replicative fitness) [39,44–46]. To be infectious, the virus must uncoat and release its genome, express its genes, and replicate. Nascent viral progeny must move cell-to-cell either as the naked genome or as encapsidated particles while evading host defenses [39,47]. The ability of PVY^{NTN} and $PVY^{N\text{-}Wi}$ to outcompete PVY^{O} can at least partially be explained by the ability of the *Ny* gene to interfere with at least one of these stages in the virus life cycle, thereby conferring resistance. This presumably favors the replication and/or spread of PVY^{NTN} and $PVY^{N\text{-}Wi}$.

Potato seed tubers are traditionally stored at low temperatures, usually between 3 °C and 5 °C [48], and then used in the next crop. Therefore, we investigated the effect of low storage temperature on the replication and cell-to-cell movement of PVY^{NTN} and PVY^{O} at 4 °C and 22 °C in tubers of cultivars 'Desiree', 'Ranger Russet', and 'Eva'. The results showed similar levels of viral RNA at both temperature regimes. Thus, unlike the pathogenicity of tuber infecting bacterial and fungal diseases, which tend to decrease at low temperatures, PVY replication and cell-to-cell movement appear not to be affected by low storage temperatures. Epidemiologically, this indicates that, even for low virus titers, at harvest, PVY can continue to replicate and move cell-to-cell in the tuber during storage. It is important to note that others have noted an unsubstantial increase in PVY levels during storage [49]. Furthermore, the fact that the higher levels of PVY^{NTN} over PVY^{O} were more significant at 4 °C than at 22 °C indicates that PVY^{NTN} is much fitter during seed tuber storage and may at least partially explain its gradual replacement of PVY^{O} in North America, where seed tubers are traditionally stored at low temperatures.

One of the most limiting factors in studying virus pathogenicity in tubers is the difficulty of carrying out reproducible transmission in tubers. Thus, here, we developed a corkborer method and used it to infect tubers of all three potato cultivars efficiently and reproducibly.

Taken together, this study indicates that the current increase in PVY^{NTN} and $PVY^{N\text{-}Wi}$ incidences in potato fields, especially in North America, may be due, at least partially, to their ability to antagonize and/or outcompete other strains, especially including the prevalent PVY^{O} strain with the increasing incorporation of the *Ny* resistance gene into the potato cultivars, since the interaction between PVY strains is variety-dependent [36].

Supplementary Materials: The following supporting information can be downloaded at: https://www.mdpi.com/article/10.3390/ijpb15020032/s1, Figure S1. Amplification efficiency validation of primers used to amplify housekeeping gene and PVY strains PVY^{O}, PVY^{NTN}, $PVY^{N:O}$, and $PVY^{N\text{-}Wi}$. The Ct values were plotted against log of four different ten-fold template RNA dilutions (1, 10, 100, and 1000). This series of four dilutions was measured in two replicates per dilution. Figure S2. Interactions between PVY strains PVY^{O}, PVY^{NTN}, $PVY^{N:O}$, and $PVY^{N\text{-}Wi}$ in potato cultivar 'Ranger Russet'. singly and doubly infected (A) Interactions between PVY strains in potato cultivar 'Ranger Russet' at 10 dpi. (B) Interactions between PVY strains in potato cultivar 'Ranger Russet' at 30 dpi. Figure S3. Confirmation of specificity of PVY^{NTN}, PVY^{O}, $PVY^{N:O}$, and PVY^{O} primers utilized in this study as indicated by an RT-PCR amplification products run on an agarose gel. Figure S4. Typical ethidium bromide-stained Tris-borate-EDTA (TBE) agarose gel showing total potato tuber RNA extracted with our (A) guanidine hydrochloride-based buffer, and with (B) guanidine thiocyanate-based extraction buffer. The ratio of absorbance at 260 nm and 280 nm ranged from 1.90 to 2.13. Figure S5. Antagonism and synergistic interactions between PVY strains in potato tubers. PVY strains were singly- and co-inoculated to determine the effect of mixed infection on replication, which was determined using RT-qPCR. Relative viral RNA levels were determined as described in Materials and Methods. Analysis with universal PVY primers shows that except for the interaction between $PVY^{N\text{-}Wi}$ and PVY^{NTN}. Figure S6. Interaction between PVY strains in tubers of 'Russet Norkotah', which does not carry the *Ny* gene. PVY strains were singly- and co-inoculated to determine the effect of mixed infection on replication, which was determined using RT-qPCR. Relative viral RNA levels were determined as described in Materials and Methods. There were significantly lower levels of PVY^{NTN} in leaves co-inoculated with PVY^{O}. Levels of $PVY^{N:O}$ were also repressed by PVY^{NTN} and $PVY^{N\text{-}Wi}$, but not by PVY^{O}. Mean differences were determined using a *t*-test. Figure S7. RT-PCR test targeting the negative sense ssRNA [(−)ssRNA] of PVY^{NTN} and PVY^{O} Hc-Pro RNA sequence. Samples collected at 6, 24, 48, and 96 hpi and used for reverse transcription and PCR amplification. Samples from inoculated tubers showed PCR product corresponding to Hc-Pro cistron from inoculated tissues (lanes 1 to 4) of (A) PVY^{NTN}, and (B) PVY^{O}; lanes 5 to 8 are from samples inoculated with the phosphate buffer.

Author Contributions: Conceptualization, V.N.F. and P.M.N.; Methodology, V.N.F., P.M.N., J.A.C., H.A.C. and D.S.G.; Validation, V.N.F.; Investigation, V.N.F.; Visualization, V.N.F.; Writing, V.N.F.; Original draft, V.N.F.; Funding acquisition, V.N.F. and B.C.M.; Data Analysis V.N.F. and P.B.; Review and editing, P.B., B.C.M. and V.N.F. All authors have read and agreed to the published version of the manuscript.

Funding: This work was supported by the USDA National Institute of Food and Agriculture, Plant Biotic Interactions project number 2020-67018-31180.

Institutional Review Board Statement: Not applicable.

Informed Consent Statement: Not applicable.

Data Availability Statement: The data are contained within the article.

Acknowledgments: We thank Joanna Friesner for the critical review of the manuscript.

Conflicts of Interest: The authors declare no conflict of interest.

References

1. Scholthof, K.-B.G.; Adkins, S.; Czosnek, H.; Palukaitis, P.; Jacquot, E.; Hohn, T.; Hohn, B.; Saunders, K.; Candresse, T.; Ahlquist, P.; et al. Top 10 Plant Viruses in Molecular Plant Pathology. *Mol. Plant Pathol.* **2011**, *12*, 938–954. [CrossRef] [PubMed]
2. Karasev, A.V.; Gray, S.M. Continuous and Emerging Challenges of Potato Virus Y in Potato. *Annu. Rev. Phytopathol.* **2013**, *51*, 571–586. [CrossRef] [PubMed]
3. Gray, S.; De Boer, S.; Lorenzen, J.; Karasev, A.; Whitworth, J.; Nolte, P.; Singh, R.; Boucher, A.; Xu, H. Potato Virus Y: An Evolving Concern for Potato Crops in the United States and Canada. *Plant Dis.* **2010**, *94*, 1384–1397. [CrossRef]
4. Karasev, A.V.; Hu, X.; Brown, C.J.; Kerlan, C.; Nikolaeva, O.V.; Crosslin, J.M.; Gray, S.M. Genetic Diversity of the Ordinary Strain of Potato Virus Y (PVY) and Origin of Recombinant PVY Strains. *Phytopathology* **2011**, *101*, 778–785. [CrossRef]
5. Green, K.J.; Brown, C.J.; Gray, S.M.; Karasev, A.V. Phylogenetic Study of Recombinant Strains of Potato Virus Y. *Virology* **2017**, *507*, 40–52. [CrossRef] [PubMed]
6. MacKenzie, T.D.B.; Nie, X.; Bisht, V.; Singh, M. Proliferation of Recombinant PVY Strains in Two Potato-Producing Regions of Canada, and Symptom Expression in 30 Important Potato Varieties with Different PVY Strains. *Plant Dis.* **2019**, *103*, 2221–2230. [CrossRef]
7. Hu, X.; He, C.; Xiao, Y.; Xiong, X.; Nie, X. Molecular Characterization and Detection of Recombinant Isolates of Potato Virus Y from China. *Arch. Virol.* **2009**, *154*, 1303–1312. [CrossRef] [PubMed]
8. Rowley, J.S.; Gray, S.M.; Karasev, A.V. Screening Potato Cultivars for New Sources of Resistance to Potato Virus Y. *Am. J. Potato Res.* **2015**, *92*, 38–48. [CrossRef]
9. Carroll, J.E.; Smith, D.M.; Gray, S.M. Preferential Acquisition and Inoculation of PVYNTN over PVYO in Potato by the Green Peach Aphid Myzus Persicae (Sulzer). *J. Gen. Virol.* **2016**, *97*, 797–802. [CrossRef]
10. Boonham, N.; Walsh, K.; Preston, S.; North, J.; Smith, P.; Barker, I. The Detection of Tuber Necrotic Isolates of Potato Virus Y, and the Accurate Discrimination of PVY(O), PVY(N) and PVY(C) Strains Using RT-PCR. *J. Virol. Methods* **2002**, *102*, 103–112. [CrossRef]
11. Nie, B.; Singh, M.; Murphy, A.; Sullivan, A.; Xie, C.; Nie, X. Response of Potato Cultivars to Five Isolates Belonging to Four Strains of Potato Virus Y. *Plant Dis.* **2012**, *96*, 1422–1429. [CrossRef]
12. Gebhardt, C.; Valkonen, J.P. Organization of Genes Controlling Disease Resistance in the Potato Genome. *Annu. Rev. Phytopathol.* **2001**, *39*, 79–102. [CrossRef] [PubMed]
13. Mestre, P.; Brigneti, G.; Baulcombe, D.C. An Ry-Mediated Resistance Response in Potato Requires the Intact Active Site of the NIa Proteinase from Potato Virus Y. *Plant J.* **2000**, *23*, 653–661. [CrossRef] [PubMed]
14. Dupuis, B.; Bragard, C.; Schumpp, O. Resistance of Potato Cultivars as a Determinant Factor of Potato Virus Y (PVY) Epidemiology. *Potato Res.* **2019**, *62*, 123–138. [CrossRef]
15. Funke, C.N.; Nikolaeva, O.V.; Green, K.J.; Tran, L.T.; Chikh-Ali, M.; Quintero-Ferrer, A.; Cating, R.A.; Frost, K.E.; Hamm, P.B.; Olsen, N.; et al. Strain-Specific Resistance to Potato Virus Y (PVY) in Potato and Its Effect on the Relative Abundance of PVY Strains in Commercial Potato Fields. *Plant Dis.* **2017**, *101*, 20–28. [CrossRef] [PubMed]
16. Jones, R.A.C.; Barbetti, M.J.; Fox, A.; Adams, I.P. Potato Virus Y Biological Strain Group YD: Hypersensitive Resistance Genes Elicited and Phylogenetic Placement. *Plant Dis.* **2021**, *105*, 3600–3609. [CrossRef] [PubMed]
17. Jones, R.a.C. Strain Group Specific and Virus Specific Hypersensitive Reactions to Infection with Potyviruses in Potato Cultivars. *Ann. Appl. Biol.* **1990**, *117*, 93–105. [CrossRef]

18. Mondal, S.; Lin, Y.-H.; Carroll, J.E.; Wenninger, E.J.; Bosque-Pérez, N.A.; Whitworth, J.L.; Hutchinson, P.; Eigenbrode, S.; Gray, S.M. Potato Virus Y Transmission Efficiency from Potato Infected with Single or Multiple Virus Strains. *Phytopathology* **2017**, *107*, 491–498. [CrossRef]
19. Hühnlein, A.; Drechsler, N.; Steinbach, P.; Thieme, T.; Schubert, J. Comparison of Three Methods for the Detection of Potato Virus Y in Seed Potato Certification. *J. Plant Dis. Prot.* **2013**, *120*, 57–69. [CrossRef]
20. Taylor, S.; Wakem, M.; Dijkman, G.; Alsarraj, M.; Nguyen, M. A Practical Approach to RT-qPCR-Publishing Data That Conform to the MIQE Guidelines. *Methods* **2010**, *50*, S1–S5. [CrossRef]
21. Wilhelm, J.; Pingoud, A.; Hahn, M. Validation of an Algorithm for Automatic Quantification of Nucleic Acid Copy Numbers by Real-Time Polymerase Chain Reaction. *Anal. Biochem.* **2003**, *317*, 218–225. [CrossRef] [PubMed]
22. Livak, K.J.; Schmittgen, T.D. Analysis of Relative Gene Expression Data Using Real-Time Quantitative PCR and the $2^{-\Delta\Delta CT}$ Method. *Methods* **2001**, *25*, 402–408. [CrossRef]
23. Jones, R.A.C.; Vincent, S.J. Strain-Specific Hypersensitive and Extreme Resistance Phenotypes Elicited by *Potato Virus Y* Among 39 Potato Cultivars Released in Three World Regions Over a 117-Year Period. *Plant Disease* **2018**, *102*, 185–196. [CrossRef] [PubMed]
24. Caplan, J.L.; Mamillapalli, P.; Burch-Smith, T.M.; Czymmek, K.; Dinesh-Kumar, S.P. Chloroplastic Protein NRIP1 Mediates Innate Immune Receptor Recognition of a Viral Effector. *Cell* **2008**, *132*, 449–462. [CrossRef]
25. Tian, Y.-P.; Valkonen, J.P.T. Recombination of Strain O Segments to HCpro-Encoding Sequence of Strain N of Potato Virus Y Modulates Necrosis Induced in Tobacco and in Potatoes Carrying Resistance Genes Ny or Nc. *Mol. Plant Pathol.* **2015**, *16*, 735–747. [CrossRef] [PubMed]
26. Jin, H.; Liu, Y.; Yang, K.-Y.; Kim, C.Y.; Baker, B.; Zhang, S. Function of a Mitogen-Activated Protein Kinase Pathway in N Gene-Mediated Resistance in Tobacco. *Plant J.* **2003**, *33*, 719–731. [CrossRef] [PubMed]
27. Komatsu, K.; Hashimoto, M.; Ozeki, J.; Yamaji, Y.; Maejima, K.; Senshu, H.; Himeno, M.; Okano, Y.; Kagiwada, S.; Namba, S. Viral-Induced Systemic Necrosis in Plants Involves Both Programmed Cell Death and the Inhibition of Viral Multiplication, Which Are Regulated by Independent Pathways. *Mol. Plant Microbe Interact.* **2010**, *23*, 283–293. [CrossRef]
28. Valkonen, J.P.; Rokka, V.M.; Watanabe, K.N. Examination of the Leaf-Drop Symptom of Virus-Infected Potato Using Anther Culture-Derived Haploids. *Phytopathology* **1998**, *88*, 1073–1077. [CrossRef] [PubMed]
29. Moury, B.; Caromel, B.; Johansen, E.; Simon, V.; Chauvin, L.; Jacquot, E.; Kerlan, C.; Lefebvre, V. The Helper Component Proteinase Cistron of Potato Virus Y Induces Hypersensitivity and Resistance in Potato Genotypes Carrying Dominant Resistance Genes on Chromosome IV. *Mol. Plant Microbe Interact.* **2011**, *24*, 787–797. [CrossRef]
30. Plaisted, R.L.; Halseth, D.E.; Brodie, B.B.; Slack, S.A.; Sieczka, J.B.; Christ, B.J.; Paddock, K.M.; Peck, M.W. Eva:A Midseason Golden Nematode- and Virus-Resistant Variety for Use as Tablestock or Chipstock. *Am. J. Pot Res.* **2001**, *78*, 65–68. [CrossRef]
31. Nazarian-Firouzabadi, F.; Visser, R.G.F. Potato Starch Synthases: Functions and Relationships. *Biochem. Biophys. Rep.* **2017**, *10*, 7–16. [CrossRef] [PubMed]
32. Kumar, G.N.M.; Iyer, S.; Knowles, N.R. Extraction of RNA from Fresh, Frozen, and Lyophilized Tuber and Root Tissues. *J. Agric. Food Chem.* **2007**, *55*, 1674–1678. [CrossRef] [PubMed]
33. Vasanthan, T.; Bergthaller, W.; Driedger, D.; Yeung, J.; Sporns, P. Starch from Alberta Potatoes: Wet-Isolation and Some Physicochemical Properties. *Food Res. Int.* **1999**, *32*, 355–365. [CrossRef]
34. Wischmann, B.; Hamborg Nielsen, T.; Lindberg Møller, B. Wischmann, null; Hamborg Nielsen T, null; Lindberg Moller B, null In Vitro Biosynthesis of Phosphorylated Starch in Intact Potato Amyloplasts. *Plant Physiol.* **1999**, *119*, 455–462. [CrossRef] [PubMed]
35. Mondal, S.; Ghanim, M.; Roberts, A.; Gray, S.M. Different Potato Virus Y Strains Frequently Co-Localize in Single Epidermal Leaf Cells and in the Aphid Stylet. *J. Gen. Virol.* **2021**, *102*, 001576. [CrossRef]
36. Singh, M.; Singh, R.P. Host Dependent Cross-Protection between PVYN, PVY°, and PVA in Potato Cultivars and Solarium Brachycarpum. *Can. J. Plant Pathol.* **1995**, *17*, 82–86. [CrossRef]
37. Szajko, K.; Strzelczyk-Żyta, D.; Marczewski, W. Ny-1 and Ny-2 Genes Conferring Hypersensitive Response to Potato Virus Y (PVY) in Cultivated Potatoes: Mapping and Marker-Assisted Selection Validation for PVY Resistance in Potato Breeding. *Mol. Breed.* **2014**, *34*, 267–271. [CrossRef] [PubMed]
38. Syller, J. Facilitative and Antagonistic Interactions between Plant Viruses in Mixed Infections. *Mol. Plant Pathol.* **2012**, *13*, 204–216. [CrossRef] [PubMed]
39. Syller, J.; Grupa, A. Antagonistic Within-Host Interactions between Plant Viruses: Molecular Basis and Impact on Viral and Host Fitness. *Mol. Plant Pathol.* **2016**, *17*, 769–782. [CrossRef] [PubMed]
40. Tatineni, S.; Riethoven, J.-J.M.; Graybosch, R.A.; French, R.; Mitra, A. Dynamics of Small RNA Profiles of Virus and Host Origin in Wheat Cultivars Synergistically Infected by Wheat Streak Mosaic Virus and Triticum Mosaic Virus: Virus Infection Caused a Drastic Shift in the Endogenous Small RNA Profile. *PLoS ONE* **2014**, *9*, e111577. [CrossRef]
41. Glais, L.; Faurez, F.; Tribodet, M.; Boulard, F.; Jacquot, E. The Amino Acid 419 in HC-Pro Is Involved in the Ability of PVY Isolate N605 to Induce Necrotic Symptoms on Potato Tubers. *Virus Res.* **2015**, *208*, 110–119. [CrossRef] [PubMed]
42. Abrams, M. Environmental Grain, Organism Fitness, and Type Fitness. In *Entangled Life*; Springer: Dordrecht, The Netherlands, 2014; pp. 127–151. [CrossRef] [PubMed]
43. Metcalf, C.J.E.; Birger, R.B.; Funk, S.; Kouyos, R.D.; Lloyd-Smith, J.O.; Jansen, V.a.A. Five Challenges in Evolution and Infectious Diseases. *Epidemics* **2015**, *10*, 40–44. [CrossRef] [PubMed]
44. Barbour, J.D.; Grant, R.M. The Role of Viral Fitness in HIV Pathogenesis. *Curr. HIV/AIDS Rep.* **2005**, *2*, 29–34. [CrossRef] [PubMed]

45. Domingo, E. Mechanisms of Viral Emergence. *Vet. Res.* **2010**, *41*, 38. [CrossRef] [PubMed]
46. Wargo, A.R.; Kurath, G. Viral Fitness: Definitions, Measurement, and Current Insights. *Curr. Opin. Virol.* **2012**, *2*, 538–545. [CrossRef] [PubMed]
47. Elena, S.F.; Lalić, J. Plant RNA Virus Fitness Predictability: Contribution of Genetic and Environmental Factors. *Plant Pathol.* **2013**, *62*, 10–18. [CrossRef]
48. Milling, A.; Meng, F.; Denny, T.P.; Allen, C. Interactions with Hosts at Cool Temperatures, Not Cold Tolerance, Explain the Unique Epidemiology of Ralstonia Solanacearum Race 3 Biovar 2. *Phytopathology* **2009**, *99*, 1127–1134. [CrossRef]
49. Fox, A.; Evans, F.; Browning, I. Direct Tuber Testing for Potato Y Potyvirus by Real-Time RT-PCR and ELISA: Reliable Options for Post-Harvest Testing?*. *EPPO Bull.* **2005**, *35*, 93–97. [CrossRef]

microorganisms

MDPI

Article

The Impact of Aboveground *Epichloë* Endophytic Fungi on the Rhizosphere Microbial Functions of the Host *Melica transsilvanica*

Chuanzhe Wang, Chong Shi *, Wei Huang, Mengmeng Zhang and Jiakun He

College of Resources and Environment, Xinjiang Agricultural University, Urumqi 830052, China; 320223547@xjau.edu.cn (C.W.); huangwei@xjau.edu.cn (W.H.); 320223517@xjau.edu.cn (M.Z.); 320233612@xjau.edu.cn (J.H.)

* Correspondence: shichong@xjau.edu.cn

Abstract: In nature, the symbiotic relationship between plants and microorganisms is crucial for ecosystem balance and plant growth. This study investigates the impact of *Epichloë* endophytic fungi, which are exclusively present aboveground, on the rhizosphere microbial functions of the host *Melica transsilvanica*. Using metagenomic methods, we analyzed the differences in microbial functional groups and functional genes in the rhizosphere soil between symbiotic (EI) and non-symbiotic (EF) plants. The results reveal that the presence of *Epichloë* altered the community structure of carbon and nitrogen cycling-related microbial populations in the host's rhizosphere, significantly increasing the abundance of the genes (*por*A, *por*G, IDH1) involved in the rTCA cycle of the carbon fixation pathway, as well as the abundance of *nxr*AB genes related to nitrification in the nitrogen-cycling pathway. Furthermore, the presence of *Epichloë* reduces the enrichment of virulence factors in the host rhizosphere microbiome, while significantly increasing the accumulation of resistance genes against heavy metals such as Zn, Sb, and Pb. This study provides new insights into the interactions among endophytic fungi, host plants, and rhizosphere microorganisms, and offers potential applications for utilizing endophytic fungi resources to improve plant growth and soil health.

Keywords: symbiosis; *Epichloë*; *Melica transsilvanica*; rhizosphere microbial functions; metagenomics

1. Introduction

Symbiosis between plants and microbes is a prevalent phenomenon in nature [1]. Microorganisms such as endophytes and rhizosphere and phyllosphere are involved in various stages of plant growth and development [2]. The symbiotic microbiota with plants can enhance the metabolic capacity of plants, improve the efficiency of nutrient uptake, and explore new nutritional pathways and defense processes for host plants [3–5]. In particular, rhizosphere microorganisms of plants are known as the second genome of plants because of their rich diversity, complex regulatory networks, and important roles in affecting plant growth and development and improving plant stress resistance [6]. Many studies have focused on the influence of environmental factors or pathogenic microorganisms on rhizosphere microorganisms [7]. Will symbiotic microorganisms, such as some endophytic fungi that are only present in the above-ground part of plants, have an impact on rhizosphere microorganisms in the subsurface part, and what impact will they have? Such research can help us better understand the interaction between endophytic fungi, host plants and rhizosphere microorganisms, which is very valuable in ecological research.

The endophytic fungi *Epichloë*, studied in this research, typically forms mutualistic symbiotic associations predominantly with cool-season grasses [8]. The host grass provides nutrients to *Epichloë*, which in turn enhances the host's resistance to biotic and abiotic stresses such as drought, herbivory, and pathogen infections [9–11]. What sets it apart is that, due to long-term coevolution with the host, this symbiotic relationship is highly stable, allowing *Epichloë* to complete part or all of its lifecycle inside the host grass and eventually vertically transmit through seeds [12]. The unique and stable symbiosis of *Epichloë*, which confers beneficial traits to the host grass, has garnered widespread scientific interest and sparked a surge of research in microbiology, genetics, plant physiology, and animal science. An aspect often overlooked is its tissue-specific distribution within the host, typically found only in aboveground parts such as stems, leaf sheaths, seed coats, and endosperm layers [13,14]. The host plant can also coexist with both endophytic fungi and rhizospheric microorganisms. Due to its vertical transmission through seeds and its distribution restricted to aboveground tissues, *Epichloë* has temporal and spatial priority over rhizospheric microorganisms compared to those associated with the host's roots [15,16]. In recent years, with the development and application of high-throughput sequencing technologies, it has been discovered that *Epichloë* fungi can influence the diversity of fungi and bacteria in the rhizosphere of grasses. For instance, endophytic fungi (*E. festucae* var. lolii strains AR1 and AR37) can alter the fungi and bacterial community structure of perennial ryegrass (*Lolium perenne* cv. Samson 11104) [17]. Casas et al. found that the endophytic fungi *E. occultans* altered the bacterial community composition of potted perennial ryegrass soil but had no significant impact on the fungi community [18]. Endophytic fungi *E. coenophiala* significantly affected the fungi community in the rhizosphere of tall fescue and increased the relative abundance of certain microbial genera [19]. This indicates that aboveground *Epichloë* endophytic fungi can impact rhizospheric microorganisms. However, studies have mainly focused on artificial grasses such as tall fescue and perennial ryegrass, and have only elucidated the effects of *Epichloë* on the structure and diversity of host rhizospheric microbial communities, while the analysis of its influence on rhizospheric microbial functional genes and functional groups is relatively lacking.

But there are also plenty of *Epichloë* fungi in native grasslands. *Melica transsilvanica*, the host of this study, is a perennial grass belonging to the gramineous family. It coexists symbiotically with *E. guerinii* [20,21]. It is extensively dispersed throughout Europe, Central Asia, Russia, Siberia, and the Caucasus; nevertheless, it is only found naturally in China, on the north slope of the Tianshan Mountain, where it is highly valuable for study due to its good palatability and soft grass [22]. As experimental technology advances, sequencing depth is growing and the metagenomic study of soil microbial function is being utilized progressively [23]. In this study, metagenomics was used to sequence the rhizosphere soil of *M. transsilvanica-E. guerinii* symbionts (EI) and non-symbionts (EF). This study aims to achieve the following: (1) to further analyze the differences in the functional genes of rhizosphere soil microorganisms, focusing on functional genes related to soil nutrient cycling, virulence factors and heavy metal resistance; and (2) to analyze the differences in the functional groups of microorganisms carrying functional genes. In order to improve the understanding of *Epichloë*—host plant—rhizosphere microbial interaction, it also provides a theoretical basis for better use of endophytic fungi resources, improve agricultural production efficiency, and promote soil health and ecological sustainable development.

2. Materials and Methods

2.1. Test Materials and Sample Collection

The mountain steppe on the north slope of the Tianshan Mountain, China (87.42° E, 43.47° N, elevation 1393 m), with an annual mean temperature of 3 °C and an annual precipitation of 208.4 mm, is where the *Melica transsilvanica* seeds were harvested in August 2020. To properly destroy the endophytic fungi in the seeds, one portion was kept at 4 °C in a refrigerator, and the other portion was exposed to dry heat at 70 °C for 25 days [24]. The treated seeds were propagated in August 2022 in order to prevent the effects of dry heat on the seeds themselves. Aniline blue staining was used to confirm that the propagated seeds had no bacteria present, and they were then seeded and proved to be non-symbiontes (EF) in further testing [25]. After being sown into sterilized vermiculite in February 2023, seeds that had not been microwave-sterilized developed into three or four leaves. The plants' sheaves of leaves were ripped, then aniline blue was added. Plants exhibiting endophytic fungi mycelia were found to be symbionts (EI).

To minimize external environmental influences, we conducted indoor potted plant experiments. Soil from the northern slope of the Tianshan Mountains (87.42° E, 43.47° N, 1393 m above sea level) was mixed with sterile vermiculite in a 3:1 volume ratio and placed into 1 L nutrient bowls. Each treatment, EI and EF, had 18 replicate bowls, totaling 36. Each bowl was planted with 5 plants of consistent growth from the sterile vermiculite substrate. Daily additions of deionized water maintained a 70% soil moisture content, and any fallen leaves or other plants were promptly removed. After 8 months of growth in a controlled environment chamber, we collected rhizosphere soil. Large pieces of soil without roots were gently shaken off, and soil adhering to fine roots was brushed off with a sterile brush. The soil was quickly passed through a 2 mm sieve to remove roots, debris, and other impurities, then stored at −80 °C for metagenomic sequencing [26]. For each treatment, rhizosphere soil from 30 plants in a total of 6 nutrient bowls was randomly mixed to create one sample, with 3 samples prepared for each group.

2.2. Soil Metagenomics Sequencing

The PowerSoil DNA Isolation Kit from Mobio Laboratories was used to extract DNA from a 0.25 g rhizosphere soil sample in accordance with the manufacturer's instructions. With the use of a Thermo Fisher Scientific NanoDrop 1000 Spectrophotometer (Waltham, MA, USA), the extracted DNA's quality and amount were evaluated. Wekemo Tech Group Co., Ltd. (Shenzhen, China) employed the Illumina NovaSeq 6000 (Illumina, San Diego, CA, USA) platform for library preparation and sequencing, adhering to the manufacturer's guidelines. After sonication to reach a 350 bp size range, end repair, adenylation, and amplification were performed on the genomic DNA according to the Illumina sequencing methodology. Using the Trimmomatic-based KneadData program (v0.10.0), quality control was carried out. The efficacy and efficiency of the quality control procedure were assessed using FastQC (v0.11.9). After removing adaptor sequences and reducing low-quality reads from the initial metagenomic sequencing data—which contained reads with N bases and a minimum quality score cutoff of 20—clean reads (Table S1) were obtained.

2.3. Metagenomic Data Processing

Functional database annotation: Initiating from the reads of quality-controlled genomic data, the HUMAnN3 software (v3.0.2), which is based on the DIAMOND algorithm, was utilized to map the reads from each sample against the KEGG gene database. The abundance of functional categories was calculated based on the sum of gene abundances corresponding to KO (Kyoto Encyclopedia of Genes and Genomes) and Pathway terms. For further annotation related to virulence factors and heavy metal resistance, the DIMOND software (v 0.8.17.79) was employed, leveraging the Virulence Factors of Bacterial Proteins Database (VFDB) and the BacMet database, which are specialized repositories for such annotations. Functional species annotation: The taxonomic composition of the samples was determined by aligning the sequences against a proprietary microbial nucleic acid database constructed by Microeco Ltd., which includes sequences from bacteria, fungi, archaea, and viruses that are present in the NCBI NT nucleic acid database and the RefSeq complete genome database. Subsequently, the Bracken software (v2.7) was used to estimate the actual abundance of species within the samples, thereby providing a comprehensive profile of both functional genes and their host taxa abundance information. This integrated approach ensures a thorough understanding of the functional capabilities and species diversity within the microbial communities under study.

2.4. Data Visualization and Statistical Analysis

By utilizing the STAMP (Version 2.1.3) software, we conducted a differential analysis of the abundance of functional genes between groups. Furthermore, we employed the Bioinformatics Cloud platform (Wekemo Bioincloud: https://www.bioincloud.tech/task-meta

accessed on 4 May 2024) to perform a principal component analysis (PCA) on the functional gene sets between the two groups. To assess the variability in microbial functional communities among different group samples, we implemented a non-metric multidimensional scaling (NMDS) analysis based on the Bray–Curtis distance metric, followed by a linear discriminant analysis (LDA) [27].

3. Results

3.1. C Cycling

Within the KEGG database, the two carbon fixation pathways are delineated by the functional pathways map00710 (carbon fixation in photosynthetic organisms) and map00720 (carbon fixation in prokaryotes), while map00680 is associated with methane metabolism. No significant differences were observed in the overall abundance of genes related to the three pathways between the EI and EF groups (Figure 1A). However, the principal component analysis (PCA) results for gene sets related to the carbon cycle revealed a clear separation between the two groups along the PC1 and PC2 axes, indicating a divergence in their functional gene compositions. Further analysis using STAMP indicated (Figure 2) that the abundance of *glpX* and *rplB* genes, which are part of the carbon fixation pathway in photosynthetic organisms, was significantly higher in the EF group ($p < 0.05$). Conversely, in the carbon fixation pathway of prokaryotes, the genes encoding isocitrate dehydrogenase (icd and IDH) were not only more abundant in the EI group but also significantly higher than in the EF group ($p < 0.05$). Notably, the *por*A, *por*C, and *por*G genes, which encode key enzymes of the reductive tricarboxylic acid cycle (rTCA cycle), pyruvate ferredoxin oxidoreductase (PFOR), were present at lower abundances in the EI group but were significantly higher than in the EF group ($p < 0.05$). This pathway, proposed by Evans in 1966 [28], is one of the few autotrophic CO_2 fixation pathways in bacteria, suggesting a stronger prokaryotic carbon fixation potential in the EI rhizosphere.

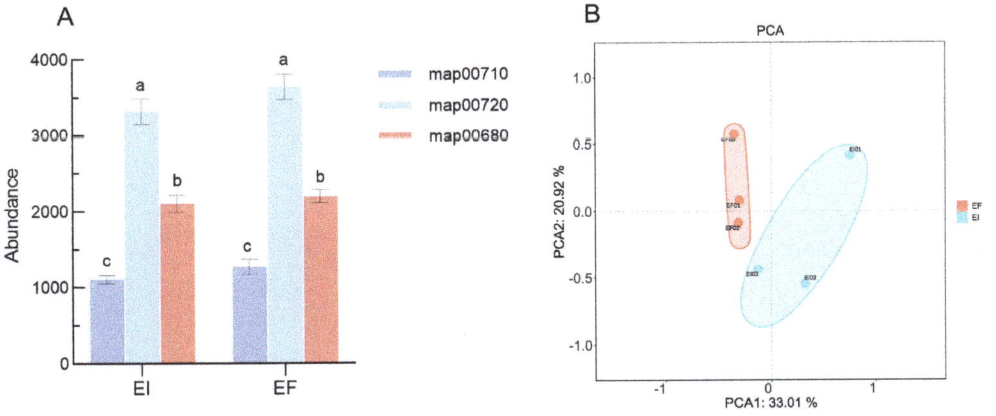

Figure 1. Total abundance of gene sets in the three main pathways of the carbon cycle (**A**); different lowercase letters above the bars indicate significant differences, as determined by one-way ANOVA ($p < 0.05$) and the principal component analysis (PCA) of carbon cycle gene sets (**B**).

We traced the microbial origins of the genes within the aforementioned pathways. Figure 3 presents the non-metric multidimensional scaling (NMDS) analysis of the three functional communities, all of which exhibit good representativeness (stress < 0.05). The results demonstrate that the community structures of the three functional groups in both the EI and EF groups are distinct. Further analysis using LEfSe to identify functionally enriched microbes with significantly altered abundance in the rhizosphere soil revealed (Figure 4) that the *Pseudomonas* genus exhibited significant differences at the genus level across all three functional groups (LDA > 2). As a typical microorganism of the reductive acetyl-CoA carbon fixation pathway [29], the higher relative abundance of *Pseudomonas* in the EI rhizosphere soil suggests a greater potential for CO_2 fixation.

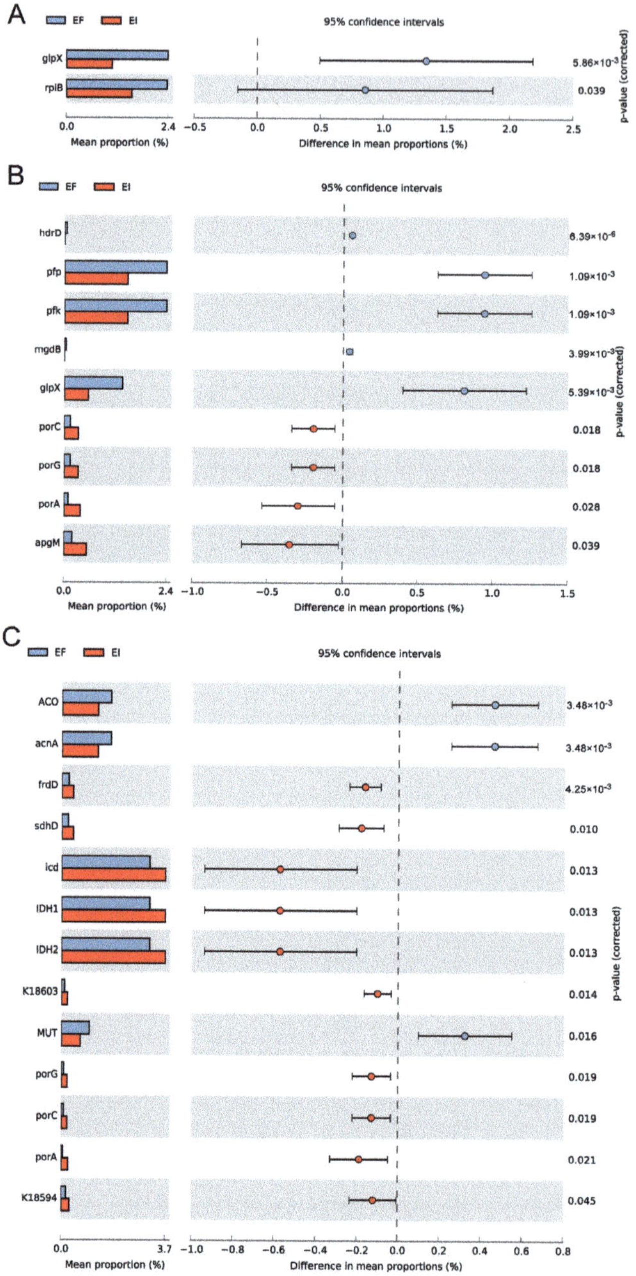

Figure 2. Intergroup variations in pathway gene sets for map00710 (**A**), map00680 (**B**), and map00720 (**C**) analysis.

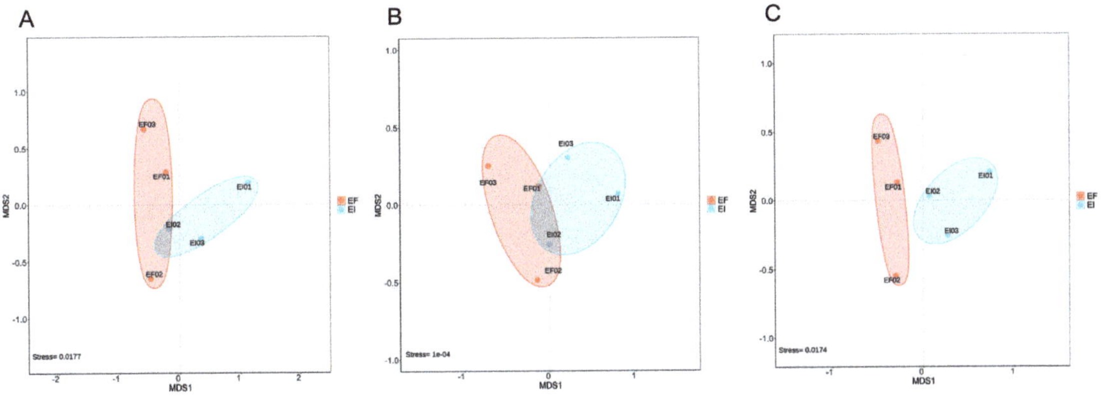

Figure 3. Nonmetric multidimensional scaling (NMDS) analysis of functional class groups by map00710 (**A**), map00720 (**B**), and map00680 (**C**).

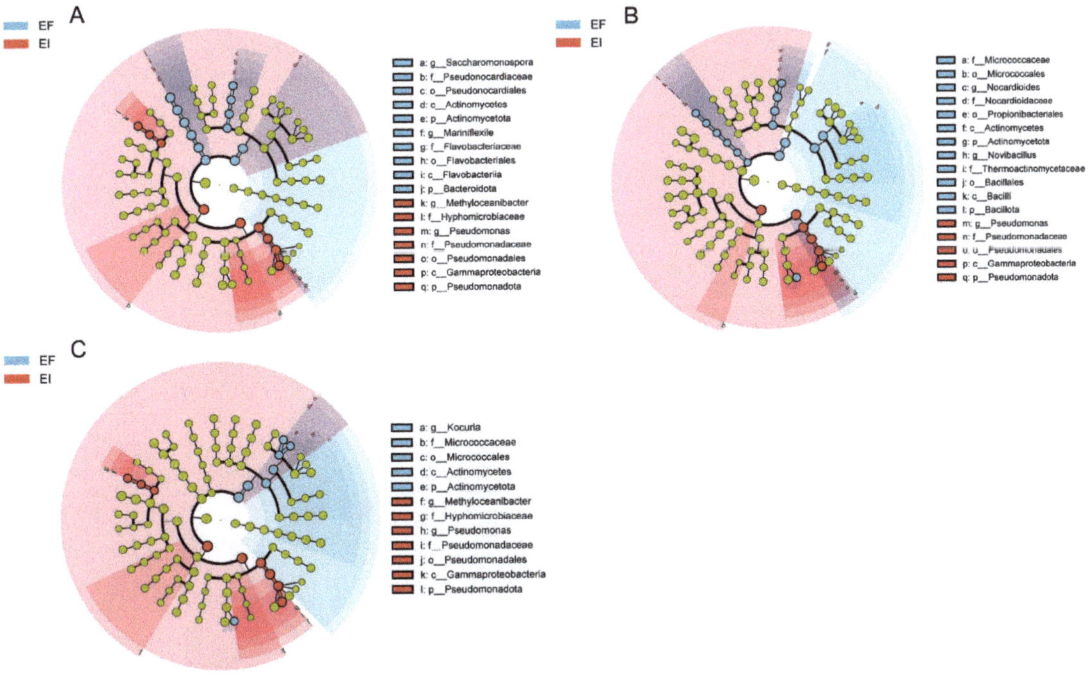

Figure 4. Variations in microorganisms' lefse analyses in the routes map00710 (**A**), map00720 (**B**), and map00680 (**C**) (LDA > 2).

3.2. N Cycling

Our principal component analysis (PCA) of the gene set of the (map00910) nitrogen cycle pathway showed that the distance between EI and EF samples was small (Figure 5A). Subsequently, we mapped the relevant genes to specific nitrogen cycle pathways (Table S2) and conducted a STAMP analysis (Figure 5C). The results indicated that the abundance of the *nxr*AB gene was significantly higher in the EI group compared to the EF group ($p < 0.05$). The *nxr*AB gene primarily participates in the nitrification process, converting nitrite to nitrate. In contrast, the abundance of *nir*BD and *nrf*AH was significantly higher in the EF group than in the EI group ($p < 0.05$), corresponding to the pathway of dissimilatory nitrate reduction to ammonia. Therefore, it can be inferred that the rhizosphere microbiota of EI plants exhibit a stronger nitrification capacity, which promotes the host's nitrogen uptake. We also traced the microbial origins of the genes within the nitrogen cycle pathways (Figure 5D). *Pseudomonas* was found to be more enriched in the rhizosphere of EI, and *Nitrosoarchaeum*, as an ammonia-oxidizing archaeon, also showed a higher abundance in EI, which may be key to the stronger nitrification capacity in the EI rhizosphere. Furthermore, NMDS analysis revealed differences in the community composition of nitrogen cycle-related functional groups between EI and EF (Figure 5B).

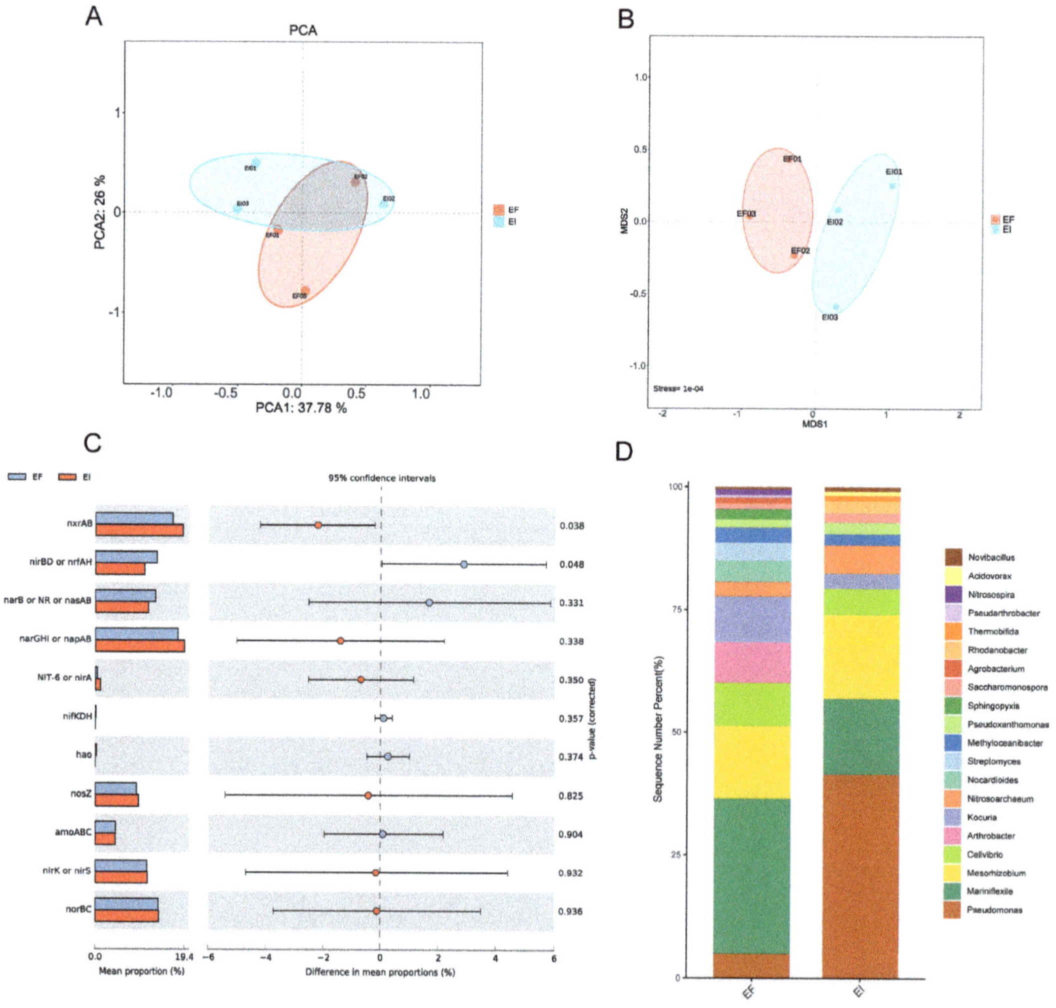

Figure 5. Principal component analysis (PCA) of the gene set associated with the nitrogen cycle (map00910) (**A**), differential gene abundance in major pathways (**C**), and the taxonomic origins of functional groups related to the nitrogen cycle (**D**) alongside a non-metric multidimensional scaling (NMDS) analysis (**B**).

3.3. P Metabolism

We separately annotated the abundance of genes related to phosphorus metabolism. PCA analysis showed clear separation and differences between the EF and EI groups in PC1 and PC2, indicating significant differences in the genes related to phosphorus metabolism between the two (Figure 6A). Classifying these functional genes into relevant pathways, it appears that the rhizosphere microbes of EF plants are more capable of promoting phosphorus cycling. Specifically, genes related to Sec-SRP (Secreted Soluble Reactive Phosphorus) involved in phosphorus transformation and cycling were significantly more abundant in EF than in EI (Figure 6B).

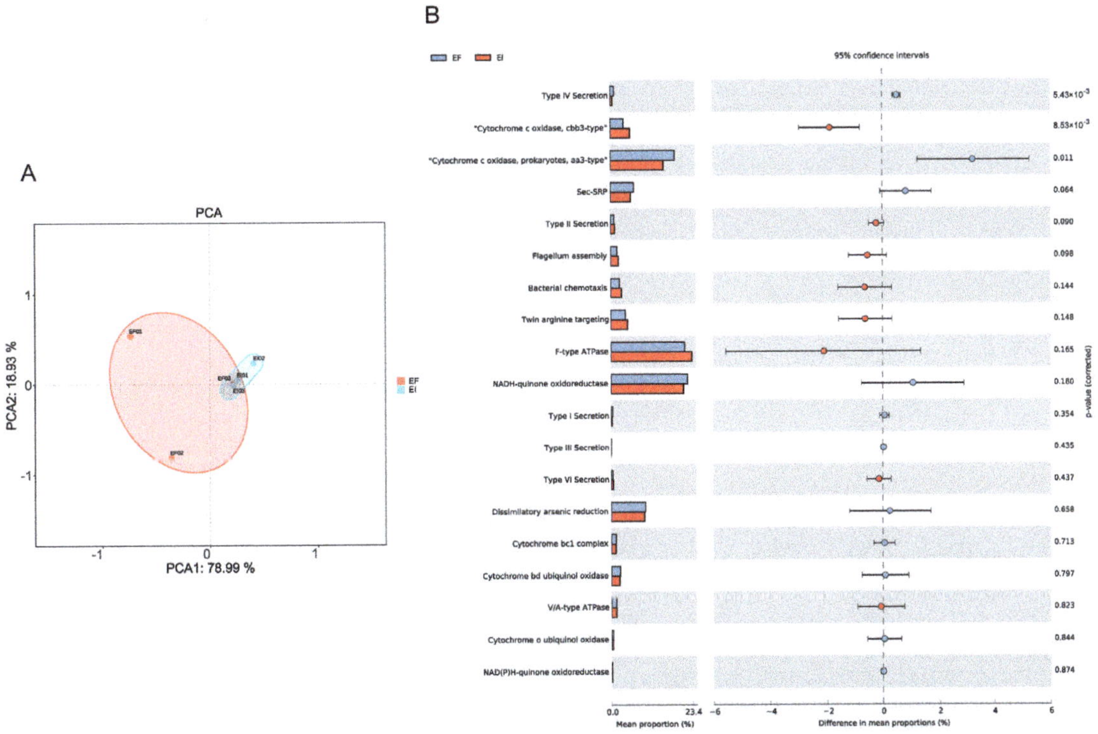

Figure 6. Principal component analysis (PCA) of the gene set associated with phosphorus metabolism (**A**), and the variation in gene abundance within key pathways (**B**).

3.4. Resistance to Heavy Metals and Virulence Factors

Through annotation with the VFDB (virulence factor) database, we found that the abundance of virulence factors in the rhizosphere of EI plants was lower, and the types were fewer compared to EF plants (Figure 7A). This indicates that the presence of the Epichloë endophytic fungi can reduce the enrichment of virulence factors in the host rhizosphere, thereby enhancing host resistance. Annotation with the BacMet (heavy metal resistance) database revealed that the rhizosphere of EI plants was enriched with more microbial functional genes resistant to Zn, Sb, and Pb (Figure 7B).

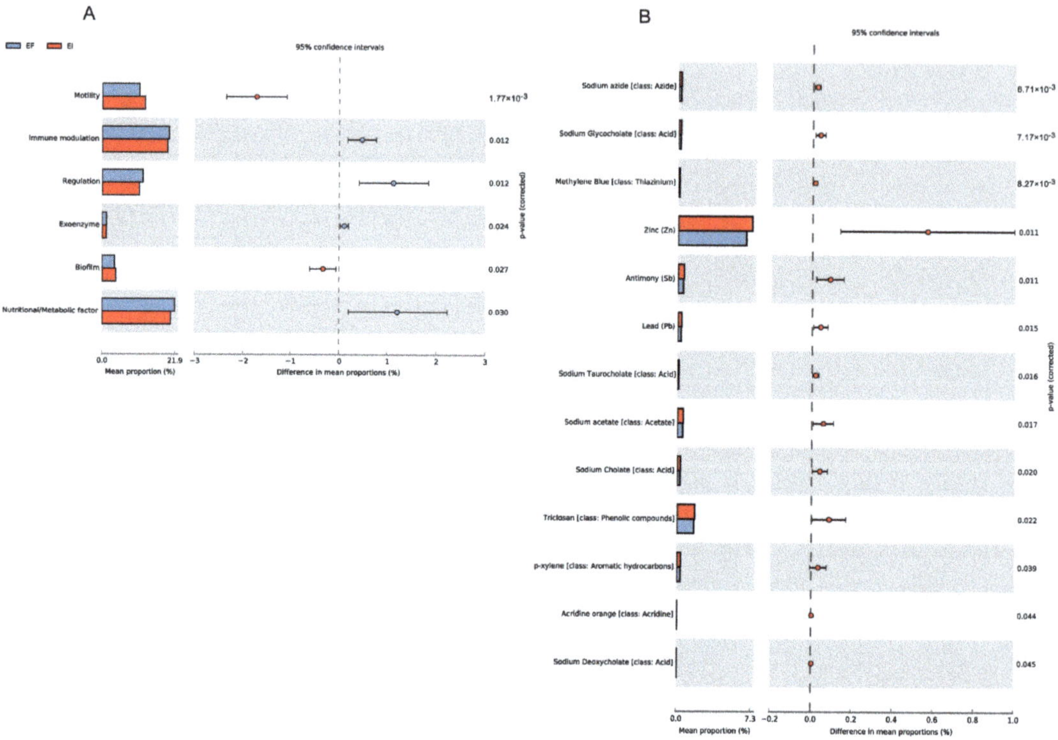

Figure 7. Variations in the rhizosphere microorganisms of EI and EF plants with respect to the virulence factor (**A**) and heavy metal resistance (**B**).

4. Discussion

In terms of carbon cycling, previous studies have shown that the host's root exudates and litter are the main sources of soil organic carbon. The infection of endophytic fungi increases the host's biomass, so that the litter into the soil increases, resulting in an increase in soil organic carbon and other substances [30]. The presence of endophytic fungi may also promote the secretion of some compounds by the host root, directly affecting the content of soil carbon [31]. Through the analysis of functional genes and related microbial groups in rhizosphere soil, this study found that the presence of *Epichloë* increased the abundance of the key gene *por*A in the host rhizosphere microorganism rTCA, and increased the abundance of Pseudomonas in the corresponding functional groups. Studies have shown that PFOR encoded by *por*A is an important enzyme in the rTCA cycle, which is involved in the transformation of pyruvate to acetyl coenzyme A. *Pseudomonas* is a typical reducing acetyl coenzyme A carbon sequestration pathway [32,33]. Yang et al. [34] found, in the carbon sequestration study of soil microorganisms in arid desert oases, that the rTCA cycle indicated by the difference in the *por*A gene is one of the most common carbon sequestration pathways. When Ren et al. [35] used the macrogenome combined with the macrotranscriptome to study the soil microbial function in Tarim Basin, he also found that the rTCA cycle driven by *por*A, *por*G and IDH1 genes is one of the carbon sequestration pathways. These two aspects indicate that the presence of *Epichloë* can enrich the host rhizosphere with more microorganisms with the ability to fix CO_2, and also increase the abundance of carbon-fixing function genes, which may be the third reason why endophytic fungi increase the organic carbon content of the host rhizosphere soil.

In terms of nitrogen cycling, endophytic fungi can reduce the impact of low nitrogen stress on the host by directly enhancing the absorption of nutrients by the host. There are also related studies that evaluate the nitrification potential of soil by measuring the nitrate nitrogen in soil; find that the presence of endophytic fungi can increase soil nitrification [36]. Endophytic fungi infection can reduce the activity of nitrification-inhibiting substances in the host rhizosphere, and then affect soil nitrification [37], which is consistent with the conclusion that the rhizosphere microorganisms of EI plants have stronger nitrification in this study. For the functional gene level of *Epichloë* affecting the nitrogen cycling of host rhizosphere microorganisms, only two articles have reported that the presence of *Epichloë festucae* affects the nitrogen cycling process of rhizosphere microorganisms of host ryegrass; under low nitrogen stress, the abundance of ammonia-oxidizing bacteria and ammonia-oxidizing archaea in the rhizosphere of the symbiotic species increased. And there are differences in the functional groups of nitrogen-fixing microorganisms between the symbiotic and non-symbiotic species [38,39]. The results of this study were more prominent than those of our study, but in the same direction as our conclusions, which may be due to the fact that the above experiments were conducted under low nitrogen conditions, and the role of endophytic fungi was highlighted.

Interestingly, the results of the comprehensive functional groups showed that *Epichloë* could recruit a large number of growth-promoting bacteria *Pseudomonas* in the host rhizosphere, which was consistent with the conclusions of previous studies, that is, *Pseudomonas* bacteria were the dominant bacterial groups widely distributed in the rhizosphere, seed and leaf of *Epichloë–Achnatherum inebrians* symbiosis [40]. *Pseudomonas* had a nitrogen fixation effect in the plant rhizosphere, as well as the role of regulating plant hormones and increasing abiotic stress tolerance [41]. As early as 1996, it was found that *Pseudomonas* was a strong candidate soil inoculant bacteria to improve crop yield [42]. Recent studies also found that inoculation of *P. aeruginosa* could promote the growth of pea (*Pisum sativum*) [43], and inoculation of *P. moraviensis* could positively promote the growth and physiology of wheat (*Triticum aestivum*), while significantly improving the soil organic matter and nitrogen availability [44]. Therefore, *Pseudomonas* play an important role in promoting plant growth and development. We concluded that the overall growth-promoting effect of *Epichloë* on the host was partially attributed to the recruitment of *Pseudomonas* by symbiosis. As for why endophytic fungi chose it as a "helper", it is also a scientific question worthy of future research.

In the context of phosphorus metabolism, this study did not observe significant differences between the EI and EF groups, and it is even possible that the potential of rhizosphere microorganisms in phosphorus metabolism for EI plants is lower than that for EF. However, some research has indicated that *E. gansuensis* improves the growth of its host *A. inebrians* under low phosphorus stress by regulating amino acid metabolic pathways, amino acid content, and organic acid content, and by increasing phosphorus utilization efficiency [45]. This phenomenon may be related to the symbiotic relationship between plants and endophytic fungi, where the endophytic fungi may enhance the host's own phosphorus absorption and reduce the host's reliance on rhizosphere microorganisms with the function of activating phosphorus or secreting organic phosphorus. Previous studies have concluded that *Epichloë* has no significant effect on the content of the available phosphorus in host rhizosphere soil under non-stress conditions, which can also confirm this view [46].

The influence of *Epichloë* on the host in producing specific metabolites to resist pathogen invasion in both aboveground and belowground parts has been extensively studied [47]. In this study, the lower abundance and diversity of virulence factors enriched in the root zone of EI plants further corroborate this finding. Additionally, *Epichloë* can enhance host tolerance to heavy metal ions such as cadmium (Cd^{2+}), aluminum (Al^{3+}), zinc (Zn^{2+}), and copper (Cu^{2+}) in soil. According to a large number of previous studies, under the stress of toxic heavy metals, symbiotic plants have higher antioxidant enzyme activity and chlorophyll a and b contents, and lower proline and malondialdehyde contents than non-symbiotic plants, indicating that the presence of *Epichloë* can better activate the antioxidant system of host plants [48–50]. The activity of antioxidant enzymes in plants was increased to reduce the ROSs (reactive oxygen species) produced under heavy metal stress. The oxidative damage of ROSs to nucleic acid, membrane and chloroplast pigment was alleviated. In addition to protecting photosynthetic pigments from destruction, *Epichloë* also enhances the activity of the enzymes related to photosynthesis and chlorophyll synthesis [51]. Other studies have shown that heavy metal stress, especially Cu^{2+}, can cause plants to produce excessive ethylene and accelerate plant senescence, while *Epichloë* can regulate ethylene synthesis and inhibit plant senescence, thereby enhancing the host's tolerance to heavy metals [52]. In conclusion, endophytic fungi enhance the host's tolerance to heavy metal ions by improving the host's antioxidant capacity and photosynthesis level, and by activating related hormone signals. In this study, it was found that the abundance of the functional genes of microorganisms resistant to Zn, Sb and Pb in the rhizosphere microorganisms of EI plants was higher than that of EF plants. To a certain extent, this indicates that Epichloë enhances the host's tolerance to heavy metal ions, which not only affects the physiological response of the host plant itself, but it also indirectly affects the resistance of rhizosphere microorganisms symbiosis with the host to heavy metals.

This study provides new insights into the interactions among *Epichloë*, host plants, and rhizosphere microorganisms, offering potential applications for utilizing these microbial relationships to improve plant growth and soil health. However, there are some limitations in the study. For example, although metagenomic approaches can reveal the functional gene composition of microbial communities, there is a lack of direct measurement of microbial activity and metabolic processes. Future research could further explore the complex interaction mechanisms between endophytic fungi and host plants by combining metagenomics, transcriptomics, and metabolomics, among other multi-omic approaches, to investigate how these interactions affect plant growth and the functional aspects of soil ecosystems.

5. Conclusions

The presence of *Epichloë* can change the community structure of the functional groups related to the carbon cycle in the rhizosphere microorganisms of the host Melica transsilvanica, and significantly increase the abundance of *por*A, *por*G and IDH1 genes in the rTCA cycle of the carbon sequestration pathway. This may be the third reason why endophytic fungi increase the organic carbon content of the host rhizosphere soil in addition to the other two main factors, namely, increasing host litter and altering host root exudates. The abundance of the *nxr*AB gene in the rhizosphere microbial nitrogen cycle pathway in EI plants was significantly higher than that in EF plants, and the community structure of related functional groups also changed. The changes in carbon and nitrogen nutrient cycling in rhizosphere microorganisms may be related to the fact that the presence of *Epichloë* recruits a large number of *Pseudomonas* bacteria in the host rhizosphere. In addition, the presence of *Epichloë* also reduces the enrichment of virulence factors by host rhizosphere microorganisms and significantly increases the accumulation of heavy metal resistance genes to Zn, Sb and Pb. To a certain extent, it indicates that *Epichloë* enhances the host's resistance to soil-borne pathogens and heavy metal ions, which not only affects the physiological response of host plants, but also indirectly affects the rhizosphere microorganisms that are symbiotic with the host and cooperate with the host to develop resistance to stress.

Supplementary Materials: The following supporting information can be downloaded at: https://www.mdpi.com/article/10.3390/microorganisms12050956/s1, Table S1: Data output quality; Table S2: Key genes corresponding to pathways associated with the nitrogen cycle; Table S3: Key genes corresponding to pathways related to phosphorus metabolism.

Author Contributions: C.S. led the study and was responsible for resources and funding; C.W. drafted the manuscript and was in charge of experiment and data analysis visualization; M.Z. and J.H. participated in the experiments; W.H. conducted research design and review. All authors have read and agreed to the published version of the manuscript.

Funding: This research was supported by the National Natural Science Foundation of China (31760704) and the Natural Science Foundation of Xinjiang Uygur Autonomous Region (2022D01A79) and special found of Xinjiang Key Laboratory of Soil and Plant Ecological Processes (23XJTRZW07).

Data Availability Statement: The sequences were submitted to the NCBI database (ID: PRJNA949598).

Acknowledgments: The authors also thank the anonymous reviewers for their constructive comments on the manuscript. Thanks to N.N. Zhang for her support of this study.

Conflicts of Interest: The authors declare no conflicts of interest.

References

1. Hassani, M.A.; Durán, P.; Hacquard, S. Microbial interactions within the plant holobiont. *Microbiome* **2018**, *6*, 58. [CrossRef] [PubMed]
2. Compant, S.; Clément, C.; Sessitsch, A. Plant growth-promoting bacteria in the rhizo-and endosphere of plants: Their role, colonization, mechanisms involved and prospects for utilization. *Soil Biol. Biochem.* **2010**, *42*, 669–678. [CrossRef]
3. Russ, D.; Fitzpatrick, C.R.; Teixeira, P.J.P.L.; Dangl, J.L. Deep discovery informs difficult deployment in plant microbiome science. *Cell* **2023**, *186*, 4496–4513. [CrossRef] [PubMed]
4. Lee, K.K.; Kim, H.; Lee, Y.H. Cross-kingdom co-occurrence networks in the plant microbiome: Importance and ecological interpretations. *Front. Microbiol.* **2022**, *13*, 953300. [CrossRef] [PubMed]
5. Zhang, J.; Cook, J.; Nearing, J.T.; Zhang, J.; Raudonis, R.; Glick, B.R.; Langille, M.G.; Cheng, Z. Harnessing the plant microbiome to promote the growth of agricultural crops. *Microbiol. Res.* **2021**, *245*, 126690. [CrossRef] [PubMed]
6. Berendsen, R.L.; Pieterse CM, J.; Bakker PA, H.M. The rhizosphere microbiome and plant health. *Trends Plant Sci.* **2012**, *17*, 478–486. [CrossRef] [PubMed]
7. Lee, J.H.; Anderson, A.J.; Kim, Y.C. Root-associated bacteria are biocontrol agents for multiple plant pests. *Microorganisms* **2022**, *10*, 1053. [CrossRef] [PubMed]
8. Clay, K.; Schardl, C. Evolutionary origins and ecological consequences of endophyte symbiosis with grasses. *Am. Nat.* **2002**, *160*, S99–S127. [CrossRef] [PubMed]
9. Mosaddeghi, M.R.; Hosseini, F.; Hajabbasi, M.A.; Sabzalian, M.R.; Sepehri, M. *Epichloë* spp. and Serendipita indica endophytic fungi: Functions in plant-soil relations. *Adv. Agron.* **2021**, *165*, 59–113. [CrossRef]
10. Waqar, S.; Bhat, A.A.; Khan, A.A. Endophytic fungi: Unravelling plant endophyte interaction and the multifaceted role of fungi endophytes in stress amelioration. *Plant Physiol. Biochem.* **2023**, *206*, 108174. [CrossRef]
11. Fardella, P.A.; Clarke, B.B.; Belanger, F.C. The *Epichloë festucae* antifungi protein Efe-AfpA has activity against numerous plant pathogens. *Microorganisms* **2023**, *11*, 828. [CrossRef]
12. Tadych, M.; Bergen, M.S.; White, J.F., Jr. *Epichloë* spp. associated with grasses: New insights on life cycles, dissemination and evolution. *Mycologia* **2014**, *106*, 181–201. [CrossRef] [PubMed]
13. Deng, Y.; Yu, X.; Yin, J.; Chen, L.; Zhao, N.; Gao, Y.; Ren, A. *Epichloë* Endophyte Enhanced Insect Resistance of Host Grass *Leymus Chinensis* by Affecting Volatile Organic Compound Emissions. *J. Chem. Ecol.* **2023**, 1–10. [CrossRef]
14. Scott, B.; Green, K.; Berry, D. The fine balance between mutualism and antagonism in the *Epichloë festucae*–grass symbiotic interaction. *Curr. Opin. Plant Biol.* **2018**, *44*, 32–38. [CrossRef] [PubMed]
15. Saikkonen, K.; Young, C.A.; Helander, M.; Schardl, C.L. Endophytic *Epichloë* species and their grass hosts: From evolution to applications. *Plant Mol. Biol.* **2016**, *90*, 665–675. [CrossRef]
16. Mack, K.M.L.; Rudgers, J.A. Balancing multiple mutualists: Asymmetric interactions among plants, arbuscular mycorrhizal fungi, and fungi endophytes. *Oikos* **2008**, *117*, 310–320. [CrossRef]
17. Wakelin, S.; Harrison, S.; Mander, C.; Dignam, B.; Rasmussen, S.; Monk, S.; Fraser, K.; O'Callaghan, M. Impacts of endophyte infection of ryegrass on rhizosphere metabolome and microbial community. *Crop Pasture Sci.* **2015**, *66*, 1049–1057. [CrossRef]
18. Rojas, X.; Guo, J.; Leff, J.W.; McNear, D.H.; Fierer, N.; McCulley, R.L. Infection with a shoot-specific fungi endophyte (*Epichloë*) alters tall fescue soil microbial communities. *Microb. Ecol.* **2016**, *72*, 197–206. [CrossRef] [PubMed]
19. Mahmud, K.; Lee, K.; Hill, N.S.; Mergoum, A.; Missaoui, A. Influence of Tall Fescue *E. pichloë* Endophytes on Rhizosphere Soil Microbiome. *Microorganisms* **2021**, *9*, 1843. [CrossRef] [PubMed]
20. Moon, C.D.; Guillaumin, J.J.; Ravel, C.; Li, C.; Craven, K.D.; Schardl, C.L. New *Neotyphodium* endophyte species from the grass tribes Stipeae and *Meliceae*. *Mycologia* **2007**, *99*, 895–905. [CrossRef]
21. Zhang, X.; Ren, A.Z.; Wei, Y.K.; Lin, F.; Li, C.; Liu, Z.-J.; Gao, Y.-B. Taxonomy, diversity and origins of symbiotic endophytes of *Achnatherum sibiricum* in the Inner Mongolia Steppe of China. *FEMS Microbiol. Lett.* **2009**, *301*, 12–20. [CrossRef] [PubMed]
22. Castro, S.; Muratet, A.; Szczepaniak, M.; Nguefack, J.; Hardion, L. RAD sequencing, morphometry and synecology clarify the taxonomy of the *Melica ciliata* (*Poaceae*) complex in France and Poland. *J. Syst. Evol.* **2023**, *61*, 764–775. [CrossRef]
23. Qian, X.; Gunturu, S.; Guo, J.; Chai, B.; Cole, J.R.; Gu, J.; Tiedje, J.M. Metagenomic analysis reveals the shared and distinct features of the soil resistome across tundra, temperate prairie, and tropical ecosystems. *Microbiome* **2021**, *9*, 108. [CrossRef] [PubMed]
24. Shi, C.; Lu, Y.; Huang, C.; Zhang, R.S.; An, S.Z. Using 70 °C dry treatment to make endophyte-free seeds of *Melica transsilvanica*. *Acta Agrestia Sin.* **2016**, *24*, 1318–1322. [CrossRef]

25. Becker, M.; Becker, Y.; Green, K.; Scott, B. The endophytic symbiont *Epichloë festucae* establishes an epiphyllous net on the surface of *Lolium perenne* leaves by development of an expressorium, an appressorium-like leaf exit structure. *New Phytol.* **2016**, *211*, 240–254. [CrossRef] [PubMed]
26. Bulgarelli, D.; Rott, M.; Schlaeppi, K.; Ver Loren van Themaat, E.; Ahmadinejad, N.; Assenza, F.; Rauf, P.; Huettel, B.; Reinhardt, R.; Schmelzer, E.; et al. Revealing structure and assembly cues for Arabidopsis root-inhabiting bacterial microbiota. *Nature* **2012**, *488*, 91–95. [CrossRef] [PubMed]
27. Gao, Y.; Zhang, G.; Jiang, S.; Liu, Y. Wekemo Bioincloud: A user-friendly platform for meta-omics data analyses. *iMeta* **2024**, *3*, e175. [CrossRef]
28. Fuchs, G. Alternative pathways of carbon dioxide fixation: Insights into the early evolution of life? *Annu. Rev. Microbiol.* **2011**, *65*, 631–658. [CrossRef] [PubMed]
29. Kozaeva, E.; Volkova, S.; Matos, M.R.A.; Mezzina, M.P.; Wulff, T.; Volke, D.C.; Nielsen, L.K.; Nikel, P.I. Model-guided dynamic control of essential metabolic nodes boosts acetyl-coenzyme A–dependent bioproduction in rewired Pseudomonas putida. *Metab. Eng.* **2021**, *67*, 373–386. [CrossRef] [PubMed]
30. Wu, L.; Chen, X.; Wei, W.; Liu, Y.; Wang, D.; Ni, B.J. A critical review on nitrous oxide production by ammonia-oxidizing archaea. *Environ. Sci. Technol.* **2020**, *54*, 9175–9190. [CrossRef]
31. Panchal, P.; Preece, C.; Peñuelas, J.; Giri, J. Soil carbon sequestration by root exudates. *Trends Plant Sci.* **2022**, *27*, 749–757. [CrossRef] [PubMed]
32. Iqbal, J.; Siegrist, J.A.; Nelson, J.A.; McCulley, R.L. Fungi endophyte infection increases carbon sequestration potential of southeastern USA tall fescue stands. *Soil Biol. Biochem.* **2012**, *44*, 81–92. [CrossRef]
33. Tang, K.H.; Blankenship, R.E. Both forward and reverse TCA cycles operate in green sulfur bacteria. *J. Biol. Chem.* **2010**, *285*, 35848–35854. [CrossRef] [PubMed]
34. Yang, Y.; Yang, X.; Gong, L.; Ding, Z.; Zhu, H.; Tang, J.; Li, X. Changes in soil microbial carbon fixation pathways along the oasification process in arid desert region: A confirmation based on metagenome analysis. *CATENA* **2024**, *239*, 107955. [CrossRef]
35. Ren, M.; Zhang, Z.; Wang, X.; Zhou, Z.; Chen, D.; Zeng, H.; Zhao, S.; Chen, L.; Hu, Y.; Zhang, C.; et al. Diversity and contributions to nitrogen cycling and carbon fixation of soil salinity shaped microbial communities in Tarim Basin. *Front. Microbiol.* **2018**, *9*, 431. [CrossRef] [PubMed]
36. Handayani, I.P.; Coyne, M.S.; Phillips, T.D. Soil organic carbon fractions differ in two contrasting tall fescue systems. *Plant Soil* **2011**, *338*, 43–50. [CrossRef]
37. Guo, J.; McCulley, R.L.; Phillips, T.D.; McNear, D.H., Jr. Fungi endophyte and tall fescue cultivar interact to differentially affect bulk and rhizosphere soil processes governing C and N cycling. *Soil Biol. Biochem.* **2016**, *101*, 165–174. [CrossRef]
38. Chen, Z.; Jin, Y.; Yao, X.; Wei, X.; Li, X.; Li, C.; White, J.F.; Nan, Z. Gene analysis reveals that leaf litter from *Epichloë* endophyte-infected perennial ryegrass alters diversity and abundance of soil microbes involved in nitrification and denitrification. *Soil Biol. Biochem.* **2021**, *154*, 108123. [CrossRef]
39. Chen, Z.; White, J.F.; Malik, K.; Chen, H.; Jin, Y.; Yao, X.; Wei, X.; Li, C.; Nan, Z. Soil nutrient dynamics relate to *Epichloë* endophyte mutualism and nitrogen turnover in a low nitrogen environment. *Soil Biol. Biochem.* **2022**, *174*, 108832. [CrossRef]
40. Ju, Y.; Kou, M.; Zhong, R.; Christensen, M.J.; Zhang, X. Alleviating salt stress on seedings using plant growth promoting rhizobacteria isolated from the rhizosphere soil of *Achnatherum inebrians* infected with *Epichloë gansuensis* endophyte. *Plant Soil* **2021**, *465*, 349–366. [CrossRef]
41. Richardson, A.E.; Barea, J.M.; McNeill, A.M.; Prigent-Combaret, C. Acquisition of phosphorus and nitrogen in the rhizosphere and plant growth promotion by microorganisms. *Plant Soil* **2009**, *321*, 305–339. [CrossRef]
42. Xie, H.; Pasternak, J.J.; Glick, B.R. Isolation and characterization of mutants of the plant growth-promoting rhizobacterium *Pseudomonas* putida GR12-2 that overproduce indoleacetic acid. *Curr. Microbiol.* **1996**, *32*, 67–71. [CrossRef]
43. Sasirekha, B.; Shivakumar, S. Statistical optimization for improved indole-3-acetic acid (iaa) production by *Pseudomonas* aeruginosa and demonstration of enhanced plant growth promotion. *J. Soil Sci. Plant Nutr.* **2012**, *12*, 863–873. [CrossRef]
44. Ul Hassan, T.; Bano, A. The stimulatory effects of L-tryptophan and plant growth promoting rhizobacteria (PGPR) on soil health and physiology of wheat. *J. Soil Sci. Plant Nutr.* **2015**, *15*, 190–201. [CrossRef]
45. Liu, Y.; Hou, W.; Jin, J.; Christensen, M.J.; Gu, L.; Cheng, C.; Wang, J. *Epichloë gansuensis* increases the tolerance of *Achnatherum inebrians* to low-P stress by modulating amino acids metabolism and phosphorus utilization efficiency. *J. Fungi* **2021**, *7*, 390. [CrossRef]
46. Chen, J.; Deng, Y.; Yu, X.; Wu, G.; Gao, Y.; Ren, A. *Epichloë* endophyte infection changes the root endosphere microbial community composition of *Leymus chinensis* under both potted and field growth conditions. *Microb. Ecol.* **2023**, *85*, 604–616. [CrossRef] [PubMed]
47. Fernando, K.; Reddy, P.; Guthridge, K.M.; Spangenberg, G.C.; Rochfort, S.J. A metabolomic study of *Epichloë* endophytes for screening antifungi metabolites. *Metabolites* **2022**, *12*, 37. [CrossRef] [PubMed]
48. Zhai, Y.; Chen, Z.; Malik, K.; Wei, X.; Li, C. Effect of fungi endophyte *Epichloë bromicola* infection on Cd tolerance in Wild Barley (*Hordeum brevisubulatum*). *J. Fungi* **2022**, *8*, 366. [CrossRef] [PubMed]
49. Wiewióra, B.; Żurek, G. The response of the associations of grass and *Epichloë* endophytes to the increased content of heavy metals in the soil. *Plants* **2021**, *10*, 429. [CrossRef]

50. Lee, K.; Missaoui, A.; Mahmud, K.; Presley, H.; Lonnee, M. Interaction between grasses and *Epichloë* endophytes and its significance to biotic and abiotic stress tolerance and the rhizosphere. *Microorganisms* **2021**, *9*, 2186. [CrossRef]
51. Żurek, G.; Wiewióra, B.; Rybka, K.; Prokopiuk, K. Different response of perennial ryegrass—*Epichloë* endophyte symbiota to the elevated concentration of heavy metals in soil. *J. Appl. Genet.* **2022**, *63*, 47–59. [CrossRef] [PubMed]
52. Wang, J.; Hou, W.; Christensen, M.J.; Li, X.; Xia, C.; Li, C.; Nan, Z. Role of *Epichloë* endophytes in improving host grass resistance ability and soil properties. *J. Agric. Food Chem.* **2020**, *68*, 6944–6955. [CrossRef] [PubMed]

microorganisms

Article

Plant Growth-Promoting Bacteria Influence Microbial Community Composition and Metabolic Function to Enhance the Efficiency of *Hybrid pennisetum* Remediation in Cadmium-Contaminated Soil

Zhao-Jin Chen, Meng-Lu Li, Shan-Shan Gao, Yu-Bo Sun, Hui Han, Bai-Lian Li and Yu-Ying Li *

Overseas Expertise Introduction Center for Discipline Innovation of Watershed Ecological Security in the Water Source Area of the Mid-line Project of South-to-North Water Diversion, College of Water Resource and Environment Engineering, Nanyang Normal University, Nanyang 473061, China; zhaojin_chen@163.com (Z.-J.C.)
* Correspondence: liyuying@nynu.edu.cn

Abstract: The green and efficient remediation of soil cadmium (Cd) is an urgent task, and plant-microbial joint remediation has become a research hotspot due to its advantages. High-throughput sequencing and metabolomics have technical advantages in analyzing the microbiological mechanism of plant growth-promoting bacteria in improving phytoremediation of soil heavy metal pollution. In this experiment, a pot trial was conducted to investigate the effects of inoculating the plant growth-promoting bacterium *Enterobacter* sp. VY on the growth and Cd remediation efficiency of the energy plant *Hybrid pennisetum*. The test strain VY-1 was analyzed using high-throughput sequencing and metabolomics to assess its effects on microbial community composition and metabolic function. The results demonstrated that *Enterobacter* sp. VY-1 effectively mitigated Cd stress on *Hybrid pennisetum*, resulting in increased plant biomass, Cd accumulation, and translocation factor, thereby enhancing phytoremediation efficiency. Analysis of soil physical-chemical properties revealed that strain VY-1 could increase soil total nitrogen, total phosphorus, available phosphorus, and available potassium content. Principal coordinate analysis (PCoA) indicated that strain VY-1 significantly influenced bacterial community composition, with Proteobacteria, Firmicutes, Chloroflexi, among others, being the main differential taxa. Redundancy analysis (RDA) revealed that available phosphorus, available potassium, and pH were the primary factors affecting bacterial communities. Partial Least Squares Discriminant Analysis (PLS-DA) demonstrated that strain VY-1 modulated the metabolite profile of *Hybrid pennisetum* rhizosphere soil, with 27 differential metabolites showing significant differences, including 19 up-regulated and eight down-regulated expressions. These differentially expressed metabolites were primarily involved in metabolism and environmental information processing, encompassing pathways such as glutamine and glutamate metabolism, α-linolenic acid metabolism, pyrimidine metabolism, and purine metabolism. This study utilized 16S rRNA high-throughput sequencing and metabolomics technology to investigate the impact of the plant growth-promoting bacterium *Enterobacter* sp. VY-1 on the growth and Cd enrichment of *Hybrid pennisetum*, providing insights into the regulatory role of plant growth-promoting bacteria in microbial community structure and metabolic function, thereby improving the microbiological mechanisms of phytoremediation.

Keywords: cadmium (Cd); phytoremediation; bacterial community and function; high-throughput sequencing; non-targeted metabolomic

1. Introduction

With the rapid development of China's industry, soil heavy metal pollution has become increasingly severe and widespread, with cadmium (Cd) being the most severely polluted among all elements [1,2]. Cd is highly ecotoxic and can accumulate in various plants, posing a serious threat to crop productivity and safety. It is considered a significant pollutant in agricultural soil [3].

In the field of heavy metal-contaminated soil remediation, phytoremediation has emerged as a green and sustainable method because it does not harm the soil ecological environment and avoids secondary pollution [4,5]. *Hybrid pennisetum* has become an excellent cellulosic energy plant due to its rapid growth and strong stress resistance. It exhibits a robust ability to absorb various heavy metals and is considered one of the most promising energy plants in China, having been extensively researched for the remediation of heavy metal-contaminated soil [6–8]. Choosing *Hybrid pennisetum* as a remediation plant not only allows for the remediation of heavy metal-contaminated farmland and abandoned mining areas but also enables the stems of the revitalized plant to be used for fermentation to produce fuel ethanol. This approach realizes both soil remediation and subsequent resource processing [9]. However, remediation plants like *Hybrid pennisetum* often exhibit slow growth, reduced biomass, and low heavy metal transfer rates under heavy metal stress conditions, severely limiting phytoremediation efficiency.

Plant growth-promoting bacteria (PGPB), which inhabit the rhizosphere soil, can mitigate the toxicity of heavy metal ions to plants by leveraging their own resistance systems [10]. They can also alter the form and mobility of heavy metals, directly or indirectly promoting plant growth and enhancing plant resistance to heavy metals. Consequently, PGPB play a crucial role in strengthening the process of phytoremediation of soil heavy metal pollution [11]. Plant-microbial joint remediation has garnered significant attention in the field of heavy metal-contaminated soil remediation as it leverages the complementary strengths of plants and microorganisms [10,12].

Soil microorganisms are integral to the soil environment, participating in complex biological and ecological processes and exerting a significant influence on plant growth and phytoremediation efficiency [13]. The effectiveness of plant-microbe joint remediation largely hinges on the adaptability of inoculated PGPB to environmental changes and their impact on the composition and metabolic functions of indigenous microbial communities. In recent years, omics technologies (such as high-throughput sequencing and metabolomics) have been employed in plant-microbial joint remediation, providing valuable insights into the mechanisms underlying PGPB-mediated improvements in phytoremediation [14–16]. Therefore, this study utilized 16S rRNA high-throughput sequencing and metabolomics technology to investigate the effects of the plant growth-promoting bacterium *Enterobacter* sp. VY-1 on the microbiological mechanism of phytoremediation enhancement from the perspectives of diversity and metabolic function.

2. Materials and Methods

2.1. Test Materials

The test strain was selected from plant growth-promoting bacteria preserved in the laboratory, identified as belonging to the genus *Enterobacter*, and named *Enterobacter* sp. VY-1. *Enterobacter* sp. VY-1 exhibits plant growth-promoting properties such as phosphorus solubilization, potassium solubilization, siderophore production, and IAA production, along with good tolerance to Cd. For detailed information, refer to Liu et al. [17]. *Hybrid pennisetum* seeds were procured from Shuyang County Tuojing Horticulture Co., Ltd. (Suqian, China). The potting soil was obtained from non-polluted soil in the pomegranate garden of Nanyang Normal University. The basic physical and chemical properties of the soil can be found in Liu et al. [17]. The soil was air-dried naturally after removing impurities such as leaves and stones, and then passed through a 20-mesh sieve for later use. $CdSO_4 \cdot 8H_2O$ was added in proportion to achieve a Cd ion content of 20 $mg \cdot kg^{-1}$ in the soil, thoroughly mixed, and allowed to equilibrate for 30 days.

2.2. Pot Experiment

The soil was filled into plastic flowerpots, with each pot containing 0.75 kg of soil. *Hybrid pennisetum* seeds of uniform size were sown in the plastic flowerpots. For greenhouse cultivation, the potted plants were randomly placed to ensure consistent growing conditions. Thinning was initiated when the *Hybrid pennisetum* seedlings reached a height of 5 cm, leaving three seedlings in each pot. *Enterobacter* sp. VY-1 was inoculated into liquid LB medium, cultured to the logarithmic growth phase, centrifuged to collect the cells, and then resuspended in sterile deionized water to achieve a cell concentration of 1×10^8 cfu·mL^{-1}. The test strains were inoculated into the pots on the 30th, 45th, and 60th days after *Hybrid pennisetum* germination, with an inoculation volume of 10 mL per strain, while the control group was inoculated with an equal amount of sterile water. After 90 days, the *Hybrid pennisetum* plants were harvested, and rhizosphere soil samples were collected.

2.3. Determination of Cd Content and Analysis of Soil Physical-Chemical Properties

The collected plants were dried at 80 °C until a constant weight was achieved, and the dry weight was measured. The sample was ground with a mortar, and 0.1 g (±0.0002 g) of the plant sample was accurately weighed into a polytetrafluoroethylene crucible. The hydrochloric acid-nitric acid-hydrofluoric acid-perchloric acid method was used for microwave digestion, followed by filtration, and the Cd content was determined by inductively coupled plasma optical emission spectrometer (ICP-OES). Additionally, 5.0 g (±0.05) of soil sample was accurately weighed into a 50 mL centrifuge tube, and DTPA (pH = 7.30 ± 0.05) extract was added. After shaking and centrifugation, the supernatant was filtered with a 0.22 μm filter, and ICP-OES was used to measure the content of DTPA-Cd. Soil pH was measured using a potentiometric method at a soil-to-water ratio of 1:5. The available phosphorus content in soil samples was determined using the 0.5 mol L^{-1} $NaHCO_3$ leaching-molybdenum antimony spectrophotometry method, while flame photometry was employed to determine the available potassium content. The hydrolyzed nitrogen content in soil samples was determined using the alkaline hydrolysis diffusion method.

2.4. High-Throughput Sequencing

A 0.5 g sample of rhizosphere soil was accurately weighed, and total DNA was extracted using the FastDNA SPIN Kit soil DNA extraction kit (MP Biochemicals, Solon, OH, USA). The DNA concentration and purity were determined using a micro-UV spectrophotometer. PCR amplification of the V3~V4 conserved region of the bacterial 16S rRNA gene was performed using universal primers 338F and 338R [18]. Sequence determination was carried out using the Illumina MiSeq PE300 high-throughput sequencer (Illumina, San Diego, CA, USA), and the PE reads obtained by Illumina sequencing were spliced based on the overlap relationship. Sequence quality control and filtering were implemented, followed by OTU cluster analysis and species taxonomy analysis. Diversity index calculation and sequencing depth detection were performed using Mothur 1.30.2, Qiime 1.9.1, etc., based on OTU. Statistical analysis of community structure at various classification levels was conducted based on taxonomic information.

2.5. Metabolomic Analysis

A 1 g soil sample was accurately weighed, 20 μL of internal standard and 1 mL of 50% methanol solution were added, and the sample was ground, vortexed, and centrifuged to collect the supernatant. It was then filtered using a 0.22 μm organic phase pinhole filter, and stored at −80 °C for subsequent analysis. Non-targeted metabolome analysis was conducted using an Agilent 7890B Infinity gas chromatograph (Agilent Co., Santa Clara, CA, USA) coupled with an Agilent 5977A GC-MS. The measurement conditions followed those described by Zhang et al. [19]. Principal component analysis (PCA) was utilized to observe the clustering patterns of metabolites in all samples and the repeatability of samples within the group. Pairwise orthogonal partial least squares-discriminant analysis (OPLS-DA) was employed to enhance the discrimination between groups and screen metabolic markers. Metabolites with VIP values > 1 and $p < 0.05$ were identified as differential metabolites between groups. The databases METLIN (https://metlin.scripps.edu/,accessed on 20

April 2023) and HMDB (http://www.hmdb.ca/, accessed on 20 April 2023) were consulted to identify these differential metabolites.

2.6. Data Analysis

Data analysis of soil physical-chemical properties, heavy metal content, high-throughput sequencing, and other data was performed using SPSS 17.0. Differences between groups were assessed using *t*-tests and one-way analysis of variance.

3. Results

3.1. Effects of Tested Strains on Growth and Cd Accumulation of *Hybrid pennisetum*

As shown in Table 1, compared to the uninoculated control, inoculation with *Enterobacter* sp. VY-1 increased the dry weight of the shoots and roots by 25.23% and 32.91%, respectively, indicating that *Enterobacter* sp. VY-1 effectively alleviates Cd-induced stress on the growth of *Hybrid pennisetum*. The Cd content in the shoots and roots correlated with the dry weight results. Inoculation with *Enterobacter* VY-1 led to a 35.48% and 24.25% increase in the Cd content in the shoots and roots, respectively. Based on the dry weight and Cd content in different tissues, *Enterobacter* sp. VY-1 increased the Cd accumulation in the shoots and roots by 64.35% and 51.66%, respectively, thereby enhancing phytoremediation efficiency. The transfer factor (TF) represents the ratio of the heavy metal content in the above-ground parts of the plant to that in the underground part. It reflects the plant's ability to transfer heavy metals from roots to above-ground parts after absorption. The transfer factor increased from 0.15 to 0.16 after inoculation with *Enterobacter* sp. VY-1, a 6.67% increase.

Table 1. Effects of inoculation *Enterobacter* sp. VY-1 on biomass, Cd content, accumulation, and TF (translocation factor) in different parts of *Hybrid pennisetum*.

Treatment	Aboveground Biomass (g)	Root Biomass (g)	Aboveground Cd Content (mg·kg^{-1})	Root Cd Content (mg·kg^{-1})	Aboveground Cd Accumulation (μg)	Root Cd Accumulation (μg)	TF
CK	5.39 ± 0.85	0.79 ± 0.15	2.48 ± 0.37	16.74 ± 1.83	13.38 ± 2.77	13.28 ± 2.80	0.15
VY-1	6.75 ± 0.93	1.05 ± 0.08	3.36 ± 1.03	20.80 ± 2.60	21.99 ± 4.08	20.14 ± 1.62	0.16

3.2. Effects of Tested Strains on Soil Physicalchemical Properties

Plant growth-promoting bacteria can enhance the content of mineral nutrients in the soil, thus improving the absorption of these nutrients by crops. Inoculation with *Enterobacter* sp. VY-1 increased soil total nitrogen, total phosphorus, available phosphorus, and available potassium contents by 3.95%, 4.33%, 19.23%, and 2.31%, respectively (Table 2).

Table 2. Effects of inoculation *Enterobacter* sp. VY-1 on the physicochemical properties of *Hybrid pennisetum* rhizosphere soil.

Treatment	pH	Total Nitrogen (mg·kg^{-1})	Total Phosphorus (mg·kg^{-1})	Available Phosphorus (mg·kg^{-1})	Available Potassium (mg·kg^{-1})
CK	7.09 ± 0.12	1.77 ± 0.10	243.07 ± 9.34	0.026 ± 0.001	108.25 ± 2.5
VY-1	7.13 ± 0.09	1.87 ± 0.19	253.60 ± 19.57	0.031 ± 0.006	110.75 ± 2.7

3.3. Effect of Tested Strains on Bacterial Community Composition

The high-throughput sequencing results revealed 14,951 OTUs in the two sample groups, with 6461 OTUs shared between CK and VY-1, 1218 OTUs unique to CK, and 811 OTUs unique to VY-1. To investigate the impact of *Enterobacter* sp. VY-1 on the bacterial community structure diversity of *Hybrid pennisetum* rhizosphere soil under Cd stress, principal coordinate analysis (PCoA) was employed to assess the dissimilarities between the two sample sets. As depicted in Figure 1, PC1 contributed to 16.54% of the variance, while PC2 contributed to 13.65%. The CK sample clustered on the left side of the PCoA plot, whereas the *Enterobacter* sp. VY-1 inoculated sample clustered on the right side, indicating that the bacterial treatment influenced the bacterial community composition. Subsequent ANOSIM analysis (Analysis of Similarities) revealed a significant intergroup difference with a p-value of 0.003.

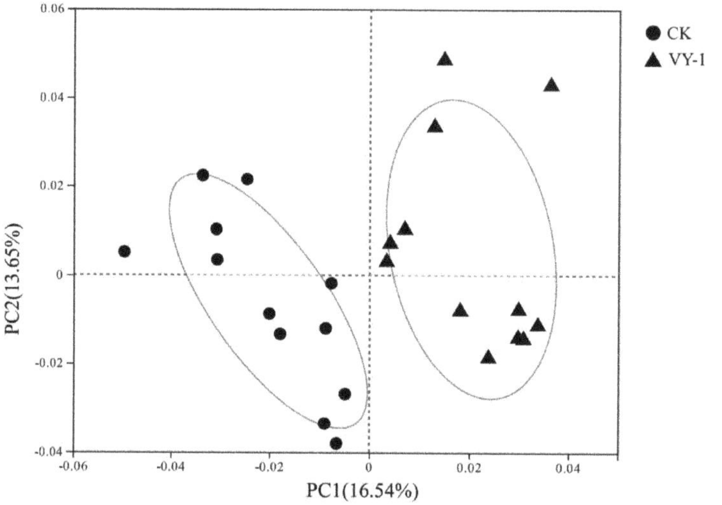

Figure 1. PCoA results of bacterial community diversity.

LefSe (Linear discriminant analysis Effect Size) software (version 1.0) was utilized to examine variations in rhizosphere soil bacterial community composition across different treatment levels. As illustrated in Figure 2, 33 taxa exhibited significant differences in relative abundance with an LDA score ≥ 3. At the phylum level, significant differences were observed in the CK treatment group for Proteobacteria, Myxococcota, Patescibacteria, and Firmicutes, while Chloroflexi and Bdellovibrionota exhibited significant differences following inoculation with *Enterobacter* sp. VY-1. At the order level, there were significant differences among the CK treatment groups in Diplorickettsiales, Thermoanaerobaculales, Myxococcales, and Bacillales. In the treatment with *Enterobacter* sp. VY-1, significant differences were noted in Thermomicrobiales, KD4-96, Propionibacteriales, and Bdellovibrionales. At the genus level, significant differences were observed among the CK treatment groups for *Bacillus* and *Subgroup-10*, while *Bdellovibrio*, *KD4-96*, *JG30-KF-CM45* exhibited significant differences among the *Enterobacter* sp. VY-1 treatment groups.

Redundancy analysis (RDA) was employed to assess the relationship between bacterial flora and environmental factors. RDA1 explained 17.40%, and RDA2 explained 5.07% of the variance (Figure 3). Utilizing the envfit function to test the significance of environmental factors, it was determined that AK ($r^2 = 0.342$, $p = 0.013$) and DTPA-Cd ($r^2 = 0.254$, $p = 0.044$) significantly influenced the composition of the rhizosphere soil bacterial community of *Hybrid pennisetum* ($p < 0.05$). The order of influence of soil physical and chemical factors on microbial community composition is AK > DTPA-Cd > AP > pH > TN > TP.

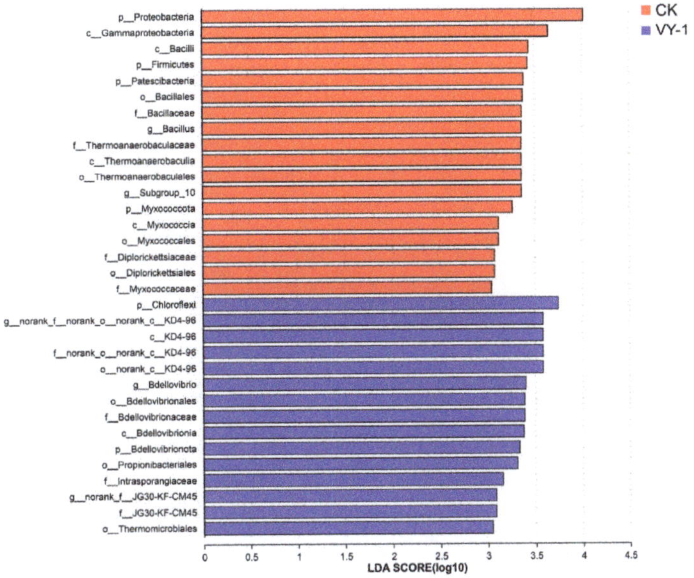

Figure 2. LefSe analysis identifying the most differentially abundant taxa.

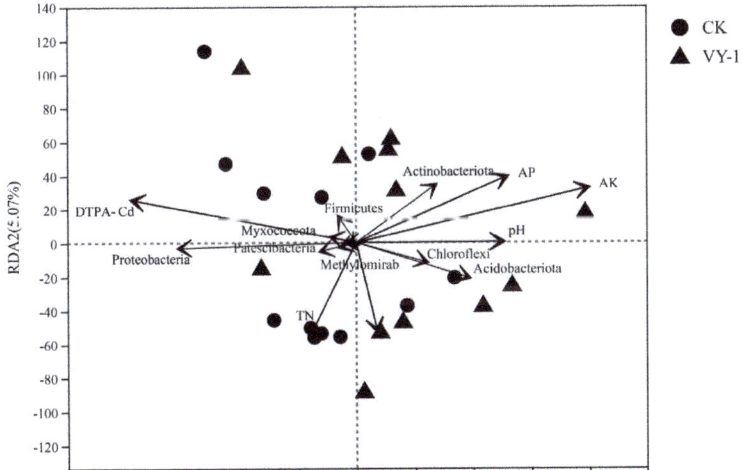

Figure 3. RDA ordination biplot between bacteria communities and environmental factors.

3.4. Effect of Test Strains on Metabolic Function

To explore the alterations in soil metabolites in the rhizosphere of *Hybrid pennisetum* under Cd stress induced by inoculation with the plant growth-promoting bacterium VY-1, Partial Least Squares Discriminant Analysis (PLS-DA) was conducted. In the PLS-DA diagram, the abscissa and ordinate represent the explanatory power of the first (Component1) and second (Component2) principal components, respectively. The PLS-DA results reveal that the metabolites in positive and negative ion modes of both the CK and *Enterobacter* sp. VY-1-inoculated groups are distinctly segregated on opposite sides of the PLS-DA diagram, indicating a significant impact of the inoculation treatment on the metabolite profile of

Hybrid pennisetum rhizosphere soil (Figure 4). The verification results of the PLS-DA model demonstrate that R^2 is greater than Q^2 in both positive and negative ion modes, with the intercepts of the Q^2 regression line and the Y vertical axis being −0.2420 and −0.2205, respectively. This indicates a good fit of the model, high predictability, and suitability for subsequent data analysis.

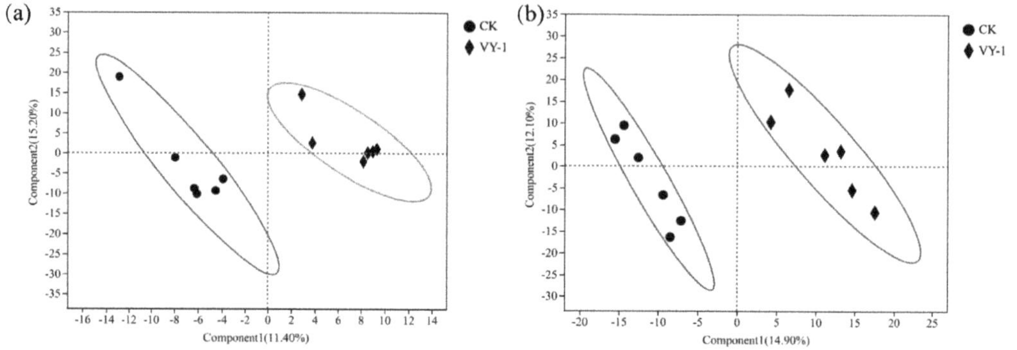

Figure 4. PLS-DA analysis chart of cationic (**a**) and anionic (**b**) metabolites in the rhizosphere soil of *Hybrid pennisetum*.

The variable weight value (variable important in projection, VIP) extracted from the OPLS-DA analysis was utilized, with VIP > 1 considered significant. Additionally, a fold change (FC) value greater than 1 or less than 1 was used as a screening criterion. The T test standard of $p < 0.05$ was employed to identify differential metabolites. The results are depicted in Figure 5. In comparison with the control group, a total of 27 differential metabolites were identified in the *Enterobacter* sp. VY-1 treatment group. Among these, metabolites exhibiting up-regulated expression differences mainly included Poly-g-D-glutamate, Methyl 9,10-epoxy-12,15-octadecadienoate, and 1-Formylneogrifolin, among others, totaling 19 metabolites. Conversely, metabolites with down-regulated expression differences mainly comprised Petasitenine, 3-Hydroxy-10′-apo-b,y-carotenal, (3beta, 5alpha, 6alpha, 22E, 24R)-Ergosta-7,9(11),22-triene-3,5,6-triol, and 8 others. These findings indicate that inoculation with *Enterobacter* sp. VY-1 regulated the metabolites in Pennisetum rhizosphere soil to varying degrees under Cd stress.

Analysis of the metabolic functional pathways of the differentially expressed metabolites using KEGG revealed that inoculation with *Enterobacter* sp. VY-1 primarily affected primary metabolic pathways, particularly those related to Metabolism and Environmental Information Processing. Among these, the metabolic pathway was the most prominently impacted. The secondary classification of KEGG metabolic pathways included membrane transport, lipid metabolism, amino acid metabolism, metabolism of terpenoids and polyketides, and nucleotide metabolism.

Fisher's exact test was employed to conduct enrichment analysis of the 27 significantly different metabolites. The Benjamini and Hochberg (BH) method was utilized for multiple correction of the p values, with a threshold of 0.05 for the corrected p value. KEGG pathways meeting this criterion were considered significantly enriched in the metabolic set. The results revealed that the 27 differential metabolites were mainly enriched in 6 metabolic pathways, including glutamine and glutamate metabolism, alpha-linolenic acid metabolism, pyrimidine metabolism, purine metabolism, zeatin biosynthesis, and ABC transporters (Figure 6). Notably, four pathways were significantly enriched ($p < 0.05$), namely glutamine and glutamate metabolism, alpha-linolenic acid metabolism, pyrimidine metabolism, and purine metabolism. Glutamine and glutamate metabolism exhibited the highest enrichment factor among these pathways.

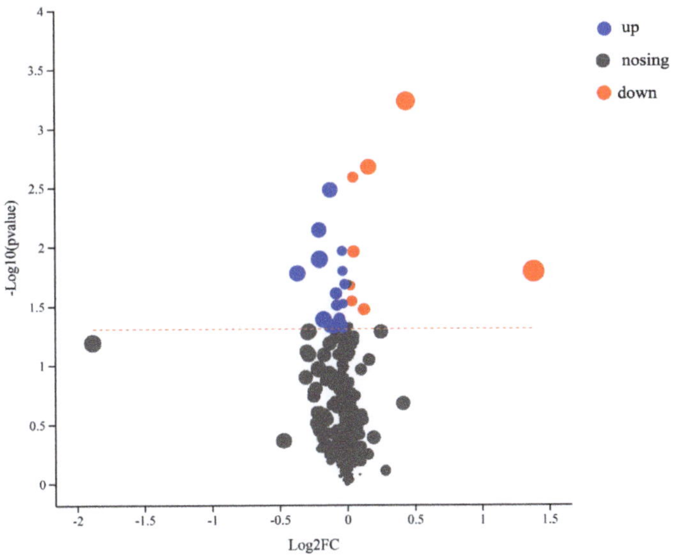

Figure 5. Volcano plot showing differential metabolites between groups.

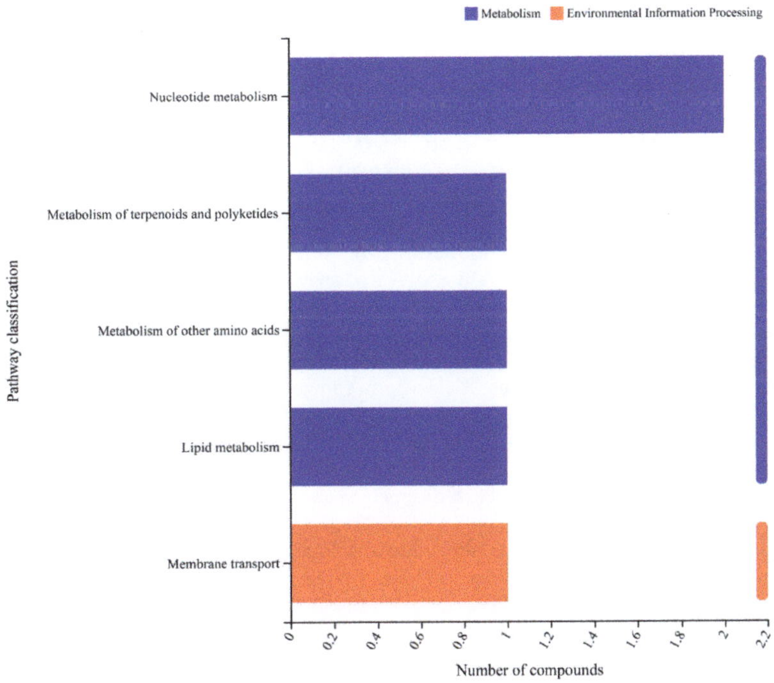

Figure 6. Analysis of KEGG metabolic functional pathways of differentially expressed metabolites.

4. Discussion

As a high-biomass energy plant, *Hybrid pennisetum* has been extensively studied for its ability to remediate heavy metal-contaminated soil. Hou et al. [20] conducted a comparative analysis of the potential of four herbaceous energy plants, namely switchgrass, barley, arundodis, and *Hybrid pennisetum*, for remediating soil heavy metals. They found that *Hybrid pennisetum* exhibited the greatest remediation potential. However, under the stress of heavy metals, the biomass of *Hybrid pennisetum* decreases, slowing down its growth and reducing its efficiency in remediation [6,8]. Plant growth-promoting bacteria are commonly employed in various phytoremediation systems to aid in enhancing phytoremediation [9,11]. Wu et al. [7] isolated the strain *Bacillus megaterium* BM18-2 from *Hybrid pennisetum* plants. The plant growth-promoting characteristics of this strain, including its ability to produce IAA and dissolve phosphorus, among other growth-promoting abilities, resulted in an increase in the biomass of *Hybrid pennisetum* by 83.1% and Cd accumulation by 28.6%. This indicates that plant growth-promoting bacteria have a positive effect on remediating heavy metal pollution in *Hybrid pennisetum*. Similarly, in this experiment, *Enterobacter* sp. VY-1, which exhibits strong growth-promoting ability, was inoculated into *Hybrid pennisetum*. The pot experiment demonstrated that the inoculation treatment increased plant biomass and Cd accumulation by 25.23% to 32.91%, respectively, and 64.35% to 51.66%, effectively alleviating Cd-induced stress and improving phytoremediation efficiency. Moreover, the transfer factor (TF) increased from 0.15 to 0.16, indicating that *Enterobacter* sp. VY-1 enhanced the ability of *Hybrid pennisetum* to transfer Cd absorbed from the roots to the shoots [21]. Heavy metals can significantly interfere with the utilization efficiency of plant nutrients such as N, P, and S, thereby limiting the growth of remediation plants. Plant growth-promoting bacteria activate the content of effective mineral elements in the soil, which is an important mechanism for enhancing plant growth and phytoremediation efficiency [9,11]. The test strains isolated in this experiment all possess the ability to produce siderophores, dissolve phosphorus, and dissolve potassium. Pot experiments revealed that these strains increased the total nitrogen, total phosphorus, available phosphorus, and available potassium content in the rhizosphere soil of *Hybrid pennisetum* by 3.95%, 4.33%, 19.23%, and 2.31%, respectively. This finding aligns with the study by Kamal et al. [6], which concluded that the plant growth-promoting bacterium *Bacillus megaterium* BM18-2 primarily acts as a biofertilizer to enhance the Cd remediation efficiency of *Hybrid pennisetum*.

Soil microbial communities constitute an integral part of the soil ecosystem, exerting a significant influence on plant growth, development, and the efficiency of Cd remediation by plants [22,23]. In the process of plant-microbial joint remediation, inoculation with plant growth-promoting bacteria can impact not only plant growth directly but also the soil microbial community, thereby influencing the function of the rhizosphere microbiome [24,25]. In this experiment, PLS-DA analysis of *Hybrid pennisetum* rhizosphere soil samples revealed significant changes in the bacterial community following inoculation with the VY-1 strain. This finding is consistent with the findings of Kong et al. [24], suggesting that plant growth-promoting bacteria can directly or indirectly impact the soil microbial community structure. Analysis of differential bacteria showed that at the phylum level, the dominant phyla were Proteobacteria, Myxococcus, Patellae, Firmicutes, and Chloroflexi. Bourceret et al. [26] also demonstrated that Proteobacteria, Acidobacteria, and Bacteroidetes are the predominant bacterial phyla in heavy metal-contaminated soil. Liu et al. [27] observed that inoculation of rhizosphere bacteria, such as *Rhodococcus erythropolis*-NSX2, not only enhances the growth of *Sedum sedum* and its Cd accumulation but also leads to an increase in the numbers of indigenous heavy metal-tolerant bacteria like actinomycetes and Bradyrhizobia. Following inoculation with the VY-1 strain, the relative abundance of Actinobacteria, Acidobacteria, Chloroflexi, and Bacteroidetes increased. Numerous studies have highlighted the growth-promoting properties associated with most members of these bacterial phyla [28–30]. For instance, Bacteroidetes, an important indicator for assessing soil ecosystem health, exhibits potential biological control functions [31]. Many genera of Actinobacteria possess plant growth-promoting abilities and heavy metal resistance, thereby promoting plant growth and enhancing plant resistance to peroxidative damage under metal stress [28,32]. Moreover, the majority of Acidobacteria phylum members play crucial roles in soil organic-matter recycling and ecological environment construction, exhibiting significant growth-promoting effects [33]. Therefore, inoculation with plant growth-promoting bacteria in this study may enhance the relative abundance of beneficial microorganisms in the rhizosphere soil of *Hybrid pennisetum* [21,34]. Analyzing the factors influencing the composition of the bacterial community, it was found that AK, AP, DTPA-Cd, and pH are the physical-chemical factors that predominantly contribute to the community structure. This finding aligns with the study by Zhao et al. [35], indicating that plant growth-promoting bacteria can influence soil bacterial community composition by releasing organic acids and dissolving phosphorus and potassium.

The addition of exogenous microorganisms can influence the metabolic activities of soil microorganisms. Utilizing LC-MS non-targeted metabolomics technology to investigate changes in soil metabolites under heavy metal stress induced by inoculation of plant growth-promoting bacteria is crucial for analyzing the joint remediation mechanism of plants and microorganisms [14,36]. Zuluaga et al. [37] demonstrated that inoculation with *Enterobacteriaceae* leads to specific alterations in the metabolism and function of tomato rhizosphere soil under adverse stress conditions. By enhancing the catabolism of amino acids by rhizosphere microorganisms, it can enhance the tolerance of tomato plants and promote plant growth. Similarly, our study employed PLS-DA analysis to illustrate that inoculation with the functional strain VY-1 can impact metabolites in the rhizosphere soil of *Pennisetum* spp., resulting in the identification of 27 differential metabolites, among which 19 were up-regulated and eight were down-regulated. This indicates that inoculating the VY-1 strain in the rhizosphere of *Hybrid pennisetum* enhances the activity of surrounding microorganisms and induces certain changes in soil metabolites, thereby mitigating the toxic effects of heavy metals on microorganisms and *Hybrid pennisetum* [38]. Cluster analysis of these differential metabolites revealed an increase in lipids, amino acids, carbohydrates, and other metabolites following inoculation with the VY-1 strain. Zhang et al. [39] utilized non-targeted metabolomics technology to analyze the metabolites of *Ganoderma lucidum* under different concentrations of Cd stress and found that *Ganoderma lucidum* can alleviate the toxic effects of Cd by enriching lipid metabolites. Wang et al. [40] demonstrated that amino acids can play a functional role in reducing oxidative damage, regulating cell penetration, and maintaining plant growth. Carbohydrates serve as the primary source of energy for plants, and changes in their content can serve as an important indicator of plant physiological conditions. An increase in carbohydrate content indicates favorable plant growth conditions, and its augmentation significantly enhances the biomass of burdock roots under Cu stress [41]. Through KEGG pathway enrichment analysis of significantly different metabolites, it was observed that metabolic pathways were predominantly affected by inoculation with the VY-1 strain among the primary metabolic pathways. After inoculation with the VY-1 strain, six main metabolic pathways were affected, with four pathways being significantly enriched ($p < 0.05$): glutamine and glutamate metabolism, α-linolenic acid metabolism, pyrimidine metabolism, and purine metabolism. Limami et al. [42] found that aspartate, alanine, and glutamate metabolism can promote plant growth by regulating nitrogen flux and increasing carbon availability.

5. Conclusions

Research on the joint remediation mechanism of plant-microorganism systems is crucial for enhancing the practical application of remediating heavy metal-contaminated soil. This study utilized high-throughput sequencing combined with metabolomics omics technology to investigate the microbiological mechanism of *Enterobacter* sp. VY-1, a plant growth-promoting bacterium, in remediating Cd pollution in *Hybrid pennisetum*. The findings revealed that *Enterobacter* sp. VY-1 effectively mitigates Cd-induced stress, increases *Hybrid pennisetum* biomass, Cd accumulation, and transfer factor (TF), thus enhancing phytoremediation efficiency. The primary mechanism involves activating soil mineral nutrients to increase soil total nitrogen, total phosphorus, available phosphorus, and available potassium content. Moreover, it influences the composition of the *Hybrid pennisetum* rhizosphere soil bacterial community, impacting phyla such as Proteobacteria, Myxococcus, Patellae, and Firmicutes. Furthermore, it modulates the metabolite profile of *Hybrid pennisetum* rhizosphere soil, particularly affecting primary metabolic pathways like metabolism and environmental information processing as well as specific pathways such as glutamine and glutamate metabolism, α-linolenic acid metabolism, pyrimidine metabolism, and purine metabolism. Hence, plant growth-promoting bacteria can enhance the efficiency of phytoremediation in heavy metal-contaminated soil by influencing the composition and metabolic functions of microbial communities.

Author Contributions: Z.-J.C. and Y.-Y.L., designed the experiments; Z.-J.C., M.-L.L. and S.-S.G., participated in writing the paper; Z.-J.C., M.-L.L. and Y.-B.S., performed the experiments and analyzed the data; Z.-J.C., H.H., B.-L.L. and Y.-Y.L., reviewed the manuscript. All authors have read and agreed to the published version of the manuscript.

Funding: This research was funded by the National Natural Science Foundation of China (Grant Nos. U2004145, 42377039), and the Program for Science & Technology Innovation Talents in Universities of Henan Province (No. 23HASTIT018), the Key Research and Development Projects of Henan Province (Grant No. 221111520600, and 231111113000) and the Key Scientific and Technological Project of Henan Province (Grant No. 242102521067, and 232102111024).

Institutional Review Board Statement: Not applicable.

Informed Consent Statement: Not applicable.

Data Availability Statement: The data that support the findings of this study are available on request from the corresponding author.

Conflicts of Interest: The authors declare no conflicts of interest.

References

1. Qin, G.; Niu, Z.; Yu, J.; Li, Z.; Ma, J.; Xiang, P. Soil heavy metal pollution and food safety in China: Effects, sources and removing technology. *Chemosphere* **2021**, *267*, 129205. [CrossRef] [PubMed]
2. Cheng, S. Heavy metal pollution in China: Origin, pattern and control. *Environ. Sci. Pollut. Res.* **2003**, *10*, 192–198. [CrossRef] [PubMed]
3. Wen, M.; Ma, Z.; Gingerich, D.B.; Zhao, X.; Zhao, D. Heavy metals in agricultural soil in China: A systematic review and meta-analysis. *Eco-Environ. Health* **2022**, *1*, 219–228. [CrossRef]
4. Pilon-Smits, E. phytoremediation. *Annu. Rev. Plant Biol.* **2005**, *56*, 15–39. [CrossRef]
5. Gavrilescu, M. Enhancing phytoremediation of soils polluted with heavy metals. *Curr. Opin. Biotechnol.* **2022**, *74*, 21–31. [CrossRef]
6. Kamal, N.; Liu, Z.; Qian, C.; Wu, J.; Zhong, X. Improving hybrid Pennisetum growth and cadmium phytoremediation potential by using *Bacillus megaterium* BM18-2 spores as biofertilizer. *Microbiol. Res.* **2021**, *242*, 126594. [CrossRef] [PubMed]
7. Wu, J.; Kamal, N.; Hao, H.; Qian, C.; Liu, Z.; Shao, Y.; Zhong, X.; Xu, B. Endophytic *Bacillus megaterium* BM18-2 mutated for cadmium accumulation and improving plant growth in *Hybrid pennisetum*. *Biotechnol. Rep.* **2019**, *24*, e00374. [CrossRef]
8. He, L.; Zhu, Q.; Wang, Y.; Chen, C.; He, M.; Tan, F. Irrigating digestate to improve cadmium phytoremediation potential of *Pennisetum hybridum*. *Chemosphere* **2021**, *279*, 130592. [CrossRef]
9. Weyens, N.; van der Lelie, D.; Taghavi, S.; Newman, L.; Vangronsveld, J. Exploiting plant-microbe partnerships to improve biomass production and remediation. *Trends Biotechnol.* **2009**, *27*, 591–598. [CrossRef]
10. Kisvarga, S.; Hamar-Farkas, D.; Ördögh, M.; Horotán, K.; Neményi, A.; Kovács, D.; Orlóci, L. The Role of the Plant–Soil Relationship in Agricultural Production—With Particular Regard to PGPB Application and Phytoremediation. *Microorganisms* **2023**, *11*, 1616. [CrossRef]
11. Alves, A.R.A.; Yin, Q.; Oliveira, R.S.; Silva, E.F.; Novo, L.A.B. Plant growth-promoting bacteria in phytoremediation of metal-polluted soils: Current knowledge and future directions. *Sci. Total Environ.* **2022**, *838*, 156435. [CrossRef]
12. Ren, X.-M.; Guo, S.-J.; Tian, W.; Chen, Y.; Han, H.; Chen, E.; Li, B.-L.; Li, Y.-Y.; Chen, Z.-J. Effects of Plant Growth-Promoting Bacteria (PGPB) Inoculation on the Growth, Antioxidant Activity, Cu Uptake, and Bacterial Community Structure of Rape (*Brassica napus* L.) Grown in Cu-Contaminated Agricultural Soil. *Front. Microbiol.* **2019**, *10*, 1455. [CrossRef] [PubMed]
13. Deng, S.; Zhang, X.; Zhu, Y.; Zhuo, R. Recent advances in phyto-combined remediation of heavy metal pollution in soil. *Biotechnol. Adv.* **2024**, *72*, 108337. [CrossRef]
14. Yadav, R.; Singh, G.; Santal, A.R.; Singh, N.P. Omics approaches in effective selection and generation of potential plants for phytoremediation of heavy metal from contaminated resources. *J. Environ. Manag.* **2023**, *336*, 117730. [CrossRef]
15. Li, X.-Q.; Liu, Y.-Q.; Li, Y.-J.; Han, H.; Zhang, H.; Ji, M.-F.; Chen, Z.-J. Enhancing Mechanisms of the Plant Growth-Promoting Bacterial Strain *Brevibacillus* sp. SR-9 on Cadmium Enrichment in Sweet Sorghum by Metagenomic and Transcriptomic Analysis. *Int. J. Environ. Res. Public Health* **2022**, *19*, 16309. [CrossRef] [PubMed]

16. Rane, N.R.; Tapase, S.; Kanojia, A.; Watharkar, A.; Salama, E.-S.; Jang, M.; Kumar Yadav, K.; Amin, M.A.; Cabral-Pinto, M.M.S.; Jadhav, J.P.; et al. Molecular insights into plant–microbe interactions for sustainable remediation of contaminated environment. *Bioresour. Technol.* **2022**, *344*, 126246. [CrossRef] [PubMed]
17. Liu, Y.-Q.; Chen, Y.; Li, Y.-Y.; Ding, C.-Y.; Li, B.-L.; Han, H.; Chen, Z.-J. Plant growth-promoting bacteria improve the Cd phytoremediation efficiency of soils contaminated with PE–Cd complex pollution by influencing the rhizosphere microbiome of sorghum. *J. Hazard. Mater.* **2024**, *469*, 134085. [CrossRef]
18. Chen, Y.; Tian, W.; Shao, Y.; Li, Y.-J.; Lin, L.-A.; Zhang, Y.-J.; Han, H.; Chen, Z.-J. *Miscanthus* cultivation shapes rhizosphere microbial community structure and function as assessed by Illumina MiSeq sequencing combined with PICRUSt and FUNGUIld analyses. *Arch. Microbiol.* **2020**, *202*, 1157–1171. [CrossRef]
19. Zhang, Z.; Tang, S.; Liu, X.; Ren, X.; Wang, S.; Gao, Z. The Effects of *Trichoderma viride* T23 on Rhizosphere Soil Microbial Communities and the Metabolomics of Muskmelon under Continuous Cropping. *Agronomy* **2023**, *13*, 1092. [CrossRef]
20. Hou, X.C.; Fan, X.F.; Wu, J.Y.; Zhu, Y.; Zhang, Y.X. Potentiality of herbaceous bioenergy plants of remediation of soil contaminated by heavy metals soi. *Chin. J. Grassl.* **2012**, *34*, 59–76.
21. Chen, Z.-J.; Tian, W.; Li, Y.-J.; Sun, L.-N.; Chen, Y.; Zhang, H.; Li, Y.-Y.; Han, H. Responses of rhizosphere bacterial communities, their functions and their network interactions to Cd stress under phytostabilization by *Miscanthus* spp. *Environ. Pollut.* **2021**, *287*, 117663. [CrossRef] [PubMed]
22. Chen, Z.J.; Zheng, Y.; Ding, C.Y.; Ren, X.M.; Yuan, J.; Sun, F.; Li, Y.Y. Integrated metagenomics and molecular ecological network analysis of bacterial community composition during the phytoremediation of cadmium-contaminated soils by bioenergy crops. *Ecotoxicol. Environ. Saf.* **2017**, *145*, 111–118. [CrossRef] [PubMed]
23. Song, J.; Brookes, P.C.; Shan, S.; Xu, J.; Liu, X. Effects of remediation agents on microbial community structure and function in soil aggregates contaminated with heavy metals. *Geoderma* **2022**, *425*, 116030. [CrossRef]
24. Kong, Z.; Wu, Z.; Glick, B.R.; He, S.; Huang, C.; Wu, L. Co-occurrence patterns of microbial communities affected by inoculants of plant growth-promoting bacteria during phytoremediation of heavy metal-contaminated soils. *Ecotoxicol. Environ. Saf.* **2019**, *183*, 109504. [CrossRef]
25. Wang, Q.; Zhou, Q.; Huang, L.; Xu, S.a.; Fu, Y.; Hou, D.; Feng, Y.; Yang, X. Cadmium phytoextraction through *Brassica juncea* L. under different consortia of plant growth-promoting bacteria from different ecological niches. *Ecotoxicol. Environ. Saf.* **2022**, *237*, 113541. [CrossRef] [PubMed]
26. Bourceret, A.; Cébron, A.; Tisserant, E.; Poupin, P.; Bauda, P.; Beguiristain, T.; Leyval, C. The Bacterial and Fungal Diversity of an Aged PAH- and Heavy Metal-Contaminated Soil is Affected by Plant Cover and Edaphic Parameters. *Microb. Ecol.* **2016**, *71*, 711–724. [CrossRef] [PubMed]
27. Liu, W.; Wang, Q.; Wang, B.; Hou, J.; Luo, Y.; Tang, C.; Franks, A.E. Plant growth-promoting rhizobacteria enhance the growth and Cd uptake of *Sedum plumbizincicola* in a Cd-contaminated soil. *J. Soils Sediments* **2015**, *15*, 1191–1199. [CrossRef]
28. Boukhatem, Z.F.; Merabet, C.; Tsaki, H. Plant Growth Promoting Actinobacteria, the Most Promising Candidates as Bioinoculants? *Front. Agron.* **2022**, *4*, 849911. [CrossRef]
29. Compant, S.; Samad, A.; Faist, H.; Sessitsch, A. A review on the plant microbiome: Ecology, functions, and emerging trends in microbial application. *J. Adv. Res.* **2019**, *19*, 29–37. [CrossRef]
30. Chen, Y.; Ding, Q.; Chao, Y.; Wei, X.; Wang, S.; Qiu, R. Structural development and assembly patterns of the root-associated microbiomes during phytoremediation. *Sci. Total Environ.* **2018**, *644*, 1591–1601. [CrossRef]
31. Zheng, X.; Zhu, Y.; Wang, Z.; Zhang, H.; Chen, M.; Chen, Y.; Wang, J.; Liu, B. Effects of a novel bio-organic fertilizer on the composition of rhizobacterial communities and bacterial wilt outbreak in a continuously mono-cropped tomato field. *Appl. Soil Ecol.* **2020**, *156*, 103717. [CrossRef]
32. Soumare, A.; Boubekri, K.; Lyamlouli, K.; Hafidi, M.; Ouhdouch, Y.; Kouisni, L. Efficacy of phosphate solubilizing Actinobacteria to improve rock phosphate agronomic effectiveness and plant growth promotion. *Rhizosphere* **2021**, *17*, 100284. [CrossRef]
33. Kalam, S.; Basu, A.; Ahmad, I.; Sayyed, R.Z.; El-Enshasy, H.A.; Dailin, D.J.; Suriani, N.L. Recent Understanding of Soil Acidobacteria and Their Ecological Significance: A Critical Review. *Front. Microbiol.* **2020**, *11*, 580024. [CrossRef] [PubMed]
34. Lin, H.; Liu, C.; Li, B.; Dong, Y. *Trifolium repens* L. regulated phytoremediation of heavy metal contaminated soil by promoting soil enzyme activities and beneficial rhizosphere associated microorganisms. *J. Hazard. Mater.* **2021**, *402*, 123829. [CrossRef]
35. Zhao, Y.; Zhang, M.; Yang, W.; Di, H.J.; Ma, L.; Liu, W.; Li, B. Effects of microbial inoculants on phosphorus and potassium availability, bacterial community composition, and chili pepper growth in a calcareous soil: A greenhouse study. *J. Soils Sediments* **2019**, *19*, 3597–3607. [CrossRef]
36. Bell, T.H.; Joly, S.; Pitre, F.E.; Yergeau, E. Increasing phytoremediation efficiency and reliability using novel omics approaches. *Trends Biotechnol.* **2014**, *32*, 271–280. [CrossRef]
37. Alzate Zuluaga, M.Y.; Milani, K.M.L.; Miras-Moreno, M.B.; Lucini, L.; Valentinuzzi, F.; Mimmo, T.; Pii, Y.; Cesco, S.; Rodrigues, E.P.; de Oliveira, A.L.M. The adaptive metabolomic profile and functional activity of tomato rhizosphere are revealed upon PGPB inoculation under saline stress. *Environ. Exp. Bot.* **2021**, *189*, 104552. [CrossRef]
38. Kaur, S.; Suseela, V. Unraveling Arbuscular Mycorrhiza-Induced Changes in Plant Primary and Secondary Metabolome. *Metabolites* **2020**, *10*, 335. [CrossRef] [PubMed]
39. Zhang, B.; Zhou, J.; Li, Q.; Gan, B.; Peng, W.; Zhang, X.; Tan, W.; Jiang, L.; Li, X. Manganese affects the growth and metabolism of Ganoderma lucidum based on LC-MS analysis. *PeerJ* **2019**, *7*, e6846. [CrossRef]

40. Wang, Y.; Ren, W.; Li, Y.; Xu, Y.; Teng, Y.; Christie, P.; Luo, Y. Nontargeted metabolomic analysis to unravel the impact of di (2-ethylhexyl) phthalate stress on root exudates of alfalfa (*Medicago sativa*). *Sci. Total Environ.* **2019**, *646*, 212–219. [CrossRef]
41. Jung, Y.; Ha, M.; Lee, J.; Ahn, Y.G.; Kwak, J.H.; Ryu, D.H.; Hwang, G.-S. Metabolite Profiling of the Response of Burdock Roots to Copper Stress. *J. Agric. Food Chem.* **2015**, *63*, 1309–1317. [CrossRef] [PubMed]
42. Limami, A.M.; Glévarec, G.; Ricoult, C.; Cliquet, J.-B.; Planchet, E. Concerted modulation of alanine and glutamate metabolism in young Medicago truncatula seedlings under hypoxic stress. *J. Exp. Bot.* **2008**, *59*, 2325–2335. [CrossRef] [PubMed]

International Journal of
Plant Biology

MDPI

Article

Alteration of Photosynthetic and Antioxidant Gene Expression in Sugarcane Infected by Multiple Mosaic Viruses

Intan Ria Neliana [1,2], Wardatus Soleha [1], Suherman [3], Nurmalasari Darsono [4], Rikno Harmoko [4], Widhi Dyah Sawitri [5] and Bambang Sugiharto [1,2,6,*]

1 Laboratory of Molecular Biology and Biotechnology, Center for Development of Advanced Science and Technology (CDAST), University of Jember, Jl. Kalimantan No. 37, Kampus Tegalboto, Jember 68121, Indonesia; nelianaintanria@gmail.com (I.R.N.); wardatussholeha91@gmail.com (W.S.)
2 Postgraduate Program in Biotechnology, University of Jember, Jl. Kalimantan No. 37, Kampus Tegalboto, Jember 68121, Indonesia
3 Faculty of Agriculture, Moch. Sroedji University, Jl. Sriwijaya No. 32, Jember 68124, Indonesia; suherman_16@unej.ac.id
4 Research Center for Genetic Engineering, National Research and Innovation Agency, Jl. Raya Jakarta-Bogor, Cibinong, Bogor 16911, Indonesia; nurmalasaridarsono@gmail.com (N.D.); rknharmoko@gmail.com (R.H.)
5 Department of Agronomy, Faculty of Agriculture, Gadjah Mada University, Yogyakarta 55281, Indonesia; widhi.d.s@ugm.ac.id
6 Department of Biology, Faculty of Mathematic and Natural Science, University of Jember, Jl. Kalimantan No. 37, Kampus Tegalboto, Jember 68121, Indonesia
* Correspondence: sugiharto.fmipa@unej.ac.id; Tel.: +62-811-350314

Abstract: Sugarcane mosaic virus (SCMV), sugarcane streak mosaic virus (SCSMV), and sorghum mosaic virus (SrMV) are the causative pathogens of mosaic disease. This study aimed to identify mosaic virus infection and its impact on photosynthetic and antioxidant gene expression in eight commercial sugarcane cultivars grown on sugarcane plantations in East Java, Indonesia. The disease incidence and severity were observed in symptomatic leave samples, and then the virus was identified. A high incidence and severity of mosaic symptoms were observed in the PS881 and NX04 cultivars compared with the other cultivars. RT-PCR analysis detected SCSMV infection in all cultivars; double infections with SCSMV and SCMV in the PS881, PS882, and Cening cultivars; and triple infections with SCSMV, SCMV, and SrMV in the PS881 cultivar. Ascorbate peroxidase (*Apx*) expression was upregulated in all virus-infected cultivars and significantly increased in the triple-infected PS881 cultivar. However, catalase (*Cat*) expression was only slightly increased in the PS881 cultivar. The chlorophyll content was reduced, and the *PsaA* gene was downregulated in all cultivars. The expression of *PsaA*, *RbcS*, and *Sps* was significantly suppressed in the triple-infected PS881 cultivar. Moreover, the downregulation of both the *RbcS* and *Pepc* genes was concomitant with that of their protein levels.

Keywords: sugarcane cultivars; multiple mosaic virus infections; virus identification; RT-PCR analysis; expression of antioxidant and photosynthetic genes

1. Introduction

Sugarcane (*Saccharum* spp.) is a major crop used for the production of sugar, biofuel, and other products, such as animal feed, paper, organic fertilizer, and industrial enzymes [1,2], and it is mainly propagated using lateral buds, which are easily transmitted and infected by sugarcane mosaic disease [3,4] and yellow leaf disease [5]. Mosaic virus is one of the most severe diseases in sugarcane and leads to the development of light and dark green irregular patterns, yellow patches, or streaks [6,7]. Multiple infections by two or more viruses frequently occur in plants and result in severe symptoms [8]. Mosaic diseases of sugarcane with multiple infections of sugarcane mosaic virus (SCMV), sugarcane streak mosaic virus (SCSMV), and sorghum mosaic virus (SrMV) have been reported in

China [8,9]; SCMV and SrMV have been reported in the USA [10]; and SCMV and SCSMV have reported in India [11].

Reactive oxygen species (ROS) accumulate in response to biotic stress caused by viral infection. Mosaic virus infection significantly upregulates ROS-producing genes, such as NAD oxidase, malate dehydrogenase, and flavin-binding monooxygenase [12], and the production of ROS can potentially damage cellular components and disrupt programmed cell death [13,14]. However, accumulated ROS can be detoxified by non-enzymatic and enzymatic antioxidants, such as ascorbate peroxidase (Apx), catalase (Cat), superoxide dismutase, and glutathione reductase [15–17]. This antioxidant defense mechanism for scavenging ROS protects cells from oxygen radicals [18].

Under light illumination, SCMV infection stimulates malate synthesis by enhancing pyruvate orthophosphate dikinase (PPDK) activity, leading to ROS accumulation [19]. As a C4 plant, sugarcane exhibits high rates of photosynthesis under high light intensity. Phosphoenolpyruvate carboxylase (PEPC) captures atmospheric CO_2 and converts it to oxaloacetate and malate using phosphoenolpyruvate (PEP) as a substrate, and then malate is transported into the bundle sheath cell and decarboxylated to produce CO_2 molecules, which are re-fixed by ribulose-1,5-bisphosphate carboxylase (Rubisco) to form sucrose and starch [20]. To maintain the C4 cycle, PEP is regenerated from pyruvate using PPDK and then used by PEPC to produce malate. Light is well known to regulate the activity of several carbon-assimilating enzymes, including PPDK, PEPC, and sucrose-phosphate synthase (SPS) [21,22]. However, the effects of mosaic virus infections on photosynthetic activity remain poorly understood.

Mosaic diseases significantly reduce the chlorophyll content, photosynthetic efficiency, yield, and quality of sugarcane [6,23,24]. The expression of light-harvesting proteins and primary carbon-assimilating enzymes is high in both resistant cultivars [12] and non-infected sugarcane [24]. Moreover, NADPH, Rubisco, and photosystem I (PSI) reaction center expression, including the *PsaA* gene, is upregulated in mosaic-resistant sugarcane compared to susceptible sugarcane. Protein analysis confirmed that PEPC and Rubisco activity is decreased in virus-infected sugarcane seedlings [23,24]. Alterations in photosynthesis-related activities slow sugarcane growth [17] and reduce the sucrose content and sugarcane yield by up to 75% [6]. In addition, mosaic virus infection reduces the chlorophyll content in cassava [25] and poplar [26] and perturbs pigment biosynthesis in tomato [27]. These results indicate that mosaic virus infection downregulates the expression of photosynthesis-related genes [26,28,29].

Mosaic symptoms in sugarcane are frequently associated with several different viral infections. Multiple or mixed viral infections have also been reported in sweet potatoes in Uganda [29] and cotton plants in China [9]. Multiple infections from two or more viral interactions are categorized as synergistic, antagonistic, or neutral interactions [8,30] that induce a decline in plant vigor and productivity [31]. Combined infection by viruses that act synergistically exacerbates symptom severity and leads to plant death or severely reduced yields [32], as reported in maize [33].

In this study, mosaic viruses and their distribution were determined in sugarcane grown at four locations in East Java, Indonesia. Symptomatic sugarcane leaves were collected from the field, and SCSMV, SCMV, and SrMV were identified in the leaf samples using specific primer pairs for RT-PCR analysis. The effects of viral infection on the expression of antioxidant and photosynthetic genes were determined by RT-PCR and immunoblot analysis.

2. Materials and Methods

2.1. Observation Area and Leaf Sample Collection

Mosaic symptoms were observed in commercial sugarcane cultivars grown in four districts of East Java Province: Kediri, Lumajang, Jember, and Bondowoso. This province is the primary location of sugarcane production in Indonesia. Observations were conducted in plants at three to four months of age in sugarcane plantations under the authority of PT. Perkebunan Nusantara X and PT. Perkebunan Nusantara XI. Symptomatic leaf samples were collected from fields and assessed for disease incidence and severity. Disease incidence was determined as the percentage of sugarcane mosaic symptoms relative to the total sugarcane observed in the field, while disease severity was estimated using a previously described scoring system for young sugarcane leaf area in which symptoms at the first and third leaves from the green top are recorded [3]. The percentage severity was calculated using the following equation [34]:

$$\text{Percent age of disease severity index } (\%) = \frac{\sum(\text{class frequency} \times \text{score of rating class})}{(\text{total number of plants} \times (\text{maximal disease index})} \times 100$$

Disease incidence and severity were observed over approximately 10,000 m^2 for each sugarcane cultivar. Within the observed areas, a total of 61 sugarcane leaf samples with mosaic symptoms were selected, disinfected with 75% alcohol, and immersed in liquid nitrogen for further laboratory analysis.

2.2. RNA Extraction and Mosaic Virus Determination

For mosaic virus detection, RNA was extracted from sugarcane leaf samples using an RNAprep Pure Plant Plus Kit (Tiangen, Beijing, China) [35], and the concentration was quantified using a NanoVuePlus-UV Spectrophotometer (GE Healthcare, Chicago, IL, USA). The three major mosaic viruses, SCSMV, SCMV, and SrMV, were identified by RT-PCR using primer pairs specific for the CP gene, as shown in Table 1. Total RNA (1 µg) was converted to cDNA using a ReverTra AceTM Kit (Toyobo, Osaka, Japan) and then subjected to a PCR reaction using a KOD-Plus-Neo Kit according to the manual instruction (Toyobo, Japan). The PCR analysis was performed at 95 °C for 5 min, followed by 35 cycles of denaturation at 95 °C for 30 s; annealing at 53 °C (SCSMV), 58 °C (SCMV), or 52 °C (SrMV) for 30 s; and extension at 72 °C for 1 min—followed by a final extension at 72 °C for 5 min using a T100TM Thermal Cycler (Bio-Rad, Hercules, CA, USA). PCR products were separated by agarose gel electrophoresis containing ethidium bromide and visualized using a GelDoc system (Major Science, Saratoga, CA, USA). The corresponding DNA bands were excised from the agarose gel, purified using a GenepHlow Gel/PCR Kit (Geneaid, Taipei, Taiwan), and used for nucleotide sequence determination.

Table 1. Description of primers used in this study.

Primer	Sequence Information (5′-3′)	Amplicon Length (bp)	Annealing (°C)	Reference
CP-SCSMV-F	5′-MTCTTCATCRGCCGCMTCRATAC-3′	335 bp	53	This study
CP-SCSMV-R	5′-AGAACTGAACCCACTTGTACGCC-3′			
CP-SCMV-F	5′-GACATATGGATGTAGATGCTGGTACGACA-3′	735 bp	58	[35]
CP-SCMV-R	5′-ATGGATCCTAGTGGTGCTGCTGCACTCCC-3′			
CP-SrMV-F	5′-GCAGATGCTGATGCGAAA-3′	480 bp	52	This study
CP-SrMV-R	5′-ACGCTTCAGCTGCATCAC-3′			
PEPC-F	5′-TGGGTGGTGACCGTGATGG-3′	129 bp	60	This study
PEPC-R	5′-GCAGCGCCACATAGAGAGC-3′			
PsaA-F	5′-AGGGGCTTATACCCTCAG-3′	121 bp	52	This study
PsaA-R	5′-GGATTAGGTGCCTAACGGAC-3′			
RbcS-F	5′-CACTAGCTTCGCCAAAGT-3′	205 bp	55	This study
RbcS-R	5′-AAGCCTTCCTTGCTGAAC-3′			
SPS-F	5′-GGGTCTCCATAGGACCATTA-3′	110 bp	55	This study
SPS-R	5′-GGGGTGTTATTGTGTGAGTA-3′			

Table 1. *Cont.*

Primer	Sequence Information (5′-3′)	Amplicon Length (bp)	Annealing (°C)	Reference
CatF	5′-GCTCAGTTCGACAGGGAACG-3′	208 bp	55	[16]
CatR	5′-CACGTGGATCCCTCAAGGTC-3′			
ApxF	5′-GATTTGATTGCCGTGGCTGG-3′	134 bp	55	[36]
ApxR	5′-TCTTCAGGAAGTTTGCCAGTTG-3′			
β-tubulinF	5′-CATCTTCGTGTGGAATCCT-3′	129 bp	55	This study
β-tubulinR	5′-AACTCCAGAACAGACCGTA-3′			

2.3. DNA Sequencing and Phylogenetic Analysis

Nucleotide sequencing of the isolated PCR products was performed by a DNA sequencing provider (1st BASE, Singapore). The obtained nucleotide sequences were compared with the sequences in the NCBI virus database using BLAST/N (https://blast.ncbi.nlm.nih.gov (accessed on 8 January 2024)) to confirm the degree of sequence similarity. A phylogenetic tree was constructed from the SCSMV, SCMV, and SrMV nucleotide sequences using the neighbor-joining (NJ) method (1000 bootstrap replicates) and MEGA 11 software version 11.0.13 [36].

2.4. Quantitative Real-Time PCR (RT-qPCR)

RT-qPCR was performed using a CFX connectTM Real-Time System (Bio-Rad, USA) to analyze the expression levels of photosynthesis and antioxidant ROS-scavenging genes. The 25 μL reaction system contained 12.5 μL of 2× SYBR Green master mix (ThunderbirdTM SYBR qPCR Mix, Toyobo, Japan), 1 μL of 10 μmol/L of primers, and 1.5 μL of 50 ng cDNA, with nuclease-free water added to reach the reaction volume. The gene-specific primer pairs are listed in Table 1. The RT-qPCR program was 95 °C for 30 s, followed by 40 cycles at 95 °C for 15 s, primer-dependent annealing temperatures (Table 1) for 15 s, elongation at 72 °C for 15 s, and final extension at 72 °C for 60 s. Three independent biological replicates were performed, each sample was analyzed in duplicate, and expression was normalized based on β-tubulin expression. The relative gene expression was calculated via Livak's method using the formula $2^{-\Delta\Delta CT}$ [37].

2.5. Protein Extraction and Immunoblotting

One gram of frozen sugarcane leaf sample was ground in liquid nitrogen, and the proteins were extracted according to the method previously described [35]. The soluble protein concentration was determined using a Bradford reagent (Bio-Rad, USA). The levels of Rubisco and PEPC proteins were determined by immunoblotting using specific polyclonal antibodies against PEPC and Rubisco, with 15 μg and 10 μg soluble proteins for PEPC and Rubisco, respectively. The immunoblotting analysis was conducted according to the method previously described [38].

2.6. Analysis of the Total Chlorophyll of Leaves

The total chlorophyll content was determined using an optical method [39] with slight modifications. Briefly, symptomatic sugarcane leaves (0.1 g) were finely cut and incubated with 5 mL of a solution containing a mixture of acetone and ethanol (2:1, *v*/*v*) in the dark for 24 h. Three independent biological replicates were performed, and each sample was analyzed in triplicate. The absorbance of the solution was measured at 645 and 663 nm using a spectrophotometer (UV-VIS double-beam spectrophotometer (Hitachi, Tokyo, Japan). Solution without the sample was used as the blank. The total chlorophyll concentration was estimated using the following equation [40]:

$$\text{mg total chlorophyll/g tissue} = 20.2 \times (A645) + 8.02 \times (A663) \times V \times W$$

where:

A645: absorbance at 645 nm;
A663: absorbance at 663 nm;
V: final volume of chlorophyll extract;
W: fresh weight of leaf extract.

2.7. *Statistical Analysis*

Statistical analyses were performed in triplicates and represented as the mean ± standard errors. Statistical significance was calculated using Dunnett's test with an ANOVA to determine significant differences. A *p*-value of 0.05 was considered statistically significant.

3. Results

3.1. *Mosaic Symptom, Incidence, and Severity*

Mosaic symptoms were observed over approximately 10,000 m^2 for each sugarcane cultivar, and 61 symptomatic leaf and 1 healthy non-symptomatic leaf samples were collected. Mosaic symptoms include blotchy or streaky patterns of yellow and green colors with varying lengths, sizes, and irregularities, and they were recorded to determine the incidence percentage. The incidence of symptoms observed in Kediri was higher than that in the other locations (Figure 1A). The highest incidence was observed in NX04, followed by PS881 (80 and 54%, respectively). However, the incidence of PS881 was slightly lower in Kediri compared with that reported in a previous study in the Jember area [3]. Furthermore, the percentage of severity was high at above 80% in all cultivars (Figure 1B), with the highest percentages found in PS58, PS881, and NX04.

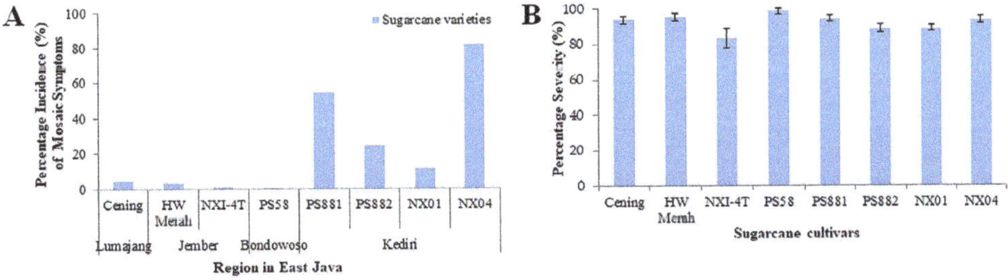

Figure 1. Percentage of mosaic disease incidence (**A**) and severity (**B**) observed on the leaves of eight sugarcane cultivars grown in the field at four locations in East Java, Indonesia: Lumajang, Jember, Bondowoso, and Kediri.

To determine whether mosaic infection of one or more viral types occurred, RT-PCR was conducted using specific primer pairs for the CP gene to detect SCMV, SCSMV, and SrMV (Table 1), which are considered the major mosaic viruses in sugarcane. The analysis showed that CP-SCSMV DNA, with a molecular size of 335 kb, was present in all sugarcane leaf samples (Figure 2A), which indicated that SCSMV was more widespread in all cultivars and regions of the experiment compared with that observed in previous studies [4]. Furthermore, SCMV with a molecular size of 735 kb was found mainly in PS881 and PS882, with only a few instances observed in the Cening cultivar; however, the corresponding DNA band was not detected in NX04, NX01, HW Merah, PS58, and NXI-4T (Figure 2B). Interestingly, a specific primer for SrMV detection, which is a rare mosaic virus infection, was found in the PS881 cultivar but not in the other cultivars (Figure 2C). These results indicate the presence of multiple or mixed mosaic virus infections, including double and triple infections, which are reported for the first time in an Indonesian sugarcane plantation. Furthermore, the mosaic patterns were similar in all sugarcane varieties; however, the intensity of the symptoms varied (Figure 2D). The intensity of the mosaic symptoms was more severe in double and triple infections than in single infections.

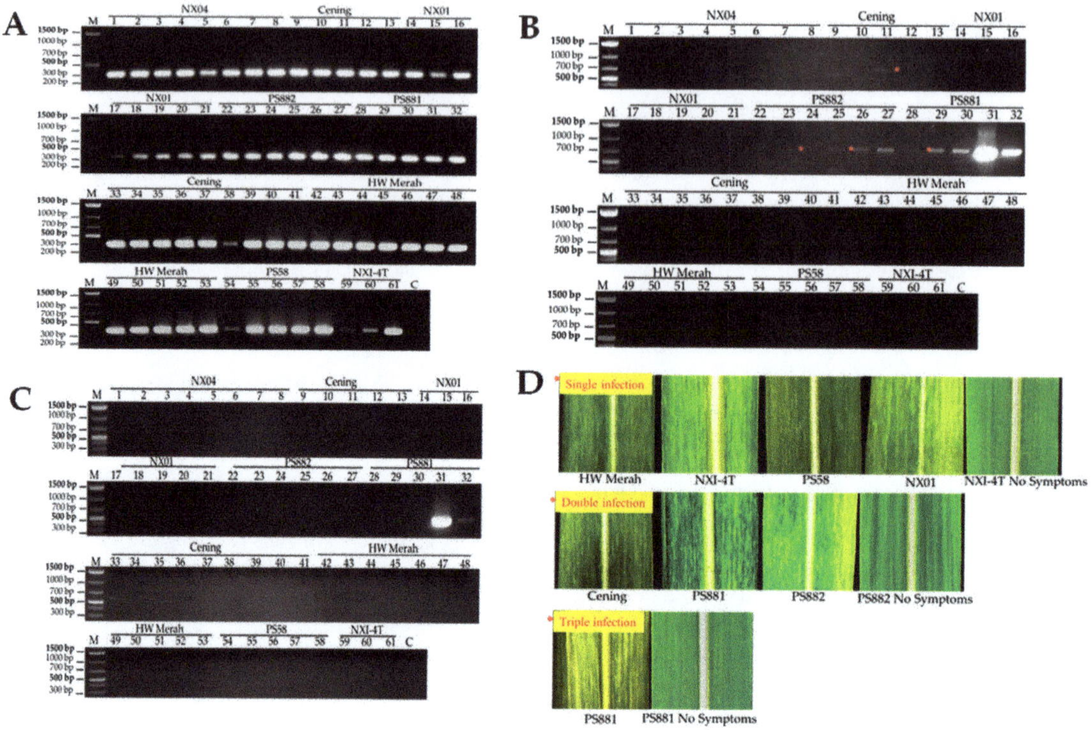

Figure 2. RT-PCR analysis of mosaic virus coat protein genes and mosaic symptom pattern variations in field-grown sugarcane. RT-PCR was conducted with RNA isolated from infected leaves of eight sugarcane cultivars and using specific primers for SCSMV (**A**), SCMV (**B**), and SrMV (**C**). The image presents the RT-PCR products from five infected samples of each sugarcane cultivar among the 61 positive mosaic virus infection samples. M, 1 kb DNA ladder (Thermo Fisher Scientific, Massachusetts, USA); C: control, non-symptomatic sugarcane. Red asterisk indicates positive CP-SCMV DNA. (**D**) Variation of mosaic symptoms intensity in the leaves of sugarcane varieties infected by SCSMV (single), SCSMV-SCMV (double), and SCSMV-SCMV-SrMV (triple) compared to non-symptomatic healthy leaf (control).

3.2. Phylogenetic Analysis of the Mosaic Virus Isolates

To confirm infection by the mosaic virus, the corresponding CP-DNAs of SCSMV, SCMV, and SrMV were determined based on their nucleotide sequences. The sequences from each mosaic virus were correctly verified using NCBI BLAST, and phylogenetic trees were constructed using MEGA 11 software. The sequence similarities between the target sequences and other viral isolates available in GenBank were 92.11% (KP987832.1), 92.58% (MT725538.1), and 97.39% (EU189037.1) for SCSMV, SCMV, and SrMV, respectively. Phylogenetic analysis grouped SCSMV with isolates from Thailand (KP987832.1); SCMV with isolates from Florida, USA (MT725538.1) and Boetzingen, Germany (X98168.1) (Figure 3A,B); and SrMV with isolates from Argentina (EU189037.1), the USA (EF078962.1), and China (DQ227695.1) (Figure 3C).

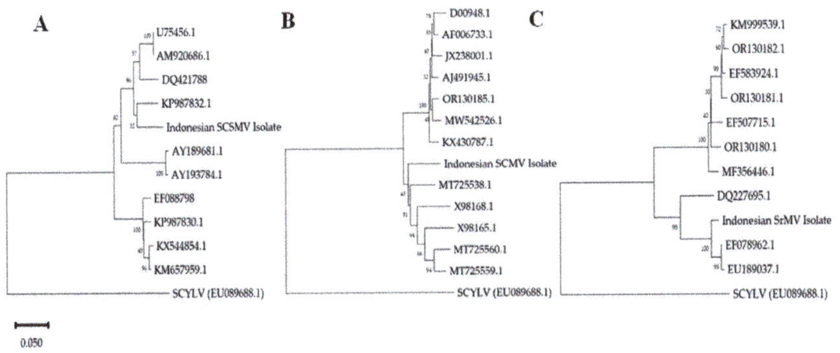

Figure 3. Phylogenetic analysis of SCSMV (**A**), SCMV (**B**), and SrMV (**C**) based on the CP nucleotide sequences along with the neighbor-joining (NJ) method with 1000 bootstrap replicates. SCYLV: Sugarcane yellow leaf virus was selected as outgroup. Sequence clusters according to the geographical origin of the virus isolates, such as Thailand (KP987832.1) for SCSMV; USA (MT725538.1) and Germany (X98168.1) for SCMV; and Argentina (EU189037.1), USA (EF078962.1), and China (DQ227695.1) for SrMV.

3.3. Expression of *Apx* and *Cat* Genes

Environmental stresses, including biotic stresses, produce ROS that regulate cellular metabolism and inhibit cell growth and development. Plants have developed antioxidant mechanisms to overcome the detrimental effects of ROS production. In this study, *Apx* expression was upregulated in infected cultivars compared with that in asymptomatic healthy controls (Figure 4A). Moreover, the expression was substantially higher in PS881 and NX04 than in the other cultivars, while only a slight increase was observed in NX4T and NX01. Changes in *Apx* expression are associated with the ROS detoxification pathway in SCMV-resistant sugarcane genotypes [12]. However, whether *Apx* expression in sugarcane indicates the degree of resistance to viral infections is not well understood.

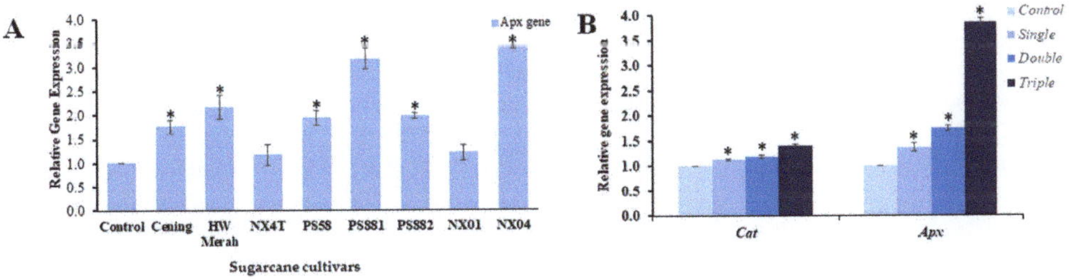

Figure 4. Expression of *Apx* and *Cat* gene in response to mosaic virus infection in sugarcane leaves. Relative expression of the *Apx* gene under single infection (**A**) and the *Cat* and *Apx* genes under single, double, and triple infection (**B**). The expression of *Cat* and *Apx* in the control, single, double, and triple infection samples was determined using the PS881 sugarcane cultivar. β-tubulin was used as the reference gene. Values are reported as the means ± SD of three biological replicates. Asterisk (*) denotes statically significant differences at *t*-test: $p \leq 0.05$.

To confirm the response of antioxidant expression to mosaic virus infection, RNA was isolated from PS881 leaves exhibiting single, double, and triple mixed infections, and the expression of the *Apx* and *Cat* genes was determined. Mixed infections were detected in the leaves of the PS881 cultivar, as previously described (Figure 2). *Apx* expression was almost upregulated in all cultivars, except in NX-4T and NX01, which remained constant compared to those healthy non-infected leaves (control) (Figure 4A). The expression of *Cat* was slightly upregulated in single infections and increased by approximately 1.4-fold in mixed triple infections with the addition of SrMV compared to that in the control (Figure 4B). However, the expression of *Apx* significantly increased in the mixed infection group, and the highest expression (approximately 3.9-fold) was found in the triple-infection group. These results indicate that *Apx* expression was more affected by mosaic infection compared with *Cat* expression.

3.4. Expression of Photosynthesis Related-Genes in Response to Mosaic Virus Infection

Mosaic virus infection disturbs the chloroplast structure, which leads to leaf chlorosis, chloroplast malfunctions, and downregulated photosystem efficiency [23,41]. To observe how the viral infections disrupted the chloroplasts, the chlorophyll content of the infected leaves was determined. The content was significantly lower in the PS58, PS881, PS882, NX04, and HW Merah cultivars (Figure 5A) but only slightly reduced in the Cening, NXI-4T, and NX01 cultivars compared with the other varieties. These results clearly indicate that mosaic virus infection causes chloroplast disruption or damage that leads to disturbances in photosynthetic activity.

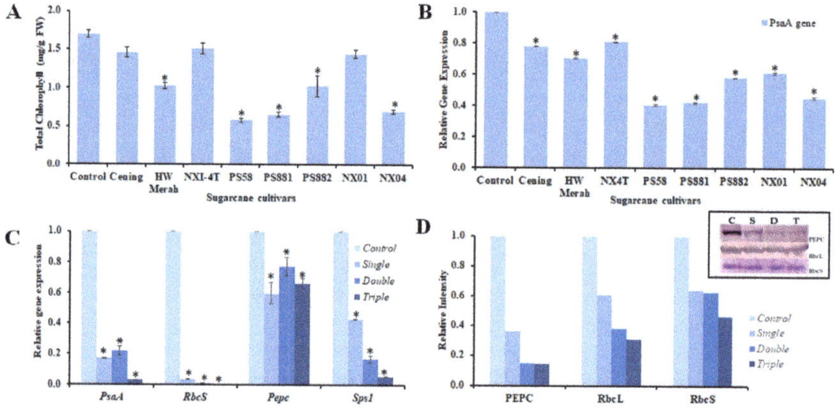

Figure 5. Chlorophyll content and gene expression of photosynthetic-related in response to mosaic virus infection in sugarcane leaves. Total chlorophyll content (**A**) and PsaA expression (**B**) were determined in single-infection sugarcane leaves. Expression of the *PsaA*, *RbcS*, *Pepc*, and *Sps* genes (**C**) and levels of PEPC and Rubisco proteins (**D**) were determined in the control and single-, double-, and triple-infection PS881 leaves. Values represent the means ± SD of three biological replicates. Asterisk (*) denotes statically significant differences at *t*-test: $p \leq 0.05$. Inset shows the PEPC and Rubisco large subunit (RbcL) and small subunit (RbcS) proteins detected by immunoblot analysis. C, S, D, and T represent the control and single, double, and triple infections, respectively. The intensities of the protein bands were quantitively expressed using ImageJ free software version 1.54h.

Quantitative RT-PCR was used to determine whether the expression of *PsaA*, located in the chloroplast genome, was affected by viral infection. The expression of *PsaA* was reduced in the single-infected leaves of all sugarcane cultivars compared to that in the healthy leaves, although the expression varied among the infected cultivars (Figure 5B). Consistent with the chlorophyll content, the expression of *PsaA* was markedly reduced in the leaves of PS58, PS881, and NX04. However, *PsaA* expression was slightly reduced in the other cultivars compared with that in the healthy plants.

This change in chlorophyll content suggests that viral infection has a significant impact on energy metabolism and photosynthetic efficiency and may contribute to alterations in the expression of photosynthetic and carbon-assimilating genes in infected plants. The expression of the photosynthetic genes for *RbcS* (Rubisco small subunit), *Pepc*, and *Sps* (sucrose-phosphate synthase) was measured along with that of the *PsaA* gene in sugarcane leaves with triple mixed infection (Figure 5C). The expression of *PsaA* gradually decreased with the degree of viral infection in single, double, and triple PS881 leaves. The expression of *RbcS* and *Sps* was gradually suppressed in single and double infections, while the expression of *RbcS*, *Sps*, and *PsaA* was almost depleted in the triple infection. Furthermore, the expression of *Pepc* was reduced by approximately 22.7–40.3% in the single-, double-, and triple-infected leaves compared with that in healthy leaves.

To ensure the consistency of gene expression under mosaic virus infection, which reduces photosynthetic activity, the protein content of Rubisco and PEPC was detected using immunoblotting. Rubisco and PEPC are key enzymes for carbon assimilation in C4 plants, such as sugarcane. Consistent with the gene expression, the PEPC, RbcL, and RbcS protein expression decreased according to the degree of single, double, and triple infections in the PS881 cultivar (Figure 5D). Because Rubisco proteins have a conservative and slow turnover [42], the inhibited proteins were still detectable and not as depleted as the level of mRNA in the triple-infected leaves.

4. Discussion

In this study, mosaic diseases were observed in four sugarcane plantations located in Lumajang, Jember, Bondowoso, and Kediri in East Java, Indonesia. Eight sugarcane cultivars were observed for mosaic symptoms and sampled for viral analysis. Mosaic diseases were found to have a higher incidence in Kediri than in the other locations, although they had a similarly high percentage of severity (above 80%). The higher incidence at this location may have been caused by differences in integrated disease management, such as heated water treatment prior to planting. Among the sugarcane cultivars, PS881 and NX04 showed the highest incidence and were categorized as susceptible cultivars [3,4]. RT-PCR analysis revealed that SCSMV was distributed in all sugarcane cultivars, while SCMV and SrMV were only observed in a few cultivars. The nucleotide sequences of the amplified *CP* gene and phylogenetic analyses confirmed the presence of SCSMV, SCMV, and SrMV in sugarcane. Interestingly, the PS882 and PS881 cells were infected with double SCSMV-SCMV and triple SCSMV-SCMV-SrMV infections, respectively. These results indicate the presence of a mixed mosaic virus infection in sugarcane, which has not been previously reported in Indonesia.

Mixed infections by two or more viruses occur naturally in plants [8,43]. Infection of sugarcane with SCSMV, SCMV, or SrMV resulted in mosaic-like symptoms, such as irregular light and dark green patterns on the leaves [7]. Infections with these three viruses in sugarcane present highly similar mosaic symptoms and are difficult to differentiate. Interactions between two or more plant viruses in mixed infections can be categorized as synergistic or antagonistic. A synergistic interaction refers to a situation in which two or more viral infections result in severe symptoms, and antagonistic interaction refers to only one of the viruses having a beneficiary and higher activity than the second virus [30]. PS881 with triple infection and PS882 with double infection showed higher disease severity (Figure 1B) and mosaic symptom intensity (Figure 2D) than the cultivars with single infection. These results indicate that mixed mosaic viral infection with SCSMV, SCMV, and SrMV results in synergistic interactions.

Infections caused by mosaic viruses induce ROS accumulation; however, plants have developed antioxidant scavenging and detoxification systems to overcome these problems. Apx is one of the most important genes for ROS scavenging [36], and its expression increased in all infected sugarcane cultivars, with the highest expression in the mixed infection of PS881 leaves (Figure 4A). Increased Apx expression can regulate ROS accumulation, which may be considered a characteristic of plant resistance to viral infections. However, elevated Apx expression may be required to balance oxidation, and antioxidation cascades to maintain redox homeostasis.

Viral infection has a significant effect on energy metabolism, respiration, and photosynthetic rates. Viral proliferation requires energy from the infected plant cells, which leads to ROS accumulation [19]. ROS play dual roles in plants. Low ROS concentrations act as a signaling molecule to promote growth and development, whereas excessive ROS accumulation triggers cellular damage, leading to cell death. Viral infection modified the chloroplast structure [25] and caused a reduction in chlorophyll content (Figure 5A) as well as the expression of *PsaA*, which is located in the chloroplast (Figure 5B). Chloroplasts play a central role in the generation of energy during photosynthesis. The disruption of chloroplasts causes a reduction in photosynthetic energy for carbon assimilation. The significant reduction in *PsaA* expression in PS881 plants with mixed infection was followed by the suppression of photosynthetic carbon-assimilating genes, such as *RbcS*, *Pepc*, and *Sps*, as well as their protein content (Figure 5C,D). These results indicate that mosaic virus infection damages the chlorophyll and subsequently reduces carbon assimilation activity. Thus, infection with mosaic viruses significantly reduces the photosynthetic efficiency, yield, and quality of sugarcane.

The development of virus-resistant sugarcane is considered the most effective approach for managing viral diseases. Pathogen-derived resistance (PDR) and RNA interference (RNAi) technologies have been previously demonstrated as powerful approaches for plant defense against sugarcane mosaic viruses [35,44]. Moreover, RNAi has been reported to effectively induce high resistance against mosaic viruses in sugarcane [45]. Mixed infections with SCSMV and SCMV, which present synergistic interactions, exacerbate sugarcane yield. A strategy to induce dual resistance to the mosaic virus can apply RNAi by assembling a hairpin element composed of CP gene sequences from SCSMV and SCMV in tandem and an intron sequence as a loop. Dual resistance to synergistically interacting viruses has been reported in transgenic orchids [46] and wheat [47]. Therefore, RNAi could be used to induce dual resistance to SCSMV and SCMV in sugarcane.

5. Conclusions

Mosaic symptoms were observed in eight commercial sugarcane varieties cultivated in four districts of East Java Province: Kediri, Lumajang, Jember, and Bondowoso. The incidence and severity of mosaic symptoms were higher in the NX04 and PS881 cultivars compared with the other cultivars. Single infection with SCSMV was found in all sugarcane cultivars, and triple infection was observed in the PS881 cultivar. Multiple infections prominently increased the expression of *Apx*, the most important antioxidant required to balance oxidation and antioxidation cascades. The chlorophyll content and chloroplast photosystem I gene expression were significantly reduced in the infected sugarcane leaves. The disruption of chloroplasts causes a reduction in photosynthetic energy for carbon assimilation. Therefore, the expression of the photosynthetic genes *RbcS*, *Pepc*, and *Sps* were significantly decreased in the PS881 sugarcane cultivar with multiple infections. Taken together, our results show that mosaic virus infection leads to oxidative stress and reduces photosynthetic activity in sugarcane, which may impair the crop's growth and yield.

Author Contributions: Conceptualization, B.S. and N.D.; methodology, I.R.N., R.H. and W.D.S.; formal analysis, I.R.N. and W.S.; investigation, I.R.N., W.S. and S.; software and statistics, I.R.N. and R.H.; validation, N.D.; data curation, R.H. and W.D.S.; writing—original draft preparation, I.R.N., W.D.S. and B.S.; writing—review editing, B.S., W.D.S. and R.H.; supervision, B.S.; project administration, N.D. and W.D.S.; funding acquisition, B.S. All authors have read and agreed to the published version of the manuscript.

Funding: This research was supported by the Indonesian Ministry of Education, Culture, Research and Technology (grant number 5077/UN25.3.1/LT/2024).

Institutional Review Board Statement: Not applicable.

Informed Consent Statement: Not applicable.

Data Availability Statement: Data is contained within the article.

Acknowledgments: The authors would like to thank Purnama Okviandari, Retnosari Apriasti, and Almira Siam Veda for their helps.

Conflicts of Interest: The authors declare no conflicts of interest.

References

1. Aono, A.H.; Pimenta, R.J.G.; Garcia, A.L.B.; Correr, F.H.; Hosaka, G.K.; Carrasco, M.M.; Cardoso-Silva, C.B.; Mancini, M.C.; Sforça, D.A.; Dos Santos, L.B.; et al. The Wild Sugarcane and Sorghum Kinomes: Insights Into Expansion, Diversification, and Expression Patterns. *Front. Plant Sci.* **2021**, *12*, 668623. [CrossRef]
2. Nguyen, T.-T.; Hoang, Q.-T.; Nguyen, T.-T.; Pham, T.-A.; Cao, A.-D.; Pham, H.-D.; Le, V.-H.; Vu, T.-T.; Pham, N.-H.; Nguyen, T.-C.; et al. Research and Development Prospects for Sugarcane Industry in Vietnam. *SugarTech* **2022**, *24*, 1330–1341. [CrossRef] [PubMed]
3. Addy, H.; Nurmalasari; Wahyudi, A.; Sholeh, A.; Anugrah, C.; Iriyanto, F.; Darmanto, W.; Sugiharto, B. Detection and Response of Sugarcane against the Infection of Sugarcane Mosaic Virus (SCMV) in Indonesia. *Agronomy* **2017**, *7*, 50. [CrossRef]
4. Putra, L.K.; Kristini, A.; Achadian, E.M.; Damayanti, T.A. *Sugarcane streak mosaic virus* in Indonesia: Distribution, Characterisation, Yield Losses and Management Approaches. *SugarTech* **2014**, *16*, 392–399. [CrossRef]
5. Bertasello, L.E.T.; Da Silva, M.F.; Pinto, L.R.; Nóbile, P.M.; Carmo-Sousa, M.; Dos Anjos, I.A.; Perecin, D.; Spotti Lopes, J.R.; Gonçalves, M.C. Yellow Leaf Disease Resistance and *Melanaphis sacchari* Preference in Commercial Sugarcane Cultivars. *Plants* **2023**, *12*, 3079. [CrossRef] [PubMed]
6. Lu, G.; Wang, Z.; Xu, F.; Pan, Y.-B.; Grisham, M.P.; Xu, L. Sugarcane Mosaic Disease: Characteristics, Identification and Control. *Microorganisms* **2021**, *9*, 1984. [CrossRef] [PubMed]
7. Vamsi Krishna, G.; Manoj Kumar, V.; Kishore Varma, P.; Bhavani, B.; Vijaya Kumar, G. Identification of Resistance to *Sugarcane mosaic virus*, *Sugarcane streak mosaic virus*, and *Sugarcane bacilliform virus* in New Elite Sugarcane Accessions in India. *Front. Microbiol.* **2023**, *14*, 1276932. [CrossRef] [PubMed]
8. Xu, Y.; Ghanim, M.; Liu, Y. Editorial: Mixed Infections of Plant Viruses in Nature and the Impact on Agriculture. *Front. Microbiol.* **2022**, *13*, 922607. [CrossRef]
9. He, E.-Q.; Bao, W.-Q.; Sun, S.-R.; Hu, C.-Y.; Chen, J.-S.; Bi, Z.-W.; Xie, Y.; Lu, J.-J.; Gao, S.-J. Incidence and Distribution of Four Viruses Causing Diverse Mosaic Diseases of Sugarcane in China. *Agronomy* **2022**, *12*, 302. [CrossRef]
10. Filloux, D.; Fernandez, E.; Comstock, J.C.; Mollov, D.; Roumagnac, P.; Rott, P. Viral Metagenomic-Based Screening of Sugarcane from Florida Reveals Occurrence of Six Sugarcane-Infecting Viruses and High Prevalence of *Sugarcane yellow leaf virus*. *Plant Dis.* **2018**, *102*, 2317–2323. [CrossRef]
11. Viswanathan, R.; Balamuralikrishnan, M.; Karuppaiah, R. Characterization and Genetic Diversity of *Sugarcane streak mosaic virus* Causing Mosaic in Sugarcane. *Virus Genes* **2008**, *36*, 553–564. [CrossRef] [PubMed]
12. Akbar, S.; Wei, Y.; Yuan, Y.; Khan, M.T.; Qin, L.; Powell, C.A.; Chen, B.; Zhang, M. Gene Expression Profiling of Reactive Oxygen Species (ROS) and Antioxidant Defense System following Sugarcane Mosaic Virus (SCMV) Infection. *BMC Plant Biol.* **2020**, *20*, 532. [CrossRef] [PubMed]
13. Sawitri, W.D. Identification of Chinese Cabbage Sentrin as a Suppressor of Bax-Induced Cell Death in Yeast. *J. Microbiol. Biotechnol.* **2012**, *22*, 600–606. [CrossRef] [PubMed]
14. Wu, J.; Sun, Y.; Zhao, Y.; Zhang, J.; Luo, L.; Li, M.; Wang, J.; Yu, H.; Liu, G.; Yang, L.; et al. Deficient Plastidic Fatty Acid Synthesis Triggers Cell Death by Modulating Mitochondrial Reactive Oxygen Species. *Cell Res.* **2015**, *25*, 621–633. [CrossRef] [PubMed]
15. Huseynova, I.M.; Mirzayeva, S.M.; Sultanova, N.F.; Aliyeva, D.R.; Mustafayev, N.S.; Aliyev, J.A. Virus-Induced Changes in Photosynthetic Parameters and Peroxidase Isoenzyme Contents in Tomato (*Solanum lycopersicum* L.) Plants. *Photosynthetica* **2018**, *56*, 841–850. [CrossRef]
16. Manimekalai, R.; Narayanan, J.; Ranjini, R.; Gokul, M.; Selvi, A.; Kumar, P.; Gomathi, R. Hydrogen Peroxide-Induced Oxidative Stress in Sugarcane and Response Expression Pattern of Stress-Responsive Genes through Quantitative RT-PCR. *SugarTech* **2018**, *20*, 681–691. [CrossRef]
17. Zandi, P.; Schnug, E. Reactive Oxygen Species, Antioxidant Responses and Implications from a Microbial Modulation Perspective. *Biology* **2022**, *11*, 155. [CrossRef] [PubMed]
18. Sahu, P.K.; Jayalakshmi, K.; Tilgam, J.; Gupta, A.; Nagaraju, Y.; Kumar, A.; Hamid, S.; Singh, H.V.; Minkina, T.; Rajput, V.D.; et al. ROS Generated from Biotic Stress: Effects on Plants and Alleviation by Endophytic Microbes. *Front. Plant Sci.* **2022**, *13*, 1042936. [CrossRef] [PubMed]
19. Jiang, T.; Du, K.; Xie, J.; Sun, G.; Wang, P.; Chen, X.; Cao, Z.; Wang, B.; Chao, Q.; Li, X.; et al. Activated Malate Circulation Contributes to the Manifestation of Light-Dependent Mosaic Symptoms. *Cell Rep.* **2023**, *42*, 112333. [CrossRef]
20. Ludwig, M. The Roles of Organic Acids in C4 Photosynthesis. *Front. Plant Sci.* **2016**, *7*, 647. [CrossRef]
21. Chastain, C.J.; Chollet, R. Regulation of Pyruvate, Orthophosphate Dikinase by ADP-/Pi-Dependent Reversible Phosphorylation in C3 and C4 Plants. *Plant Physiol. Biochem.* **2003**, *41*, 523–532. [CrossRef]
22. Huber, S.C.; Huber, J.L. Role and Regulation of Sucrose-Phosphate Synthase in Higher Plants. *Annu. Rev. Plant. Physiol. Plant. Mol. Biol.* **1996**, *47*, 431–444. [CrossRef] [PubMed]
23. Akbar, S.; Yao, W.; Yu, K.; Qin, L.; Ruan, M.; Powell, C.A.; Chen, B.; Zhang, M. Photosynthetic Characterization and Expression Profiles of Sugarcane Infected by Sugarcane Mosaic Virus (SCMV). *Photosynth. Res.* **2021**, *150*, 279–294. [CrossRef] [PubMed]
24. Wang, K.; Deng, Q.; Chen, J.; Shen, W. Physiological and Molecular Mechanisms Governing the Effect of Virus-Free Chewing Cane Seedlings on Yield and Quality. *Sci. Rep.* **2020**, *10*, 10306. [CrossRef] [PubMed]
25. Zanini, A.A.; Di Feo, L.; Luna, D.F.; Paccioretti, P.; Collavino, A.; Rodriguez, M.S. *Cassava common mosaic virus* Infection Causes Alterations in Chloroplast Ultrastructure, function, and carbohydrate metabolism of cassava plants. *Plant Pathol.* **2021**, *70*, 195–205. [CrossRef]

26. Wang, L.; Zhang, W.; Shen, W.; Li, M.; Fu, Y.; Li, Z.; Li, J.; Liu, H.; Su, X.; Zhang, B.; et al. Integrated Transcriptome and microRNA Sequencing Analyses Reveal Gene Responses in Poplar Leaves Infected by the Novel Pathogen Bean Common Mosaic Virus (BCMV). *Front. Plant Sci.* **2023**, *14*, 1163232. [CrossRef] [PubMed]
27. Hanssen, I.M.; Peter Van Esse, H.; Ballester, A.-R.; Hogewoning, S.W.; Parra, N.O.; Paeleman, A.; Lievens, B.; Bovy, A.G.; Thomma, B.P.H.J. Differential Tomato Transcriptomic Responses Induced by Pepino Mosaic Virus Isolates with Differential Aggressiveness. *Plant Physiol.* **2011**, *156*, 301–318. [CrossRef]
28. Pesti, R.; Kontra, L.; Paul, K.; Vass, I.; Csorba, T.; Havelda, Z.; Várallyay, É. Differential Gene Expression and Physiological Changes during Acute or Persistent Plant Virus Interactions May Contribute to Viral Symptom Differences. *PLoS ONE* **2019**, *14*, e0216618. [CrossRef] [PubMed]
29. Tugume, A.K.; Mukasa, S.B.; Valkonen, J.P.T. Mixed Infections of Four Viruses, the Incidence and Phylogenetic Relationships of Sweet Potato Chlorotic Fleck Virus (Betaflexiviridae) Isolates in Wild Species and Sweetpotatoes in Uganda and Evidence of Distinct Isolates in East Africa. *PLoS ONE* **2016**, *11*, e0167769. [CrossRef]
30. Syller, J. Facilitative and Antagonistic Interactions between Plant Viruses in Mixed Infections. *Mol. Plant Pathol.* **2012**, *13*, 204–216. [CrossRef]
31. Moreno, A.B.; López-Moya, J.J. When Viruses Play Team Sports: Mixed Infections in Plants. *Phytopathology* **2020**, *110*, 29–48. [CrossRef]
32. Redinbaugh, M.G.; Stewart, L.R. Maize Lethal Necrosis: An Emerging, Synergistic Viral Disease. *Annu. Rev. Virol.* **2018**, *5*, 301–322. [CrossRef]
33. Wang, Q.; Zhang, C.; Wang, C.; Qian, Y.; Li, Z.; Hong, J.; Zhou, X. Further Characterization of *Maize chlorotic mottle virus* and Its Synergistic Interaction with *Sugarcane mosaic virus* in Maize. *Sci. Rep.* **2017**, *7*, 39960. [CrossRef] [PubMed]
34. Bock, C.H.; Barbedo, J.G.A.; Del Ponte, E.M.; Bohnenkamp, D.; Mahlein, A.-K. From Visual Estimates to Fully Automated Sensor-Based Measurements of Plant Disease Severity: Status and Challenges for Improving Accuracy. *Phytopathol. Res.* **2020**, *2*, 9. [CrossRef]
35. Apriasti, R.; Widyaningrum, S.; Hidayati, W.N.; Sawitri, W.D.; Darsono, N.; Hase, T.; Sugiharto, B. Full Sequence of the Coat protein GENE is Required for the Induction of Pathogen-Derived Resistance against Sugarcane Mosaic Virus in Transgenic Sugarcane. *Mol. Biol. Rep.* **2018**, *45*, 2749–2758. [CrossRef]
36. Liu, F.; Huang, N.; Wang, L.; Ling, H.; Sun, T.; Ahmad, W.; Muhammad, K.; Guo, J.; Xu, L.; Gao, S.; et al. A Novel L-ascorbate Peroxidase 6 Gene, ScAPX6, Plays an Important Role in the Regulation of Response to Biotic and Abiotic Stresses in Sugarcane. *Front. Plant Sci.* **2018**, *8*, 2262. [CrossRef] [PubMed]
37. Livak, K.J.; Schmittgen, T.D. Analysis of Relative Gene Expression Data Using Real-Time Quantitative PCR and the $2^{-\Delta\Delta CT}$ Method. *Methods* **2001**, *25*, 402–408. [CrossRef] [PubMed]
38. Anur, R.M.; Mufithah, N.; Sawitri, W.D.; Sakakibara, H.; Sugiharto, B. Overexpression of Sucrose Phosphate Synthase Enhanced Sucrose Content and Biomass Production in Transgenic Sugarcane. *Plants* **2020**, *9*, 200. [CrossRef]
39. Pérez-Patricio, M.; Camas-Anzueto, J.; Sanchez-Alegría, A.; Aguilar-González, A.; Gutiérrez-Miceli, F.; Escobar-Gómez, E.; Voisin, Y.; Rios-Rojas, C.; Grajales-Coutiño, R. Optical Method for Estimating the Chlorophyll Contents in Plant Leaves. *Sensors* **2018**, *18*, 650. [CrossRef]
40. Fiorentini, M.; Zenobi, S.; Giorgini, E.; Basili, D.; Conti, C.; Pro, C.; Monaci, E.; Orsini, R. Nitrogen and Chlorophyll Status Determination in Durum Wheat as Influenced by Fertilization and Soil Management: Preliminary Results. *PLoS ONE* **2019**, *14*, e0225126. [CrossRef]
41. Bwalya, J.; Alazem, M.; Kim, K. Photosynthesis-Related Genes Induce Resistance against Soybean Mosaic Virus: Evidence for Involvement of the RNA Silencing Pathway. *Mol. Plant Pathol.* **2022**, *23*, 543–560. [CrossRef] [PubMed]
42. Studer, R.A.; Christin, P.-A.; Williams, M.A.; Orengo, C.A. Stability-Activity Tradeoffs Constrain the Adaptive Evolution of RubisCO. *Proc. Natl. Acad. Sci. USA* **2014**, *111*, 2223–2228. [CrossRef]
43. Komínková, M.; Ben Mansour, K.; Komínek, P.; Brožová, J.; Střalková, R. Multiple Infections with Viruses of the Family Tymoviridae in Czech Grapevines. *Viruses* **2024**, *16*, 343. [CrossRef] [PubMed]
44. Widyaningrum, S.; Pujiasih, D.R.; Sholeha, W.; Harmoko, R.; Sugiharto, B. Induction of Resistance to Sugarcane Mosaic Virus by RNA Interference Targeting Coat Protein Gene Silencing in Transgenic Sugarcane. *Mol. Biol. Rep.* **2021**, *48*, 3047–3054. [CrossRef] [PubMed]
45. Hidayati, W.N.; Apriasti, R.; Addy, H.S.; Sugiharto, B. Distinguishing Resistances of Transgenic Sugarcane Generated from RNA Interference and Pathogen-Derived Resistance Approaches to Combating Sugarcane Mosaic Virus. *Indones. J. Biotechnol.* **2021**, *26*, 107. [CrossRef]
46. Chen, T.-Y.; Pai, H.; Hou, L.-Y.; Lee, S.-C.; Lin, T.-T.; Chang, C.-H.; Hsu, F.-C.; Hsu, Y.-H.; Lin, N.-S. Dual Resistance of Transgenic Plants against *Cymbidium mosaic virus* and *Odontoglossum ringspot virus*. *Sci. Rep.* **2019**, *9*, 10230. [CrossRef]
47. Tatineni, S.; Sato, S.; Nersesian, N.; Alexander, J.; Quach, T.; Graybosch, R.A.; Clemente, T.E. Transgenic Wheat Harboring an RNAi Element Confers Dual Resistance Against Synergistically Interacting Wheat Streak Mosaic Virus and Triticum Mosaic Virus. *MPMI* **2020**, *33*, 108–122. [CrossRef]

MDPI AG
Grosspeteranlage 5
4052 Basel
Switzerland
Tel.: +41 61 683 77 34

MDPI Books Editorial Office
E-mail: books@mdpi.com
www.mdpi.com/books

www.ingramcontent.com/pod-product-compliance
Lightning Source LLC
LaVergne TN
LVHW070447061126
851858LV00011B/182
* 9 7 8 3 7 2 5 8 7 9 9 1 5 *